# STUDENT STUDY GUIDE
# &
# SELECTED SOLUTIONS MANUAL

DAVID D. REID
*UNIVERSITY OF CHICAGO*

# PHYSICS

FOURTH EDITION VOLUME I

JAMES S. WALKER

**Addison-Wesley**
San Francisco Boston New York
Cape Town Hong Kong London Madrid Mexico City
Montreal Munich Paris Singapore Sydney Tokyo Toronto

*Publisher*: Jim Smith
*Editorial Manager*: Laura Kenney
*Sr. Project Editor*: Katie Conley
*Media Producer*: David Huth
*Editorial Assistant*: Dyan Menezes
*Executive Marketing Manager*: Scott Dustan
*Managing Editor*: Corinne Benson
*Sr. Production Supervisor*: Nancy Tabor
*Production Coordinator*: Philip Minnitte
*Production Service*: Pre-Press PMG
*Project Manager*: Amanda Maynard
*Cover Design*: Seventeenth Street Studios
*Manufacturing Buyer*: Jeff Sargent
*Cover and Text Printer and Binder*: Edwards Brothers

*Cover Images*: Wind turbines with lightning: Mark Newman / Photo Researchers, Inc.; scanning electron micrograph of head of fly showing compound eye x96 (Colour Enhanced): S. Lowry/Univ Ulster (Getty Images); iceberg in the Errera Channel: Seth Resnick (Getty Images); surfer in tube wave, North Shore, Oahu, Hawaii, USA: Warren Bolster (Getty Images); solar coronal loops: Science Source; light passing through triangular prism: David Sutherland (Getty Images)

ISBN-10: 0-321-60200-5
ISBN-13: 978-0-321-60200-8



**Addison-Wesley**
is an imprint of

**www.pearsonhighered.com**

1 2 3 4 5 6 7 8 9 10—EB—12 11 10 09

To my wife, *Annie*

## PREFACE

This study guide is designed to assist you in your study of the fascinating and sometimes challenging world of physics using *Physics*, *Fourth Edition* by James S. Walker. To do this I have provided a Chapter Review, which consists of a comprehensive (but brief) review of almost every section in the text. Numerous solved examples and exercises appear throughout each Chapter Review. The examples follow the two-column format of the text, while the solutions to the exercises have a more traditional layout. Together with the Chapter Review, each chapter contains a list of objectives, a practice quiz, a glossary of key terms and phrases, a table of important formulas, and a table that reviews the units and dimensions of the new quantities introduced.

In addition to the above materials that I have provided, you will also find Warm-Up and Puzzle questions by Just-in-Time Teaching innovators Gregory Novak and Andrew Gavrin (Indiana University-Purdue University, Indianapolis), Answers to Selected Conceptual Questions and Solutions to Selected End-of-Chapter Problems and Conceptual Exercises from the *Instructor Solutions Manual*. Taken together, the information in this study guide, when used in conjunction with the main text, should enhance your ability to master the many concepts and skills needed to understand physics, and therefore, the world around you. Work hard, and most importantly, have fun doing it!

I am indebted to many for helping me to complete this work. Most directly, I thank Mr. Christian Botting of Parson Science for his continued work with me on this project. I also wish to acknowledge Dr. Anand P. Batra of Howard University. He provided an excellent review of the physics content for the first edition of this Study Guide and made countless valuable suggestions.

David D. Reid
University of Chicago
May 2006

# TABLE OF CONTENTS

# CHAPTER 1

# INTRODUCTION TO PHYSICS

## Chapter Objectives

After studying this chapter, you should

**1.** know the three most common basic physical quantities in physics and their units.

**2.** know how to determine the dimension of a quantity and perform a dimensional check on any equation.

**3.** be familiar with the most common metric prefixes.

**4.** be able to perform calculations, keeping a proper account of significant figures.

**5.** be able to convert quantities from one set of units to another.

**6.** be able to perform quick order-of-magnitude calculations.

**7.** know the difference between scalars and vectors.

## Warm-Ups

**1.** Estimate the number of seconds in a human lifetime. We'll let you choose the definition of lifetime. Do all reasonable choices of lifetime give answers that have the same order of magnitude?

**2.** Which is a faster speed, 30 mi/h or 13 m/s? Describe in words how you obtained your answer.

**3.** Estimate how many 20-cm × 20-cm tiles it would take to tile the floor and three sides of a shower stall. The stall has a 16-$ft^2$ floor and 5-ft walls.

## Chapter Review

### 1–1 – 1–2 Physics and the Laws of Nature & Units of Length, Mass, and Time

The study of **physics** deals with the fundamental laws of nature and many of their applications. These laws govern the behavior of all physical phenomena. We describe the behavior of physical systems using various quantities that we create for this purpose. To begin, we consider the three basic quantities of *length*, *mass*, and *time*. These three quantities will be used to create other quantities.

We define a system of units for these quantities so that we can specify how much length, mass, or time we have. The system of units used in this book is the SI, which stands for Système International. In this system, the unit of length is the *meter* (m), the unit of mass is the *kilogram* (kg), and the unit of time is the *second* (s). This system of units is still sometimes referred to by its former name, the mks system.

SI units are based on the metric system. An important aspect of this system is its hierarchy of prefixes used for quantities of different magnitudes. Certain of these prefixes are used very frequently in physics, so you should become familiar with them. Some of the more common ones are listed here:

| Power | Prefix | Symbol |
|---|---|---|
| $10^{-12}$ | pico | p |
| $10^{-9}$ | nano | n |
| $10^{-6}$ | micro | μ |
| $10^{-3}$ | milli | m |
| $10^{-2}$ | centi | c |
| $10^{3}$ | kilo | k |
| $10^{6}$ | mega | M |
| $10^{9}$ | giga | G |

**Exercise 1–1 Metric Prefixes** Write the following quantities using a convenient metric prefix.
**(a)** 0.00025 m **(b)** 25,000 m **(c)** 250 m **(d)** 250,000,000 m **(e)** 0.0000025 m

**Answer** **(a)** 0.25 mm **(b)** 25 km **(c)** 0.25 km **(d)** 250 Mm **(e)** 2.5 μm

## Practice Quiz

**1.** Which of the following quantities is *not* one of the three basic quantities?

**(a)** length **(b)** speed **(c)** time **(d)** mass

## 1–3 Dimensional Analysis

One important aspect of any physical quantity is its dimension. The **dimension of a quantity** tells us what *type* of quantity it is. When indicating the dimension of a quantity only, we use capital letters enclosed in brackets. Thus, the dimension of length is represented by [L], mass by [M], and time by [T].

We use many equations in physics, and these equations must be dimensionally consistent. It is extremely useful to perform a **dimensional analysis** on any equation about which you are unsure. If the equation is not dimensionally consistent, it cannot be correct. The rules are simple:

* Two quantities can be added or subtracted only if they are of the same dimension.
* Two quantities can be equal only if they are of the same dimension.

Notice that only the dimension needs to be the same, not the units. It is perfectly valid to write 12 inches = 1 foot because both quantities are lengths, [L] = [L], even though their units are different. However, it is not valid to write $x$ inches = $t$ seconds because the quantities have different dimensions: [L] ≠ [T].

---

**Example 1–2 Checking the Dimensions** Given that the quantities $x$ (m), $v$ (m/s), $a$ ($m/s^2$), and $t$ (s) are measured in the units shown in parentheses, perform a dimensional analysis on the following equations. **(a)** $x = t$ **(b)** $x = 2vt$ **(c)** $v = at + t/x$ **(d)** $x = vt + 3at^2$

**Picture the Problem** There is no picture.

**Strategy** Write each equation in terms of its dimensions and check if the equation obeys the preceding rules.

**Solution**

**Part (a)**

**1.** Write the equation with dimensions only: Because these dimensions are not the same, the equation is not valid.

$$[L] = [T]$$

**Part (b)**

**2.** Write out the dimensions of this equation: The right-hand-side dimension is equal to the dimension on the left, so the equation is dimensionally correct.

$$[L] = \frac{[L]}{[T]} \times [T] = [L]$$

**Part (c)**

**3.** Write out the dimension of this equation: The first and second terms on the right-hand side are not of equal dimension and cannot be added. This is not a valid equation.

$$\frac{[L]}{[T]} = \frac{[L]}{[T^2]} \times [T] + \frac{[T]}{[L]} = \frac{[L]}{[T]} + \frac{[T]}{[L]}$$

**Part (d)**

**4.** Write out the dimension of this equation: Here, both terms on the right have the same dimension, which is also equal to the dimension on the left. This equation is dimensionally correct.

$$[L] = \frac{[L]}{[T]} \times [T] + \frac{[L]}{[T^2]} \times [T^2] = [L] + [L]$$

**Insight** Notice that in dimensional analysis, purely numerical factors are ignored because they are **dimensionless.** Because there are dimensionless quantities, dimensional consistency does not guarantee that the equation is physically correct, but it makes for a quick and easy first check.

## Practice Quiz

**2.** Which of the following expressions is dimensionally correct?

**(a)** [L] = [M] × [T]  **(b)** [T] = [L]/[T]  **(c)** $[\mathrm{L}] = \frac{[\mathrm{L}]}{[\mathrm{T}]} \times [\mathrm{T}]$  **(d)** $[\mathrm{M}] = \frac{[\mathrm{L}^2]}{[\mathrm{T}]}$

**3.** If speed $v$ has units of m/s, distance $d$ has units of m, and time $t$ has units of s, which of the following expressions is dimensionally correct?

**(a)** $v = t/d$  **(b)** $t = vd$  **(c)** $d = v/t$  **(d)** $t = d/v$

## 1–4 Significant Figures

All measured quantities carry some uncertainty in their values. When working with the values of quantities, it is important to keep proper account of the digits that are reliably known. Such digits are called **significant figures.** The rules for working with significant figures are as follows:

* *Zeros*: Zeros that are written only to locate the decimal point (set the order of magnitude) are not significant figures. All other zeros are counted as significant figures. Some zeros can be ambiguous. This will be discussed later.
* *Multiplication and Division*: The number of significant figures in the result of a multiplication or division equals the number of significant figures in the factor containing the fewest significant figures.
* *Addition and Subtraction*: The significant figures in the result of an addition or subtraction are located only in the *places* (hundreds, ones, tenths, etc.) that are reliably known for *every* value in the sum.

**Exercise 1–3 Significant Figures** A calculation involves the addition of two measured distances $d_1$ = 1250 m, and $d_2$ = 336 m. If each measurement is given to three significant figures, what is the result of the calculation?

**Solution** Adding the two distances, we get $d_1 + d_2 = 1250\text{ m} + 336\text{ m} = 1590\text{ m}$.

The answer is not 1586 because even though the 6 is significant in 336 m, the one's place of 1250 m (the 0) is not significant, so the ones place of the result cannot be significant. You may wonder about the fact that there is no significant figure in the thousands place of 336 m because the value requires no digit there; however, because there is no digit there, we know that place with certainty.

---

**Example 1–4 Driving in a Residential Zone** On most residential streets in the United States the speed limit is 25 mi/h (= 11 m/s). If a car drives down a neighborhood side street at the legal speed limit for 120.46 s, how much distance does the car cover?

**Picture the Problem** Our sketch shows the car moving along a straight road.

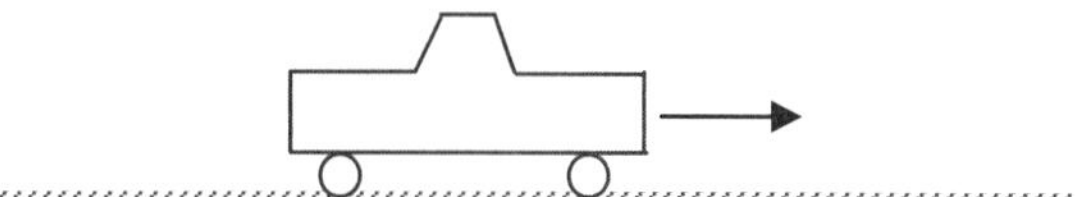

**Strategy** The distance traveled is the speed multiplied by the time of travel.

**Solution**

Multiply the speed and the time to get the distance: $d = 11 \text{ m/s} \times 120.46 \text{ s} = 1300 \text{ m}$

**Insight** There are two important things to notice about the result. First, despite the fact that the time is known to five significant figures, the speed is known only to two, and so the result has only two significant figures. Second, the final two zeros in the value 1300 are not significant. They must be written, however, to give the proper magnitude of the value. It can often be unclear whether such zeros are significant. This problem can be avoided by using scientific notation.

---

**Scientific Notation**

A very useful way of writing numerical values is to use **scientific notation.** In this notation, a value is written as a number of order unity (meaning that only one digit is left of the decimal point) times the appropriate power of 10. The value of scientific notation is that it allows for quick identification of the **order-of-magnitude** (power of 10) of a quantity, calculations are often easier to perform when the values are listed this way, and it removes any ambiguity in the number of significant figures. For example, if the number 3500 has only one significant figure, we write it as $4 \times 10^3$; if it has two, we write $3.5 \times 10^3$; if it has three, we write $3.50 \times 10^3$; and if it has four significant figures we write $3.500 \times 10^3$.

**Exercise 1–5 Scientific Notation** Write the following quantities using scientific notation assuming three significant figures.

**(a)** 0.00250 m **(b)** 12,060 m **(c)** 451 m **(d)** 8.00 m **(e)** 0.00003593 m

**Answer:** **(a)** $2.50 \times 10^{-3}$ m **(b)** $1.21 \times 10^{4}$ m **(c)** $4.51 \times 10^{2}$ m **(d)** $8.00 \times 10^{0}$ m **(e)** $3.59 \times 10^{-5}$ m

When the value is already of order unity, as with part (d), the power of 10 is often dropped. In such cases, the fact that the two zeros are written after the decimal point indicates that they are significant figures.

### Round-off Error

Be aware that to avoid excessive round-off error you should round only to the proper number of significant figures at the very end of a calculation. In Exercise 1–34, if the distance calculated is only an intermediate step in a longer calculation, then the value 1586 m should be used in the subsequent steps. In general, keep one or more additional digits for values calculated in intermediate steps. Another, even better, approach is not to calculate intermediate values numerically but to just carry through the formulas inserting numerical values only at the end.

**Example 1–6 Don't Round-off Too Soon** A cardboard box has measurements of $L$ = 1.92 m, and $W$ = 0.725 m. Its height is $H$ = 1.88 m. **(a)** Calculate the area ($A$) of the base of the box. **(b)** Calculate its volume ($V$) using the result of (a). **(c)** Calculate its volume using the formula for volume.

**Picture the Problem** The diagram shows a box representing the box whose base area and volume we wish to determine.

1.88 m
0.725 m
1.92 m

**Strategy** First calculate the area of the base.

**Solution**

**1.** Calculate the area of the base: $A = LW = 1.92\text{ m} \times 0.725\text{ m} = 1.39\text{ m}^2$

**2.** The volume of the box is area × height: $V = A \times H = 1.39\text{ m}^2 \times 1.88\text{ m} = 2.61\text{ m}^3$

**3.** The volume is length × width × height: $V = A \times H = LWH = 1.92\text{ m} \times 0.725\text{ m} \times 1.88\text{ m} = 2.62\text{ m}^3$

**Insight** The answers to parts (b) and (c) differ in the final digit. Which one is correct? Part (c) is correct because the full values were used. The round-off to three significant figures in part (a) is the reason for the difference.

## Practice Quiz

**4.** Assuming that every nonzero digit is significant, consider the following product of numbers: $1.34 \times 10.75 \times 0.042$. Which answer is correct to the proper number of significant figures?

**(a)** 6 **(b)** 0.61 **(c)** 6.05 **(d)** 6.0501

**5.** Assuming that only nonzero digits are significant, consider the following sum of numbers: 1700 + 338 + 13. Which answer is correct to the proper number of significant figures?

**(a)** 2051 **(b)** 2050 **(c)** 2100 **(d)** 2000

**6.** Consider the following expression: $(5.93) \times (8.762) + (2.116) \times (3.70)$. Which answer is correct to the proper number of significant figures?

**(a)** 59.78786 **(b)** 59.79 **(c)** 59.83 **(d)** 59.8

**7.** Assuming that only nonzero digits are significant, which of the following numbers is the proper scientific notation for 25,300?

**(a)** $2.53 \times 10^4$ **(b)** $25.3 \times 10^3$ **(c)** $2.53 \times 10^3$ **(d)** $0.253 \times 10^5$

**8.** The number $7.4 \times 10^5$ is equivalent to which of the following?

**(a)** 7.4 **(b)** 740 **(c)** 7,400 **(d)** 740,000

## 1–5 Converting Units

Even though we predominantly use SI units, it will often be necessary to convert between SI and other units. A conversion can be accomplished using a **conversion factor** that is constructed by knowing how much of a quantity in one unit equals that same quantity in another unit. A conversion factor is a ratio of equal quantities written such that, when multiplied by a quantity, the undesired unit algebraically cancels leaving only the desired unit. This concept is best illustrated by example.

**Example 1–7 Volume of a Box** A typical cardboard box provided by moving companies measures 1.50 ft × 1.50 ft × 1.33 ft. Determine the volume ($V$) of clothes that you can pack into this box in cubic meters.

**Picture the Problem** The diagram represents the box whose volume we wish to determine.

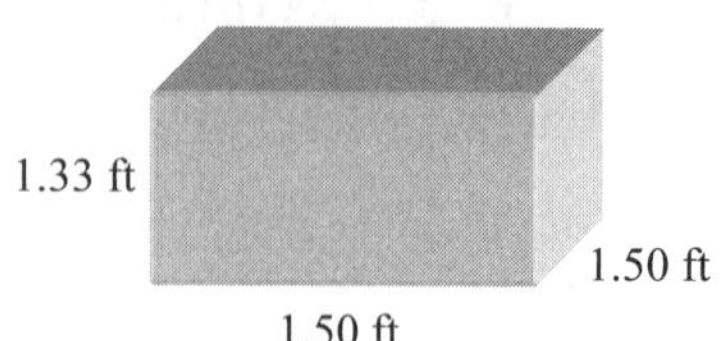


**Strategy** We first calculate the volume in the given units, determine the conversion factor, and then convert the volume to cubic meters.

**Solution**

**1.** Calculate the volume as given: $V = 1.50\text{ ft} \times 1.50\text{ ft} \times 1.33\text{ ft} = 2.993\text{ ft}^3$

**2.** Write the number of meters in a foot: $1\text{ m} = 3.281\text{ ft}$

**3.** Write the conversion factor from feet to meters: $\dfrac{1\text{ m}}{3.281\text{ ft}}$

**4.** The conversion factor from $\text{ft}^3$ to $\text{m}^3$ is: $\left(\dfrac{1\text{ m}}{3.281\text{ ft}}\right)^3 = \dfrac{1\text{ m}^3}{35.32\text{ ft}^3}$

**5.** Multiply the volume by the conversion factor: $V = 2.993\ \text{ft}^3\left(\dfrac{1\ \text{m}^3}{35.32\ \text{ft}^3}\right) = 0.0847\ \text{m}^3$

**Insight** In the final step, the unit $\text{ft}^3$ cancels just as numbers would. Setting up this cancellation is the crucial step in unit conversion. You will get plenty of practice converting units in your study of physics.

## Practice Quiz

**9.** Given that 1 in. = 2.54 cm, convert 250.0 cm to inches.

**(a)** 635 in. **(b)** 0.394 in. **(c)** 98.4 in. **(d)** 150.0 in.

**10.** Convert the speed $1.00 \times 10^2$ m/s to km/h.

**(a)** 36 km/h **(b)** 360 km/h **(c)** $3.60 \times 10^8$ km/h **(d)** 27.8 km/h

## 1–6 Order-of-Magnitude Calculations

As stated previously, the order-of-magnitude of a quantity is the power of ten for that quantity when its value is written in scientific notation; it identifies the general size of the quantity. An order-of-magnitude calculation is designed only to determine a value to within a factor of 10 (1 order of magnitude). These types of calculations provide quick estimates of values to give some indication of what to expect from a more detailed calculation. Keep in mind that a useful order-of-magnitude calculation requires that you know, or can reasonably estimate, values to about one significant figure.

---

**Exercise 1–8 Cars in the Motor City** How many automobiles are owned by the residents of Detroit?

**Solution:**

The population of the city of Detroit is about one million ($10^6$) people. People live to be about 80 years old and the driving age is 16. So, roughly 16/80 = one-fifth of these residents are expected to be of driving age. This means that there are $0.2 \times 10^6 = 2 \times 10^5$ potential drivers. Some of these residents won't own a car, but others will own more than one, averaging out to approximately one car per person of driving age. Therefore, there are roughly $10^5$ cars owned by the residents of Detroit.

---

### 1–7 Scalars and Vectors

All of the quantities you'll study in this course fall into one of two categories. Some quantities such as mass are specified by a single number and its units. Such quantities are called **scalar** quantities. Other quantities, however, require a directional specification in addition to a numerical value. Quantities of this latter type are called **vector** quantities.

### 1–8 Problem Solving in Physics

Solving physics problems is a logical and creative endeavor for which there is no set prescription; however, there are certain practices that help this creative process to flourish. First, a *careful reading* of the problem is necessary to fully grasp the question being posed and the information being given. It is often useful to separately write out all the given and required information; several of the solved examples in this study guide illustrate that approach. It is also a good practice to make a *sketch* of the problem and to *visualize* the physics that is taking place. A correct mental picture of the problem takes you a long way toward a correct solution. Next, map out your *strategy* for the solution. Here, you basically solve the problem logically before doing it mathematically. For the mathematical solution, you need to *identify and solve* the appropriate equations for the relevant physics. Finally, you should *check and explore* your result to be sure that the answer makes sense in the context of the problem.

## Reference Tools and Resources

### I. Key Terms and Phrases

**physics** the study of the fundamental laws of nature and many of their applications

**SI units** the internationally adopted standard system of units (based on meters, kilograms, and seconds) for quantitatively measuring quantities

**dimension of a quantity** the fundamental type of a quantity such as length, mass, or time

**dimensional analysis** a type of calculation that checks the dimensional consistency of an equation

**dimensionless quantity** a value that is purely numerical, or a quantity defined such that all dimensional factors cancel

**significant figures** the digits in the numerical value of a quantity that are known with certainty

**scientific notation** a method of writing numbers that consists of a number of order unity times 10 to the appropriate power

**order-of-magnitude** the power of 10 characterizing the size of a quantity

**conversion factor** a factor (equal to 1) that multiplies a quantity to convert its value to another unit

**scalar** a numerical value with appropriate units

**vector** a mathematical quantity having a numerical value and direction (with appropriate units)

## II. Tips

### Dimensional Analysis

You should be aware that, typically, arguments of mathematical functions are dimensionless. Angles, for example, are dimensionless, as can be seen by the equation for the length of a circular arc, $s = r\theta$, where $\theta$ is in radians; hence, angular measures such as radians and degrees signify only how we choose to measure the angle. The trigonometric functions, therefore, such as sine, cosine, and tangent are applied to dimensionless quantities. Other examples of dimensionless functions are $\log(x)$, $\ln(x)$, and their inverse functions $10^x$ and $\exp(x)$.

# Puzzle

### WHERE ARE YOU?

The standard geographical coordinates of Chicago are as follows:

Latitude: 41 degrees 50 minutes

Longitude: 87 degrees 45 minutes

What are the $x$, $y$, $z$ coordinates of Chicago in a coordinate system centered at the center of Earth, with the $z$-axis pointing from the South Pole to the North Pole, and the $x$-axis passing through the zero longitude meridian pointing away from Europe into space? Answer this question in words, not equations, briefly explaining how you obtained your answer.

## Answers to Selected Conceptual Questions

**2.** The quantity $T + d$ does not make sense because it adds quantities of different dimensions. The quantity $d/T$ does make sense; it could represent the distance $d$ traveled by an object in the time $T$.

**4. (a)** In one year there are $\left(6\times10^{1}\,\frac{\text{s}}{\text{min}}\right)\left(6\times10^{1}\,\frac{\text{min}}{\text{h}}\right)\left(2.4\times10^{1}\,\frac{\text{h}}{\text{d}}\right)\left(3.65\times10^{2}\,\frac{\text{d}}{\text{y}}\right)=3.15\times10^{7}\,\text{s}$. So, the order of magnitude of this is $10^7$ s.

**(b)** A baseball game lasts roughly 3 hours; this gives $3\left(6\times10^{1}\,\frac{\text{s}}{\text{min}}\right)\left(6\times10^{1}\,\frac{\text{min}}{\text{h}}\right)=1.08\times10^{4}\,\text{s}$. The order of magnitude of this $10^4$ s or 10,000 s.

**(c)** By feeling your own heartbeat you can tell that 0.1 s ($10^{-1}$s) per beat is too short and that 10 s ($10^1$ s) per beat is too long. So, the order of magnitude of a heartbeat must be 1 s ($10^0$ s).

**(d)** The earth is about 4 billion years old. Using the result from part (a) we get $\left(4\times10^{9}\,\text{y}\right)\left(3.15\times10^{7}\,\frac{\text{s}}{\text{y}}\right)=1.26\times10^{17}\,\text{s}$. The order of magnitude of this is $10^{17}$ s.

**(e)** People generally live up to about 100 years; so the age of a person is typically within the range from 0 – 100 years. Using the result from part (a) we get the order of magnitude of this range to be $10^7$ s to $10^9$ s.

## Solutions to Selected End-of-Chapter Problems and Conceptual Exercises

8. **Picture the Problem**: This is a dimensional analysis question.

**Strategy:** Manipulate the dimensions in the same manner as algebraic expressions.

**Solution:** Substitute dimensions for the variables:

$$v^2 = 2ax^p$$

$$\left(\frac{\text{m}}{\text{s}}\right)^2 = \left(\frac{\text{m}}{\text{s}^2}\right)(\text{m})^p$$

$$\text{m}^2 = \text{m}^{p+1} \quad \text{therefore } \boxed{p=1}$$

**Insight:** The number 2 does not contribute any dimensions to the problem.

16. **Picture the Problem**: The weights of the fish are added.

**Strategy:** Apply the rule for addition of numbers, which states that the number of decimal places after addition equals the smallest number of decimal places in any of the individual terms.

**Solution: 1.** Add the numbers: 2.35 + 12.1 + 12.13 lb = 26.58 lb

**2.** Round to the smallest number of decimal places in any of the individual terms: $26.58\text{ lb} \Rightarrow \boxed{26.6\text{ lb}}$

**Insight:** The 12.1 lb rock cod is the limiting figure in this case; it is only measured to within an accuracy of 0.1 lb.

28. **Picture the Problem**: This is a units conversion problem.

**Strategy:** Multiply the known quantity by appropriate conversion factors to change the units.

**Solution:** Convert m/s to miles per hour: $\left(3.00\times10^{8}\ \frac{\text{m}}{\text{s}}\right)\left(\frac{1\ \text{mi}}{1609\ \text{m}}\right)\left(\frac{3600\ \text{s}}{1\ \text{h}}\right)=\boxed{6.71\times10^{8}\ \frac{\text{mi}}{\text{h}}}$

**Insight:** Conversion factors are conceptually equal to one, even though numerically they often equal something other than one. They are often helpful in displaying a number in a convenient, useful, or easy-to-comprehend fashion.

36. **Picture the Problem**: The rows of seats are arranged into roughly a circle.

**Strategy:** Estimate that a baseball field is a circle around 300 ft in diameter, with 100 rows of seats around outside of the field, arranged in circles that have perhaps an average diameter of 500 feet. The length of each row is then the circumference of the circle, or $\pi d = \pi(500\ \text{ft})$. Suppose there is a seat every 3 feet.

**Solution:** Multiply the quantities to make an estimate: $N=(100\ \text{rows})\left(\pi 500\ \frac{\text{ft}}{\text{row}}\right)\left(\frac{1\ \text{seat}}{3\ \text{ft}}\right)=52,400\ \text{seats}\cong\boxed{10^{5}\ \text{seats}}$

**Insight:** Some college football stadiums can hold as many as 100,000 spectators, but most less than that. Still, for an order of magnitude we round to the nearest factor of ten; in this case it's $10^5$.

46. **Picture the Problem**: This is a units conversion problem.

**Strategy:** Multiply the known quantity by appropriate conversion factors to change the units.

**Solution: 1. (a)** The acceleration must be $\boxed{\text{greater than}}$ 14 ft/s$^2$ because there are about 3 ft per meter.

**2. (b)** Convert m/s$^2$ to ft/s$^2$: $\left(14\ \frac{\text{m}}{\text{s}^2}\right)\left(\frac{3.28\ \text{ft}}{\text{m}}\right)=\boxed{46\ \frac{\text{ft}}{\text{s}^2}}$

**3. (c)** Convert m/s$^2$ to km/h$^2$: $\left(14\ \frac{\text{m}}{\text{s}^2}\right)\left(\frac{1\ \text{km}}{1000\ \text{m}}\right)\left(\frac{3600\ \text{s}}{\text{h}}\right)^2=\boxed{1.8\times10^{5}\ \frac{\text{km}}{\text{h}^2}}$

**Insight:** Conversion factors are conceptually equal to one, even though numerically they often equal something other than one. They are often helpful in displaying a number in a convenient, useful, or easy-to-comprehend fashion.

51. **Picture the Problem**: This is a dimensional analysis question.

**Strategy:** Find $q$ to make the time dimensions match and then $p$ to make the distance dimensions match. Recall $L$ must have dimensions of meters and $g$ dimensions of m/s$^2$.

**Solution: 1.** Make the time dimensions match: $[\text{T}]=[\text{L}]^p\left(\frac{[\text{L}]}{[\text{T}]^2}\right)^q=[\text{L}]^p\left([\text{L}]\ [\text{T}]^{-2}\right)^q$ implies $\boxed{q=-\tfrac{1}{2}}$

**2.** Now make the distance units match: $[\text{T}]=[\text{L}]^p\left(\frac{[\text{L}]}{[\text{T}]^2}\right)^{-\frac{1}{2}}$ implies $\boxed{p=\tfrac{1}{2}}$

**Insight:** Sometimes you can determine whether you've made a mistake in your calculations simply by checking to ensure the dimensions work out correctly on both sides of your equations.

## Answers to Practice Quiz

**1.** (b) **2.** (c) **3.** (d) **4.** (b) **5.** (c) **6.** (d) **7.** (a) **8.** (d) **9.** (c) **10.** (b)

# CHAPTER 2
# ONE-DIMENSIONAL KINEMATICS

## Chapter Objectives

After studying this chapter, you should

1. know the difference between distance and displacement.
2. know the difference between speed and velocity.
3. know the difference between velocity and acceleration.
4. be able to define acceleration and give examples of both positive and negative acceleration.
5. be able to calculate displacements, velocities, and accelerations using the equations of one-dimensional motion.
6. be able to interpret $x$-versus-$t$ and $v$-versus-$t$ plots for motion with both constant velocity and constant acceleration.
7. be able to describe the motion of freely falling objects.

## Warm-Ups

1. During aerobic exercise, people often suffer injuries to knees and other joints due to high accelerations. When do these high accelerations occur?
2. Estimate the acceleration you subject yourself to if you walk into a brick wall at normal walking speed. (Make a reasonable estimate of your speed and of the time it takes you to come to a stop.)
3. A man drops a baseball from the edge of the roof of a building. At exactly the same time, another man shoots a baseball vertically up toward the man on the roof in such a way that the ball just barely reaches the roof. Does the ball from the roof reach the ground before the ball from the ground reaches the roof, or vice versa?
4. Estimate the time it takes for a free-fall drop from a height of 10 m. Also estimate the time a 10-m platform diver is in the air if he takes off straight up with a vertical speed of 2 m/s (and clears the platform of course!).

## Chapter Review

### 2–1 Position, Distance, and Displacement

Any description of motion takes place in a *coordinate system* that allows us to track the *position* of an object. *One-dimensional motion* means that objects are free to move back and forth only along a single line. As a coordinate system for one-dimensional motion, we choose this line to be the $x$-axis together with a specified origin and positive and negative directions. The location of the origin and the direction that is called positive or negative may be chosen according to convenience.

The **distance** traveled by an object that moves from one position to another is the total length of travel during the trip. This total length of travel depends on the path taken as the object moves from its initial position $x_i$ to its final position $x_f$. Distance should be distinguished from **displacement**, $\Delta x = x_f - x_i$, which is the change in position of the object regardless of the path taken. The primary difference between the two is that the distance an object travels tells us nothing about the direction of travel, whereas displacement tells us precisely how far, and in what direction from its initial position an object is located. To put it succinctly, distance is the *total* length of travel, and displacement is the *net* length of travel accounting for direction.

---

**Example 2–1 Parking in the Same Spot** From your apartment, you leave your favorite parking spot and drive 4.83 km east on Main Street to go to the grocery store. After shopping, you go back home by traveling west on Main Street and find that your favorite parking spot is still available. **(a)** What distance do you travel during this trip? **(b)** What is your displacement?

**Picture the Problem** We choose the $x$-axis to represent Main Street. The origin is placed at the initial parking spot.

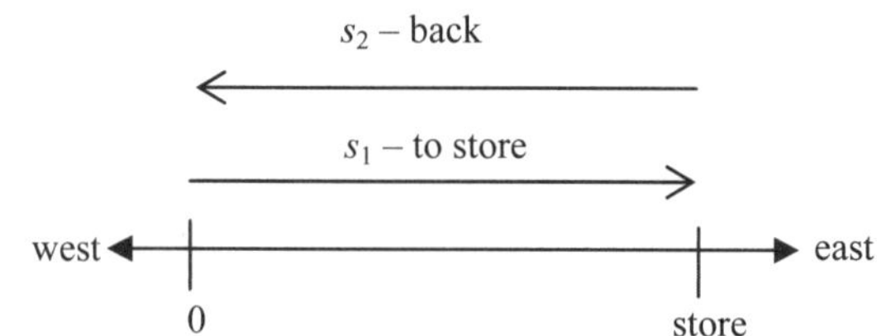


**Strategy** We must remember the distinction between distance and displacement. For distance, we consider the entire length of the path. For displacement, we focus on the initial and final positions only.

**Solution**

**Part (a)**

The trip has two length segments: $s_1$, going to the store, and $s_2$, coming back home. The total length of travel, $s$, must be the sum of these two segments.

$$s = s_1 + s_2 = 4.83 \text{ km} + 4.83 \text{ km} = 9.66 \text{ km}$$

**Part (b)**

Here we notice that the initial and final positions are the same. Subtract the initial position from the final position to get the displacement.

$$\Delta x = x_{\text{f}} - x_{\text{i}} = 0\text{ m} - 0\text{ m} = 0\text{ m}$$

**Insight** This example clearly shows how different the distance and displacement can be. It doesn't matter where the origin is placed; the results for both parts (a) and (b) would be the same. As an exercise, you should convince yourself of this fact.

## Practice Quiz

**1.** If you walk exactly four times around a quarter-mile track, what is your displacement?

**(a)** one mile **(b)** half a mile **(c)** one-quarter mile **(d)** zero

## 2–2 – 2–3 Average Speed and Velocity & Instantaneous Velocity

An important part of describing motion is to specify how rapidly an object moves. One way to do this is to quote an **average speed,**

$$\text{average speed} = \frac{\text{distance}}{\text{elapsed time}}$$

Thus, average speed is the distance traveled divided by the amount of time it took to travel that distance.

Another, sometimes more appropriate, way to describe the rate of motion is to quote an *average* **velocity,**

$$v_{\text{av}} = \frac{\text{displacement}}{\text{elapsed time}} = \frac{\Delta x}{\Delta t}$$

Thus, average velocity is the displacement divided by the amount of time it took to undergo that displacement. The difference between average speed and average velocity is that average speed tells us about the average rate of motion over the entire path taken and contains no directional information. Average velocity, in contrast, relates only to the rate at which an object goes from $x_{\text{i}}$ to $x_{\text{f}}$, regardless of the path taken, and thereby specifies direction.

**Example 2–2 Average Speed versus Average Velocity** For the trip described in Example 2–1, if it took $\Delta t_1 = 10.0\text{ min}$ to drive to the store and $\Delta t_2 = 12.0\text{ min}$ to drive back home, calculate the average speed and average velocity for the trip.

**Picture the Problem** The same picture from Example 2–1 applies here.

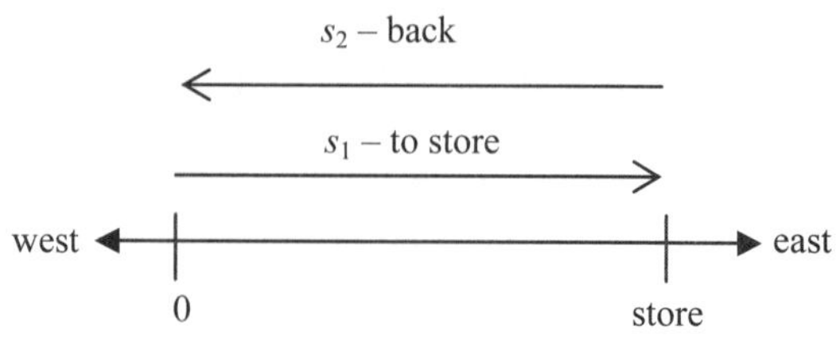


**Strategy** Knowing the definitions of average speed and average velocity, we can apply them directly using the results of Example 2–1.

**Solution**

**1.** To get the answers in SI units, let us first determine the elapsed time for the entire trip and convert it to seconds:

$$\Delta t = \Delta t_1 + \Delta t_2 = 10.0 \text{ min} + 12.0 \text{ min} = 22.0 \text{ min}$$
$$= 22.0 \text{ min}\left(\frac{60 \text{ s}}{\text{min}}\right) = 1320 \text{ s}$$

**2.** Convert the distance traveled to meters:

$$9.66 \text{ km} = 9.66 \times 10^3 \text{ m}$$

**3.** Use the definition of average speed:

$$\text{ave speed} = \frac{\text{distance}}{\Delta t} = \frac{9.66 \times 10^3 \text{ m}}{1320 \text{ s}} = 7.32 \text{ m/s}$$

**4.** Similarly, use the definition of average velocity to calculate its value:

$$\upsilon_{\text{av}} = \frac{\Delta x}{\Delta t} = \frac{0 \text{ m}}{1320 \text{ s}} = 0 \text{ m/s}$$

**Insight** Notice here that just as distance and displacement can be very different, so can average speed and average velocity. Don't confuse the terms.

---

Some situations require more than just the average rate of motion. Often, we require the velocity that an object has at a specific instant in time; this velocity is called the *instantaneous velocity*. The instantaneous velocity, $\upsilon$, can be defined in terms of the average velocity measured over an infinitesimally small elapsed time,

$$v = \lim_{\Delta t \to 0} \frac{\Delta x}{\Delta t}$$

Thus, the instantaneous velocity would be the velocity that an object has right at $t = 2.0$ s, for example, instead of the average velocity over a time period $\Delta t = 2.0$ s. The magnitude of the instantaneous velocity of an object (how fast it is going) is its *instantaneous speed.*

In general, average and instantaneous velocities will have very different values; however, the case of *constant-velocity motion* is special in that the average velocity over any time interval equals the instantaneous velocity at any time. Thus, the definition of average velocity also serves as an equation that describes constant velocity motion. For this special case, the equation is often written as

$$\Delta x = v\Delta t$$

where $v$ no longer needs to be called $v_{\text{av}}$.

---

**Example 2–3 Time the Moving Ball** How much time is required for a ball that rolls with a constant velocity of 0.64 m/s to roll across a 2.3-m-long table?

**Picture the Problem** We take the length along the tabletop as the $x$-axis with one end as the origin. The ball is taken to roll in the positive direction.

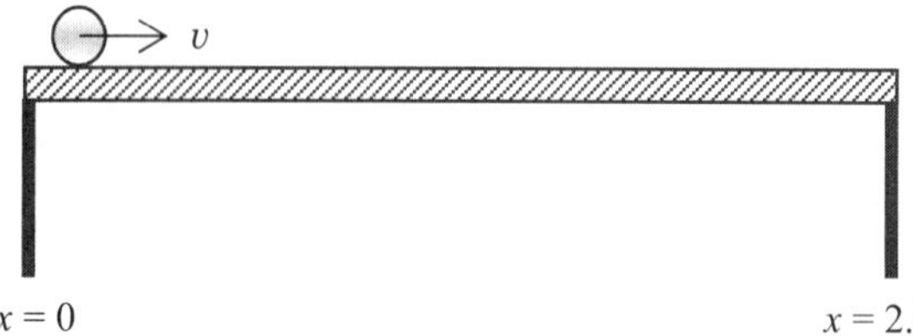


**Strategy** From the coordinate system shown, and the information given, it is clear that we know both the displacement and the velocity. Because we can take both $x_i$ and $t_i$ to be zero, we can simply write $\Delta x = x = 2.3\text{ m}$ and $\Delta t = t$.

**Solution**

**1.** Solve the constant-velocity equation for $t$: $x = vt \Rightarrow t = \dfrac{x}{v}$

**2.** Insert the values to get the numerical solution: $t = \dfrac{2.3\text{ m}}{0.64\text{ m/s}} = 3.6\text{ s}$

**Insight** Although this problem may seem short, it gets to the core of good problem solving. You should identify the information given to you, as specifically as possible, and relate it to the physics you've been learning. This is especially true for longer problems.

---

**Exercise 2–4 Playing Catch** A college athlete stands 13.2 m away from her friend and throws a ball to him at 26.8 m/s (assumed constant while in the air). If the total time, from when she throws the ball to when she hears the sound of her friend catching it, is 0.531 s, what is the speed at which the sound travels?

**Solution** Try to sketch a picture of the problem. It has two different parts, the athlete throwing the ball, with constant velocity, to her friend, followed by the sound traveling, with a different constant velocity, from the friend to the athlete. The information supplied by the problem can be listed as:

**Given** $x = 13.2$ m, $\upsilon_{\text{ball}} = 26.8$ m/s, $T_{\text{tot}} = 0.531$ s; **Find** $\upsilon_{\text{sound}}$

We first address the quantity of interest and let it guide us to the next step in the solution. We want to calculate the constant velocity of sound. From the equation for constant velocity we know that $v_{\text{sound}} = x / t_{\text{sound}}$. Because we already know $x$, we must now find $t_{\text{sound}}$. What else depends on $t_{\text{sound}}$? The total time depends on both the time for the ball to travel to the friend and the time for sound to travel back: $T_{\text{tot}} = t_{\text{ball}} + t_{\text{sound}}$. Therefore, if we can determine $t_{\text{ball}}$, we'll be able to determine $t_{\text{sound}}$. We can use the constant-velocity equation to determine the time for the ball to reach the friend:

$$t_{\text{ball}} = \frac{x}{v_{\text{ball}}} = \frac{13.2 \text{ m}}{26.8 \text{ m/s}} = 0.4925 \text{ s}$$

With this result, we now use the equations involving time to solve for $t_{\text{sound}}$:

$$t_{\text{sound}} = T_{\text{tot}} - t_{\text{ball}} = 0.531 \text{ s} - 0.4925 \text{ s} = 0.0385 \text{ s}$$

Finally, use the constant-velocity equation to solve for $v_{\text{sound}}$:

$$v_{\text{sound}} = \frac{x}{t_{\text{sound}}} = \frac{13.2 \text{ m}}{0.0385 \text{ s}} = 343 \text{ m/s}$$

There are several things to notice about this problem. At first glance it may not appear as a problem that can be solved using the constant-velocity equation, because there is clearly more than one velocity involved. However, by dividing the problem into two segments we were able to apply constant-velocity motion to each. Also, be sure you understand why the same value of $x$ was used for both the ball and the sound. Here $x$ actually represents *distance*, the distance between the athlete and her friend; the *displacements* of the ball and the sound in this problem are of opposite signs. Finally, you'll notice that the answer for $t_{\text{ball}}$ contains four digits instead of three. This is because it is only an intermediate step. As discussed in Chapter 1 of this study guide, you should retain a minimum of one extra digit in intermediate calculations to help avoid excessive round-off error.

---

## Practice Quiz

**2.** Which of the following is the correct SI unit for the product of velocity and time?

**(a)** m **(b)** m/s **(c)** s **(d)** s/m **(e)** m/s$^2$

**3.** In a coordinate system in which east is the positive direction and west is the negative direction, you take a total time of 105 seconds to walk 20 m west, then 10 m east, followed by 15 m west. With what average speed have you walked?

**(a)** 0.24 m/s **(b)** 0.43 m/s **(c)** –0.24 m/s **(d)** –0.43 m/s **(e)** 0 m/s

**4.** For the information given in question 3, with what average velocity have you walked?

**(a)** 0.24 m/s **(b)** 0.43 m/s **(c)** –0.24 m/s **(d)** –0.43 m/s **(e)** 0 m/s

**5.** How long does it take a person on a bicycle to travel exactly 1 km if she rides at a constant velocity of 20 m/s?

**(a)** 20,000 s **(b)** 0.020 s **(c)** 50 s **(d)** 20 s **(e)** 5.0 s

## 2–4 Acceleration

A very important concept in physics is the **acceleration** of an object. Acceleration is the rate at which an object's velocity is changing. This velocity can be changing because the object is slowing down, speeding up, turning around, or any combination thereof. If there is any change in the speed and/or direction of motion of an object, it is accelerating. Sometimes we need only the average rate of change of velocity, or *average acceleration*,

$$a_{\text{av}} = \frac{v_{\text{f}} - v_{\text{i}}}{t_{\text{f}} - t_{\text{i}}} = \frac{\Delta v}{\Delta t}$$

At other times we may need the acceleration at a specific instant in time, or *instantaneous acceleration*, which is defined in terms of the average acceleration measured over an infinitesimally small elapsed time,

$$a = \lim_{\Delta t \to 0} \frac{\Delta v}{\Delta t}$$

As an object accelerates from an initial velocity $v_{\text{i}}$ to a final velocity $v_{\text{f}}$, it may have many different values of instantaneous acceleration along the way. Basically, the average acceleration tells us what constant acceleration would produce the same velocity change $\Delta v$ in the same amount of elapsed time $\Delta t$.

The change in velocity that results from a particular acceleration depends on how the velocity and acceleration relate to each other. In general, when the velocity and acceleration have opposite signs, the object in question will slow down. When the velocity and acceleration have the same sign (whether both are negative or positive), the object will speed up.

**Example 2–5 Leaving a Stop Sign** After stopping at a stop sign, you begin to accelerate with an average acceleration of $-1.4\ \text{m/s}^2$. What is your speed after 2.0 s have passed?

**Picture the Problem** Here we picture your automobile accelerating along the negative *x*-axis.

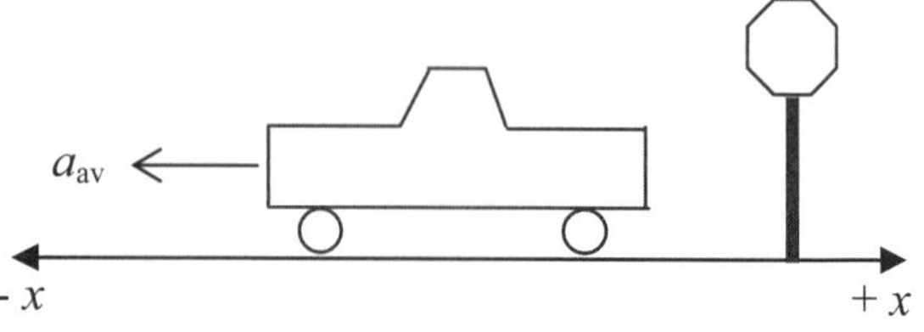


**Strategy** We first take note of the information given in the problem. Because you were at a stop sign, we know your car starts from rest, so that $v_i = 0$. We are also directly given the elapsed time and the average acceleration. Comparing these quantities with the equation for average acceleration, we know everything except $v_f$, which is closely related to the final speed.

**Solution**

**1.** Solve the average acceleration equation for $v_f$: $a_{av} = \dfrac{v_f - v_i}{\Delta t} \Rightarrow v_f = a_{av}\Delta t$

**2.** Insert the values to get the final velocity: $v_f = -1.4\ \text{m/s}^2\,(2.0\ \text{s}) = -2.8\ \text{m/s}$

**3.** Obtain the final speed as the magnitude of the final velocity: $\text{speed} = |v_f| = 2.8\ \text{m/s}$

**Insight** Notice that your car sped up even though it had a negative acceleration; it simply went faster in the negative direction. Also, note that the problem asked for the *speed* after 2.0 s, not the velocity. Always pay careful attention to precisely what is being asked in a problem; it would have been very easy to just stop at $v_f = -2.8$ m/s, but this answer cannot be correct because speed is never negative.

**Example 2–6 Average Acceleration** At a particular instant in time, a particle has a velocity of 6.00 m/s. Over the next 4.00 s it experiences an average acceleration of $-3.00\ \text{m/s}^2$. Determine its velocity at the end of this 4-s time interval.

**Picture the Problem** Our sketch shows the initial situation in which the particle has velocity in the positive direction while accelerating in the negative direction.

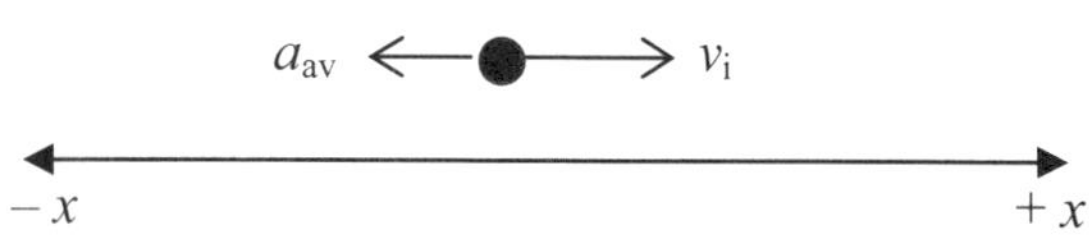


**Strategy** As in the previous example, we need to solve for the final velocity using the equation for average acceleration.

**Solution**

**1.** Solve the average acceleration equation for $v_f$: $v_f = v_i + a_{av}\Delta t$

**2.** Substitute in the values given in the problem: $v_f = 6.00\text{ m/s} + (-3.00\text{ m/s}^2)(4.00\text{ s}) = -6.00\text{ m/s}$

**Insight** Getting the numerical solution to this problem was easy; however, make sure you understand the physics just described. Because the particle's acceleration was opposite its initial velocity, it slowed down until it stopped moving in the positive direction and started moving in the negative direction. Once moving in the negative direction, the particle sped up and, at the time of interest, reached the same speed that it had initially.

## Practice Quiz

**6.** An object is moving east with a speed of 3.0 m/s. If its acceleration is westward, the object is

**(a)** speeding up. **(b)** slowing down. **(c)** maintaining constant speed.

## 2–5 – 2–6 Motion with Constant Acceleration & Applications

An important special case of accelerated motion occurs when the acceleration is constant. Conceptually, this means that the rate at which the velocity changes is the same at every instant in time during the motion. Therefore, an object undergoing constant acceleration will take the same amount of time to increase its velocity from 5 m/s to 10 m/s as it will to increase its velocity from 50 m/s to 55 m/s because in each case it has an equal change in velocity of $\Delta v = 5$ m/s.

As was the case with velocity, when the acceleration of an object is constant, the average acceleration equals the instantaneous acceleration. This means that the definition of average acceleration provides an equation that can be used to describe motion with constant acceleration, $a = \Delta v/\Delta t$. In this context it is customary to simply call the final time $t$ ($t_f = t$) and to define the initial time to be $t_i = 0$; hence,

$\Delta t = t_f - t_i = t$. With this definition, the other initial quantities, $x_i$ and $v_i$, are the values that correspond to $t = 0$ and are labeled accordingly: $v_i \rightarrow v_0$ and $x_i \rightarrow x_0$. With these modifications, we can rewrite the average velocity equation as

$$v = v_0 + at$$

Notice, however, that the preceding equation does not involve position. This suggests that to completely describe motion with constant acceleration we will need more than just this one equation. In fact, we use four equations, relating the different quantities of interest, to mathematically describe this type of motion. The other three equations are

$$x = x_0 + \tfrac{1}{2}(v_0 + v)t; \quad x = x_0 + v_0 t + \tfrac{1}{2}at^2; \text{ and } v^2 = v_0^2 + 2a(x - x_0).$$

These four equations contain all the information needed to describe motion with constant acceleration.

---

**Example 2–7 Training a Sprinter** In 2002, Tim Montgomery set the world record in the 100-m dash with a time of 9.78 s. Coming off the blocks, top sprinters have an acceleration of about 4.00 m/s². If a sprinter could train himself to maintain this acceleration for an entire 100-m dash, what would be his time?

**Picture the Problem** Our sketch shows the situation in which the sprinter has a constant forward acceleration on a 100-m track.



**Strategy** To solve this problem we focus on the three equations that contain the quantity we seek, time. We also note that $v_0 = 0$, and $x_0 = 0$; so we must select the most appropriate equation for the given information.

**Solution**

1. The best equation to use contains $t$ with all other quantities known: $x = \tfrac{1}{2}at^2$

2. Solve for $t$ and insert numerical values: $t = \sqrt{\dfrac{2x}{a}} = \sqrt{\dfrac{2(100\text{ m})}{4.00\text{ m/s}^2}} = 7.07\text{ s}$

**Insight** This solution illustrates a typical approach to problems with constant acceleration. Find an equation containing the quantity you want in which you know all the other quantities. Also, you should examine your results for their physical meaning. Considering the result of this example, do you think it is feasible to train an athlete to have a constant acceleration of 4.00 m/s$^2$ throughout a 100-m dash?

---

**Exercise 2–8 Constant Acceleration** An object moves with a constant acceleration of 1.75 m/s$^2$ and reaches a velocity of 7.80 m/s after 3.20 s. How far does it travel during those 3.20 s?

**Solution** Try to sketch a picture of the problem. The information breakdown in this problem follows:
**Given:** $a = 1.75$ m/s$^2$, $v = 7.80$ m/s, $t = 3.20$ s; **Find:** $\Delta x$

We are free to take $x_0 = 0$ and to seek only the final position $x$. A comparison of the given information with the equations for constant acceleration shows that none of them allows us to immediately solve for $x$ because we are not given $v_0$; hence, we must first solve for the initial velocity. To solve for $v_0$, we start the process all over again. Looking at the equations that contain $v_0$, we can see that the given information allows us to obtain $v_0$ from the equation

$$v = v_0 + at \;\Rightarrow\; v_0 = v - at = 7.80\text{ m/s} - \left(1.75\text{ m/s}^2\right)\left(3.20\text{ s}\right) = 2.20\text{ m/s}$$

Now that we have $v_0$, we can use any of the equations involving $x$ to calculate it.

$$x = \tfrac{1}{2}\left(v_0 + v\right)t = \tfrac{1}{2}\left(2.20\text{ m/s} + 7.80\text{ m/s}\right)\left(3.20\text{ s}\right) = 16.0\text{ m}$$

In this case, the "typical" approach of finding an equation that contains the desired unknown with all other variables known didn't work immediately. We needed an intermediate step because all the equations contain the initial velocity. See the section "Tips" for a quicker approach to this problem.

---

## Practice Quiz

**7.** Starting from rest, an object accelerates at 3.32 m/s$^2$. What will its instantaneous velocity be 3.00 s later?

**(a)** 14.9 m/s **(b)** 1.11 m/s **(c)** 0.904 m/s **(d)** 9.96 m/s

**8.** The driver of a car that is initially moving at 25.0 m/s west applies the brakes until he is going 15.0 m/s west. If the car travels 13.5 m while slowing at a constant rate, what is its acceleration and in what direction is it?

**(a)** 0.675 m/s$^2$, east **(b)** 20.0 m/s$^2$, west **(c)** 14.8 m/s$^2$, east **(d)** 22.2 m/s$^2$, west

**9.** If a car cruises at 11.3 m/s for 75.0 s, then uniformly speeds up until, after 45.0 s, it reaches a speed of 18.5 m/s, what is the car's displacement during this motion?

**(a)** 671 m **(b)** 1.52 km **(c)** 3.58 km **(d)** 848 m **(e)** 0.160 km

## 2–7 Freely Falling Objects

Constant acceleration can be applied to objects falling near Earth's surface. It is an experimental fact that when air resistance is negligible, objects near Earth's surface fall with a constant acceleration $g$ called **the acceleration of gravity.** The symbol $g$ represents the magnitude of this acceleration, which has an average value of

$$g = 9.81 \text{ m/s}^2$$

Objects undergoing this type of motion, when gravity is the only important influence, are said to be in **free fall.** The direction of this acceleration is downward. This direction is commonly taken to be the negative direction along the axis defining the coordinate system. In this case, the acceleration, $a$, is given by $a = -g$, but notice that the value of $g$ is always positive. In some cases, it may be more convenient to choose downward as the positive direction, in which case $a = +g$.

---

**Example 2–9 Learning to Juggle** An entertainer is learning to juggle balls thrown very high. One of the balls is thrown vertically upward from 1.80 m above the ground with an initial velocity of 4.92 m/s. If he fails to catch the ball and it hits the ground, how long is it in the air?

**Picture the Problem** The sketch shows the ball's upward trip on the left and downward trip on the right.

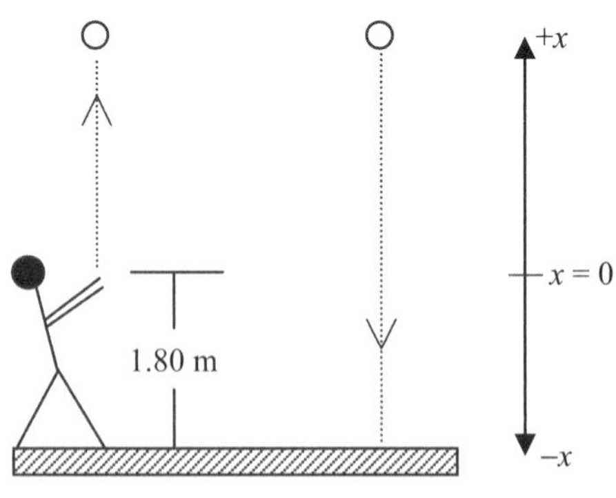


**Strategy** To solve this problem we need an expression that relates time to position, initial velocity, and acceleration. Because of how the coordinate system is chosen, $x_0 = 0$, $a = -g$, and when the ball hits the ground $x = -1.80$ m.

**Solution**

**1.** Choose the equation relating $t$ with $v_0$, $a$, and $x$. Notice that $-g$ has been used for $a$: $x = v_0 t - \frac{1}{2} g t^2$

2. Put the quadratic equation for $t$ in standard form: $\left(-\frac{g}{2}\right)t^2 + v_0 t - x = 0$

3. Apply the quadratic formula to get the solutions: $t = \frac{-v_0 \pm \sqrt{v_0^2 - 2gx}}{-g} = 0.5015\text{ s} \pm 0.7865\text{ s}$

4. The solution that gives a positive time is the correct answer: $t = 0.5015\text{ s} + 0.7865\text{ s} = 1.29\text{ s}$

**Insight** The above solution is only one approach to solving this problem. Another way might be to separately determine the times for the ball to travel upward and downward, then add them (try it). The above approach combines both motions. This can be done because the acceleration is the same going up and coming down. Make sure you can reproduce the numerical values in the solution. Be careful to properly account for all the minus signs. To what does the negative solution correspond?

---

There are a few facts concerning free-fall motion that you can utilize in analyzing situations. These facts can be deduced from the four equations for motion with constant acceleration.

* When an object launched vertically upward reaches the top of its path (its maximum height), its instantaneous velocity is zero, even though its acceleration continues to be 9.81 m/s$^2$ downward.

* An object launched upward from a given height takes an equal amount of time to reach the top of its path as it takes to fall from the top of its path back to the height from which it was launched.

* The velocity an object has at a given height, on its way up, is equal and opposite to the velocity it will have at that same height on its way back down.

---

**Example 2–10 A Child's Toy** A toy rocket is launched vertically upward from the ground. If its initial speed is 8.93 m/s, with what speed will it strike the ground?

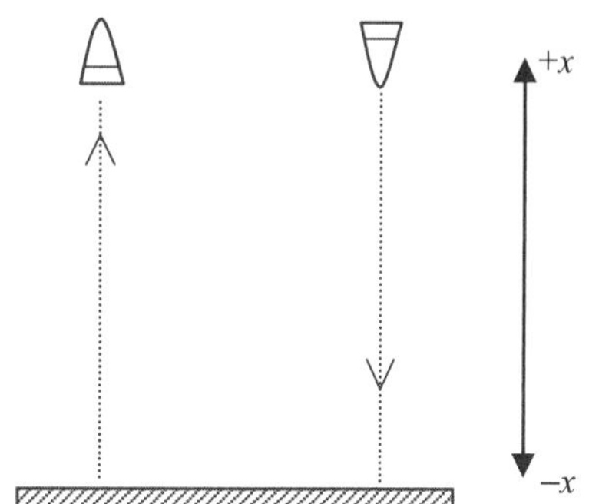


**Picture the Problem** The diagram shows the rocket's upward and downward trips.

**Strategy/Solution** Instead of using equations, we can get the solution by using the symmetry results. Because the rocket is launched upward from the ground with a certain speed, we know that when it comes back to the ground it will have the same speed (in the opposite direction). Therefore, it strikes the ground with a speed of 8.93 m/s.

**Insight** Sometimes we can use symmetry to get the answer with little or no calculation. Convince yourself of the above result by using the equation $v^2 = v_0^2 + 2a(x - x_0)$ to solve the problem.

## Practice Quiz

**10.** A ball is dropped from a height of 2.89 m above the ground. How long does it take to reach the ground?

**(a)** 0.768 s **(b)** 0.589 s **(c)** 9.81 s **(d)** 28.4 s

**11.** A ball is thrown vertically upward from a height of 2.00 m above the ground with a speed of 17.3 m/s. If the ball is caught by the same person at the same height, how long is the ball in the air?

**(a)** 1.76 s **(b)** 4.00 s **(c)** 3.53 s **(d)** 8.65 s

**12.** Can an object that is moving upward be in free fall?

**(a)** No, because an object cannot be falling if it is moving upward

**(b)** Yes, because Earth's gravity sometimes pushes objects upward

**(c)** No, because Earth's gravity never pushes objects upward

**(d)** Yes, as long as gravity is the only force acting on it

**13.** Stone A is thrown upward from the top of a high bridge while stone B is dropped from the same height. Both stones fall and strike the water below. Which stone strikes the water with greater speed?

**(a)** stone A **(b)** stone B **(c)** They strike with equal speeds.

## Graphical Analysis of Motion

The motion of an object is often analyzed graphically. Graphical analysis is useful for many things and can be used to determine what kind of motion is being observed. In order to do that, we must first know what kinds of graphs the different types of motion produces and how to obtain information from them. Specifically, we focus on graphs of position as a function of time, $x$-versus-$t$, and velocity as a function of time, $v$-versus-$t$.

**Position-versus-Time**

In general, plots of position-versus-time will be curved. Regardless of the shape of the curve, however, information about the velocity of the motion can be determined from the graph. Any two points on the graph can be connected by a straight line. The slope of this connecting line equals the average velocity of the object over the corresponding time interval. Also, for any given point on the curve there is a line called the **tangent line** that intersects the curve at that point. The slope of the tangent line at a point equals the instantaneous velocity of the object at the corresponding time.

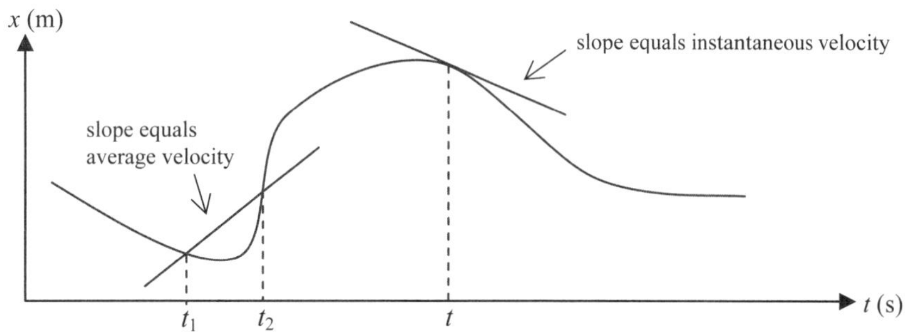


For the special case of constant-velocity motion, the equation $x = vt$ shows that we expect the $x$-versus-$t$ graph to be linear. The slope of the line will give us the velocity of the motion.

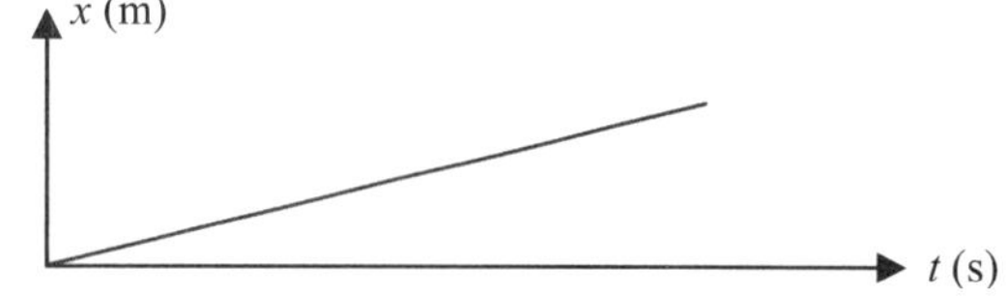


For the special case of constant acceleration, the equation $x = (1/2)at^2$ (with $v_0 = 0$) shows that we expect the $x$-versus-$t$ graph to be a parabola. From this curve we can determine the average and the instantaneous velocities.

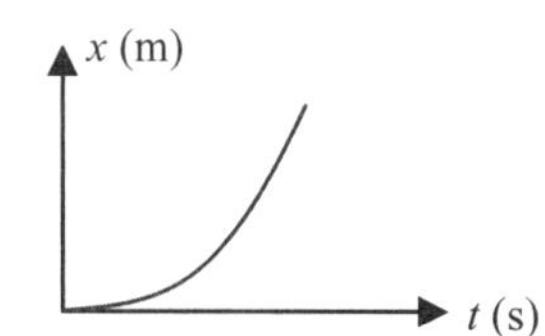


**Velocity-versus-Time**

For motion with constant acceleration, which includes constant-velocity motion ($a = 0$), the graph of velocity as a function of time is linear, as can be seen by the equation $v = v_0 + at$. The slope of the straight line equals the acceleration of the motion.

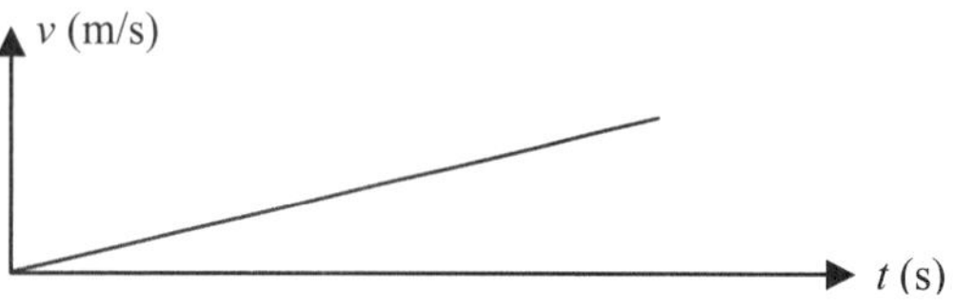

Whether the $v$-versus-$t$ curve is linear or nonlinear ($a \neq \text{constant}$), the distance traveled by an object from one time to another equals the area under the curve between those two times. For the cases of constant velocity and constant acceleration, these areas are rectangles and triangles, respectively.

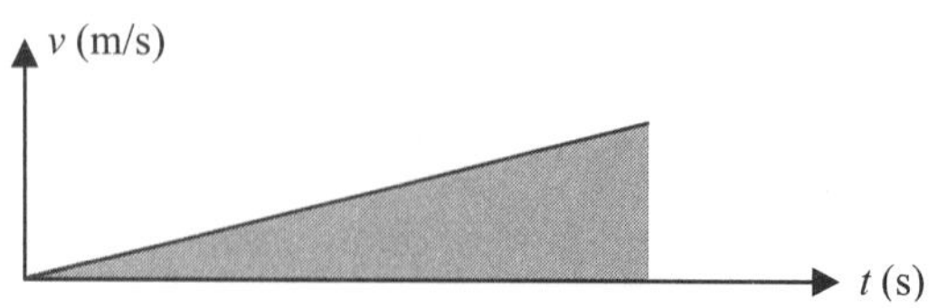


**Example 2–11 Which Type of Motion Is This?** From the following data, determine the type of motion represented. If it is constant-velocity motion, find the velocity. If it is motion with constant acceleration, find the acceleration.

| $t$ (s): | 0 | 0.50 | 1.0 | 1.5 | 2.0 | 2.5 | 3.0 |
|---|---|---|---|---|---|---|---|
| $x$ (m): | 0 | 1.0 | 2.0 | 3.0 | 4.0 | 5.0 | 6.0 |

**Picture the Problem** For the picture, we make a position-versus-time plot of the above data. The data are connected by a dashed line as a visual aid.

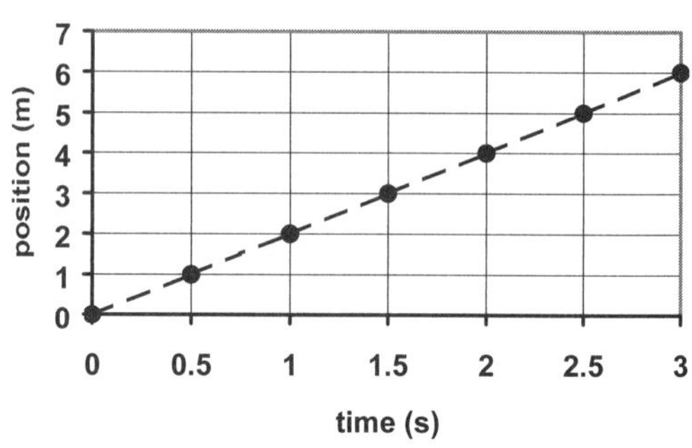


**Strategy/Solution** We examine the features of the plot. It clearly shows a linear relationship so we can conclude that the data are from constant-velocity motion. The velocity of the motion is determined by the slope of the line. To calculate the slope we select two widely spaced points on the line (1.0 m, 0.50 s) and (5.0 m, 2.5 s).

Slope equals rise divided by run:

$$v = \frac{\text{rise}}{\text{run}} = \frac{5.0\text{ m} - 1.0\text{ m}}{2.5\text{ s} - 0.50\text{ s}} = 2.0\text{ m/s}$$

**Insight** Any points along the connecting line would have given the same slope. Real-world data points would not fall so neatly on a single straight line. In such cases you would draw a best-fit line instead of a connecting line. The velocity would then equal the slope of this best-fit line.

**Example 2–12 Determine Distance from a Graph** Use the following *v*-versus-*t* graph below to find the distance traveled by the object whose motion is represented.

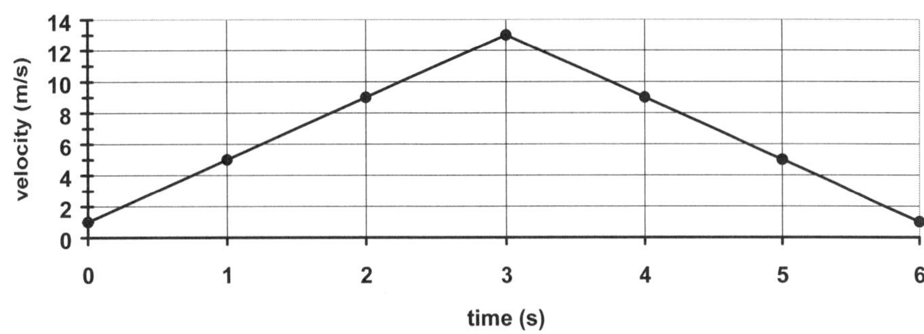


**Picture the Problem** The figure shows the graph of *v*-versus-*t* referred to in the problem.

**Strategy** To get the distance, we need the area under the curve. The diagram shows that this area can be divided into the area of the triangle bounded by the two lines plus the area of the rectangle below it.

**Solution**

| | |
|---|---|
| **1.** Use the formula for the area of a triangle to calculate one term in the sum $x_1$: | area of triangle $= \frac{1}{2} \times \text{base} \times \text{height}$<br>$\therefore \quad x_1 = \frac{1}{2} \times (6.0 \text{ s}) \times (12 \text{ m/s}) = 36 \text{ m}$ |
| **2.** Calculate the area of the rectangle beneath the triangle for the second term in the sum $x_2$: | area of rectangle $= \text{width} \times \text{height}$<br>$\therefore \quad x_2 = (6.0 \text{ s}) \times (1.0 \text{ m/s}) = 6.0 \text{ m}$ |
| **3.** The distance traveled is the sum of the two: | distance $= x_1 + x_2 = 42$ m |

**Insight** Notice that the height of the triangle is 12 m/s, not 13 m/s. This is because the motion of interest both started and ended with a velocity of 1.0 m/s.

## Practice Quiz

**14.** On an *x*-versus-*t* plot, the slope of a tangent line equals

**(a)** average velocity. **(b)** average acceleration.

**(c)** instantaneous velocity. **(d)** instantaneous acceleration.

**15.** If an object's motion is described by a *v*-versus-*t* plot that is parabolic, which kind of motion is it?

**(a)** constant velocity **(b)** constant acceleration **(c)** varying acceleration **(d)** stationary object

**16.** Given the *x*-versus-*t* graph shown on the right,

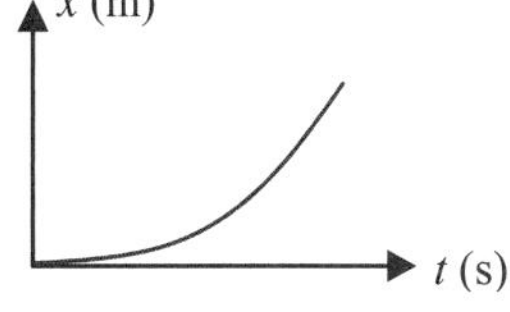

which type of motion is most likely represented?

**(a)** constant velocity motion

**(b)** accelerated motion

**(c)** motionless particle

## Reference Tools and Resources

### I. Key Terms and Phrases

**mechanics** the study of how objects move and the forces that cause motion

**kinematics** the branch of physics that describes motion

**distance** the total length of travel

**displacement** the change in position of an object

**average speed** distance divided by elapsed time

**velocity** the rate of change of displacement with time

**acceleration** the rate of change of velocity with time

**the acceleration of gravity** the acceleration that results from Earth's gravitational pull

**free fall** the motion of an object subject only to the influence of gravity

**tangent line** the straight line that intersects a curve at a point $P$ as the result of a limiting process of secant lines through points surrounding $P$

### II. Important Equations

| Name/Topic | Equation | Explanation |
|---|---|---|
| displacement | $\Delta x = x_f - x_i$ | Displacement is the change in position of an object |
| average velocity | $v_{av} = \dfrac{\Delta x}{\Delta t}$ | Average velocity is the displacement divided by the elapsed time |
| constant velocity motion | $\Delta x = v\Delta t$ | When velocity is constant, the displacement equals velocity times elapsed time. For this case $v = v_{av}$ |
| average acceleration | $a_{av} = \dfrac{\Delta v}{\Delta t}$ | Average acceleration is the velocity change divided by the elapsed time |

| | | |
|---|---|---|
| motion with constant acceleration | $v = v_0 + at$ | velocity changes linearly with time |
| | $x = x_0 + \frac{1}{2}(v_0 + v)t$ | position in terms of average velocity |
| | $x = x_0 + v_0 t + \frac{1}{2}at^2$ | position in terms of acceleration and time |
| | $v^2 = v_0^2 + 2a(x - x_0)$ | velocity squared in terms of displacement |

## III. Know Your Units

| Quantity | Dimension | SI Unit |
|---|---|---|
| displacement, distance ($x$) | [L] | m |
| velocity, speed ($v$) | [L]/[T] | m/s |
| acceleration ($a$) | $[\mathrm{L}]/[\mathrm{T}^2]$ | m/s$^2$ |

## IV. Tips

### Motion with Constant Acceleration

The four equations for constant acceleration form a set in which each equation has a key quantity missing. Taking them in the order listed in the table, "Important Equations," the first equation is missing displacement ($x - x_0$), the second is missing acceleration $a$, the third is missing final velocity $v$, and the fourth is missing time $t$. From this point of view, if an important quantity is unknown, you have an equation that does not require it. However, you may have noticed that each equation contains the initial velocity $v_0$. Show that a fifth equation missing $v_0$ is

$$x = x_0 + vt - \tfrac{1}{2}at^2$$

Use of this equation would have considerably simplified the solution to Exercise 2–8.

### Graphical Analysis

In the section on graphical analysis of accelerated motion, it was pointed out that for constant acceleration the velocity-versus-time plot is linear, and its slope equals the acceleration. Suppose, however, that you only have position-versus-time data. This plot is parabolic; is there any good way to get the acceleration from this set of data? The answer is yes if the motion starts from rest. For this special case of starting from rest, the equation for $x$ as a function of $t$ reduces to

$$x = \tfrac{1}{2}at^2$$

Notice that although $x$ is quadratic in $t$, it is linear in $t^2$. So, if you treat $t^2$ as a single variable and plot $x$-versus-$t^2$, you will get a linear curve. The slope of this line equals half of the acceleration. Just double the slope and you're done.

## Puzzle

### ROUND TRIP

An airplane flying at constant air speed due east from Chicago to Detroit in calm weather (no wind of any kind) would log the same flying time for both legs of the trip. Suppose the same trip is taken when there is wind from the west. How would the total (round trip) time in windy weather compare with the total time in calm weather? Answer this question in words, not equations, briefly explaining how you obtained your answer.

## Answers to Selected Conceptual Questions

**8.** Yes. For example, your friends might have needed to turn around at some point in the trip, which means that they had to have been moving in the opposite direction at some point.

**12.** **(a)** Yes; consider a parked car, for example. **(b)** Yes. An example would be a ball thrown straight upward; at the top of its trajectory its velocity is zero, but it has a nonzero acceleration downward.

## Solutions to Selected End-of-Chapter Problems and Conceptual Exercises

5. **Picture the Problem**: The runner moves along the oval track.

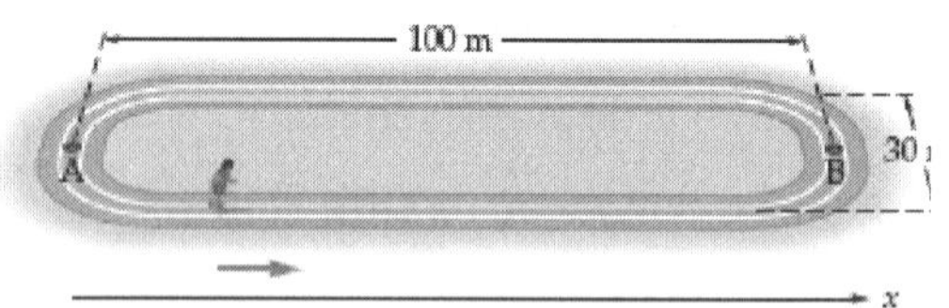


**Strategy:** The distance is the total length of travel, and the displacement is the net change in position.

**Solution: 1. (a)** Add the lengths: $(15\text{ m})+(100\text{ m})+(15\text{ m}) = \boxed{130\text{ m}}$

**2.** Subtract $x_i$ from $x_f$ to find the displacement. $\Delta x = x_f - x_i = 100 - 0\text{ m} = \boxed{100\text{ m}}$

**3. (b)** Add the lengths: $15+100+30+100+15\text{ m} = \boxed{260\text{ m}}$

**4.** Subtract $x_i$ from $x_f$ to find the displacement. $\Delta x = x_f - x_i = 0 - 0\text{ m} = \boxed{0\text{ m}}$

**Insight:** The distance traveled is always positive, but the displacement can be negative. The displacement is always zero for a complete circuit, as in this case.

17. **Picture the Problem**: The finch travels a short distance on the back of the tortoise and a longer distance through the air, with both displacements along the same direction.

**Strategy:** First find the total distance traveled by the finch and then determine the average speed by dividing by the total time elapsed.

**Solution: 1.** Determine the total distance traveled: $d = s_1\Delta t_1 + s_2\Delta t_2$

$$d = \left[(0.060 \text{ m/s})(1.2 \text{ min}) + (12 \text{ m/s})(1.2 \text{ min})\right] \times 60 \text{ s/min}$$

$$d = 870 \text{ m} = 0.87 \text{ km}$$

**2.** Divide the distance by the time elapsed: $s = \dfrac{d}{\Delta t} = \dfrac{870 \text{ m}}{2.4 \text{ min} \times 60 \text{ s/min}} = \boxed{6.0 \text{ m/s}}$

**Insight:** Most of the distance traveled by the finch occurred by air. In fact, if we neglect the 4.3 m the finch traveled while on the tortoise's back, we still get an average speed of 6.0 m/s over the 2.4 min time interval! The bird might as well have been at rest.

30. **Picture the Problem**: The given position function indicates the particle begins traveling in the negative direction but is accelerating in the positive direction.

**Strategy:** Create the *x-versus-t* plot using a spreadsheet, or calculate individual values by hand and sketch the curve using graph paper. Use the known $x$ and $t$ information to determine the average speed and velocity.

**Solution: 1. (a)** Use a spreadsheet to create the plot:

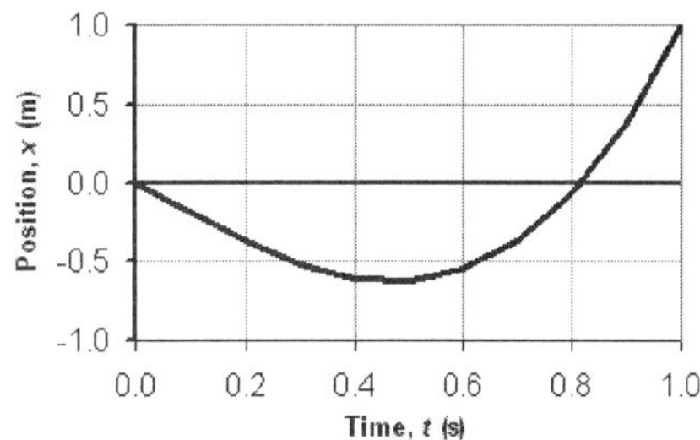


**2. (b)** Find the average velocity from $t = 0.150$ to $t = 0.250$ s:

$$v_{av} = \frac{\Delta x}{\Delta t} = \frac{\left(\left[(-2 \text{ m/s})(0.250 \text{ s}) + (3 \text{ m/s}^3)(0.250 \text{ s})^3\right] - \left[(-2 \text{ m/s})(0.150 \text{ s}) + (3 \text{ m/s}^3)(0.150 \text{ s})^3\right]\right)}{0.250 - 0.150 \text{ s}} = \boxed{-1.63 \text{ m/s}}$$

**3. (c)** Find the average velocity from $t = 0.190$ to $t = 0.210$ s:

$$v_{av} = \frac{\Delta x}{\Delta t} = \frac{\left(\left[(-2 \text{ m/s})(0.210 \text{ s}) + (3 \text{ m/s}^3)(0.210 \text{ s})^3\right] - \left[(-2 \text{ m/s})(0.190 \text{ s}) + (3 \text{ m/s}^3)(0.190 \text{ s})^3\right]\right)}{0.210 - 0.190 \text{ s}} = \boxed{-1.64 \text{ m/s}}$$

**4. (d)** The instantaneous speed at $t = 0.200$ s will be $\boxed{\text{closer to } -1.64 \text{ m/s}}$. As the time interval becomes smaller the average velocity is approaching −1.64 m/s, so we conclude the average speed over an infinitesimally small time interval will be very close to that value.

**Insight:** Note that the instantaneous velocity at 0.200 s is equal to the slope of a straight line drawn tangent to the curve at that point. Because it is difficult to accurately draw a tangent line, we usually resort to mathematical methods like those illustrated above to determine the instantaneous velocity.

48. **Picture the Problem**: The cheetah runs in a straight line with constant positive acceleration.

**Strategy:** The average velocity is simply half the sum of the initial and final velocities because the acceleration is uniform. The distance traveled is the average velocity multiplied by the time elapsed.

**Solution: 1. (a)** Calculate half the sum of the velocities: $v_{av} = \frac{1}{2}(v_0 + v) = \frac{1}{2}(0 + 25.0 \text{ m/s}) = \underline{\underline{12.5 \text{ m/s}}}$

**2.** Use the average velocity to find the distance: $d = v_{av}t = (12.5 \text{ m/s})(6.22 \text{ s}) = \boxed{77.8 \text{ m}}$

**3. (b)** For a constant acceleration the velocity varies linearly with time. Therefore we expect the velocity to be equal to $\boxed{12.5 \text{ m/s}}$ after half the time (3.11 s) has elapsed.

**4. (c)** Calculate half the sum of the velocities: $v_{av,1} = \frac{1}{2}(v_0 + v) = \frac{1}{2}(0 + 12.5 \text{ m/s}) = \boxed{6.25 \text{ m/s}}$

**5.** Calculate half the sum of the velocities: $v_{av,2} = \frac{1}{2}(v_0 + v) = \frac{1}{2}(12.5 + 25.0 \text{ m/s}) = \boxed{18.8 \text{ m/s}}$

**6. (d)** Use the average velocity to find the distance: $d_1 = v_{av,1}\, t = (6.25 \text{ m/s})(3.11 \text{ s}) = \boxed{19.4 \text{ m}}$

**7.** Use the average velocity to find the distance: $d_2 = v_{av,2}\, t = (18.8 \text{ m/s})(3.11 \text{ s}) = \boxed{58.5 \text{ m}}$

**Insight:** The distance traveled is always the average velocity multiplied by the time. This stems from the definition of average velocity.

90. **Picture the Problem**: The camera has an initial downward velocity of 2.0 m/s and accelerates straight downward before striking the ground.

**Strategy:** One way to solve this problem is to use the quadratic formula to find $t$ from the position as a function of time and acceleration equation (equation 2-11). Then the definition of acceleration can be used to find the final velocity. Here's another way: Find the final velocity from the time-free equation of motion (equation 2-12) and use the relationship between average velocity, position, and time (equation 2-10) to find the time. We'll therefore be solving this problem backwards, finding the answer to (b) first and then (a). Let upward be the positive direction, so that $v_0 = -2.0 \text{ m/s}$ and $\Delta x = x - x_0 = 0 - 45 \text{ m} = -45 \text{ m}$.

**Solution: 1. (a)** Solve equation 2-12 for $v$: $v = \sqrt{v_0^2 + 2g\Delta x} = \sqrt{(-2.0 \text{ m/s})^2 + 2(-9.81 \text{ m/s}^2)(-45 \text{ m})} = \underline{\underline{-30 \text{ m/s}}}$

**2.** Solve equation 2-10 for $t$: $t = \dfrac{\Delta x}{\frac{1}{2}(v + v_0)} = \dfrac{-45 \text{ m}}{\frac{1}{2}(-30 - 2.0 \text{ m/s})} = \boxed{2.8 \text{ s}}$

**3. (b)** We found $v$ in step 1: $v = \boxed{-30 \text{ m/s}} = -0.030 \text{ km/s}$

**Insight:** There is often more than one way to approach constant acceleration problems, some easier than others.

## Answers to Practice Quiz

**1.** (d) **2.** (a) **3.** (b) **4.** (c) **5.** (c) **6.** (b) **7.** (d) **8.** (c) **9.** (b) **10.** (a) **11.** (c) **12.** (d) **13.** (a) **14.** (c) **15.** (c) **16.** (b)

# CHAPTER 3
# VECTORS IN PHYSICS

## Chapter Objectives

After studying this chapter, you should

1. know how to represent vectors both graphically and mathematically.
2. know the difference between scalars and vectors.
3. be able to determine the magnitude and direction of a vector.
4. be able to determine the components of a vector.
5. be able to write a vector in unit vector notation.
6. know how to add and subtract vectors both graphically and algebraically.
7. be able to represent position, displacement, velocity, and acceleration as two-dimensional vectors.
8. be able to use velocity vectors to analyze constant-velocity relative motion.

## Warm-Ups

1. Which of the following are vector quantities?
   - tension in a cable
   - weight of a rock
   - volume of a barrel
   - temperature of water in a pool
   - drift of an ocean current
2. Is there a place on Earth where you can walk due south, then due east, and finally due north, and end up at the spot where you started?
3. Is it possible for the magnitude of the vector difference of two vectors to exceed the magnitude of the vector sum?
4. Reversing the algebraic sign on the velocity vector reverses the direction of motion. Is the same statement true for the acceleration vector?

## Chapter Review

In Chapter 2, you studied the one-dimensional forms of several quantities, such as displacement and velocity, that are associated with directions. In this chapter we extend that study to two dimensions. In order to extend that study to two dimensions, we must first study the mathematics needed to do so. That is, we must study vectors.

### 3–1 Scalars versus Vectors

For many quantities used in physics, such as mass and speed, a simple number together with its units suffices to specify the quantity. Such quantities are called **scalar** quantities. Other quantities, however, require a directional specification in addition to a numerical value. Such quantities are called **vector** quantities. The numerical value of a vector quantity is called its **magnitude.** A two-dimensional vector can be represented graphically by an arrow in a coordinate system.

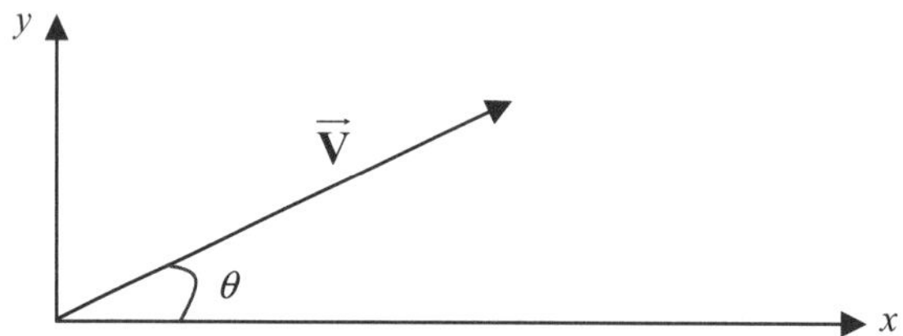


The vector quantity is represented in boldface with an arrow over it, the $\vec{\mathbf{V}}$ in the above figure. Its **direction** is specified by its orientation with respect to the axes of the coordinate system. In the figure, the $x$-axis is used. The magnitude of the vector is represented by the length of the arrow.

### 3–2 The Components of a Vector

Working with vector quantities can often be simplified by resolving them into **components.** In two dimensions, a vector has two components, one corresponding to its extent along the $x$-axis and the other corresponding to its extent along the $y$-axis. In the figure below, the vector $\vec{\mathbf{V}}$ is resolved into two *scalar components*, $V_x$ and $V_y$.

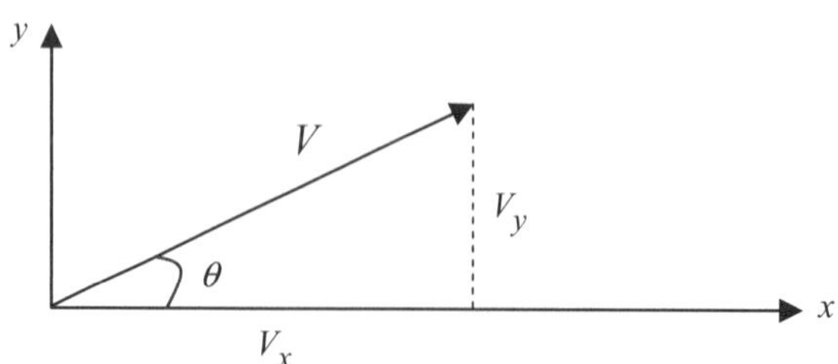

Notice that the magnitude of the vector $V$, (not boldface), and the two components form a right triangle; hence, we can use the trigonometric functions to relate all the relevant quantities

$$V_x = V\cos\theta, \quad V_y = V\sin\theta, \quad \theta = \tan^{-1}\left(\frac{V_y}{V_x}\right)$$

As you can see from these relations, the components can be positive, negative, or zero.

Be careful to note that the third of the preceding equations, for the angle, provides only a *reference angle* for the vector with respect to one of its axes. To know precisely what the direction is, you must also account for the signs of the components or, equivalently, the quadrant of the coordinate system in which the vector lies. This last point will be illustrated in solved examples. If you know the components and want the magnitude, then you can use the Pythagorean theorem

$$V = \sqrt{V_x^2 + V_y^2}$$

---

**Example 3–1 Specifying a Location** You're trying to find State Park, but you're lost. You ask a kind stranger for directions and he tells you that you can get there by first traveling 250 m west, then 310 m north. If you wish to use a vector $\vec{\mathbf{R}}$ to specify this location, **(a)** what are the components of this vector, **(b)** what is its magnitude, and **(c)** what is its direction?

**Picture the Problem** The diagram shows a coordinate system with various directions labeled. The location is indicated by the open circle, and the arrow is the vector we wish to use to specify this location. The angle $\theta$ is used to specify the direction. Try to draw $R_x$ and $R_y$ on this diagram.

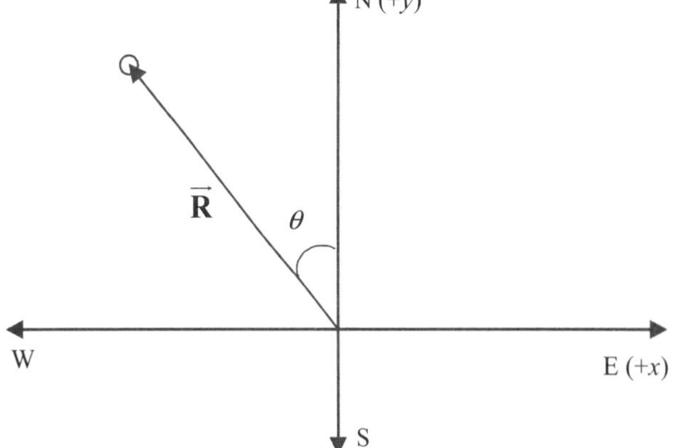


**Strategy** Because the components are given in terms of geographic directions, we need to translate that information to the $x$- and $y$-axes. Once we have the components, we can use our knowledge of right triangles to solve the rest of the problem.

**Solution**

**Part (a):** In our coordinate system, north corresponds to the $+y$-axis, and west to the $-x$-axis.

1. Write the $x$ component: $R_x = -250\text{ m}$
2. Write the $y$ component: $R_y = 310\text{ m}$

**Part (b):** Use the Pythagorean theorem to get the magnitude of $\vec{\mathbf{R}}$:

$$R = \sqrt{R_x^2 + R_y^2} = \sqrt{(-250\text{ m})^2 + (310\text{ m})^2} = 400\text{ m}$$

**Part (c):** To determine $\theta$, use tangent, showing that the direction of $\vec{\mathbf{R}}$ is 39° west of north:

$$\theta = \tan^{-1}\left(\frac{|R_x|}{R_y}\right) = \tan^{-1}\left(\frac{250\text{ m}}{310\text{ m}}\right) = 39^\circ$$

**Insight** There are several points to notice here. First, even though the problem gives only positive values, because of the coordinate system chosen, $R_x$ must be given a negative value or you will arrive at the wrong location. Also, in determining the direction, we did not just naively apply the formula $\tan\theta = R_y/R_x$. You must pay close attention to where the vector lies in the coordinate system. Thus, in part (c) the absolute value of $R_x$ is used to give a positive result for the angle.

---

**Example 3–2 Designing a Garden** When planting your garden you notice that one plant is 0.75 m away from another at approximately 60° above the horizontal. If you want to reproduce this pattern with another pair of plants, how far should you measure horizontally and vertically to determine where to dig the hole?

**Picture the Problem** The diagram shows the garden and two points that mark the locations of the plants. The vector $\vec{\mathbf{R}}$ is the position vector of one plant relative to the other. The dashed lines are the horizontal and vertical components.

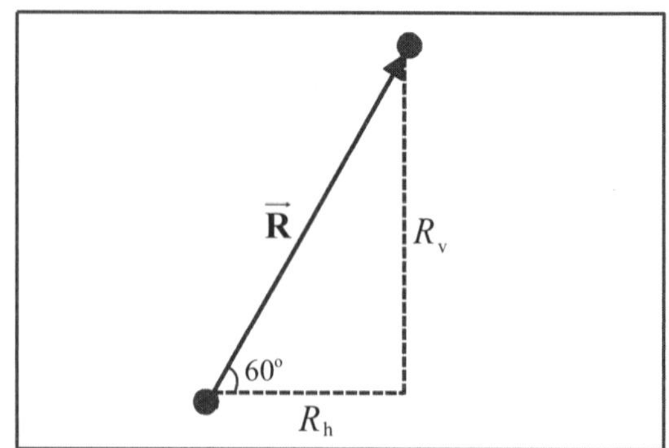


**Strategy** Comparing the information given with the picture, we see that the problem gives us the magnitude and direction of the position vector. Therefore, the distances we seek correspond to the horizontal and vertical components of this vector.

**Solution**

1. Calculate the horizontal distance, which is the horizontal component of the vector $\vec{\mathbf{R}}$: $R_h = R\cos\theta = (0.75\text{ m})\cos(60^\circ) = 0.38\text{ m}$
2. Calculate the corresponding vertical distance: $R_v = R\sin\theta = (0.75\text{ m})\sin(60^\circ) = 0.65\text{ m}$

**Insight** In this Example, unlike in Example 3–1, we started out knowing the magnitude and direction of the vector and used these to determine its components.

## Practice Quiz

1. Taking north to be the $+y$ direction and east to be $+x$, calculate the $x$ and $y$ components $(x, y)$ of a vector whose magnitude is 15 m and is directed 40° south of west.

   **(a)** (11 m, 9.6 m) **(b)** (−9.6 m, 11 m) **(c)** (−13 m, −11 m)
   **(d)** (9.6 m, 13 m) **(e)** ( −11 m , −9.6 m)

2. A certain vector has a $y$ component of 17 in arbitrary units. If its direction is 153° counterclockwise from the $+x$-direction, what is its magnitude?

   **(a)** 44 **(b)** 37 **(c)** 16 **(d)** −44 **(e)** 40

## 3–3 Adding and Subtracting Vectors

There are two approaches to vector addition and subtraction; a graphical method based on geometry, and a component method based on algebra. For precise calculations, you will predominantly use the component method, but the graphical approach can be invaluable for picturing the physical situation being described.

### Adding and Subtracting Vectors Graphically

To picture the graphical addition of two vectors $\vec{\mathbf{A}}$ and $\vec{\mathbf{B}}$ to form a third vector $\vec{\mathbf{C}}$, that is, $\vec{\mathbf{C}} = \vec{\mathbf{A}} + \vec{\mathbf{B}}$, imagine traveling along the two vectors. First you travel along $\vec{\mathbf{A}}$ from its tail to its head, and then you immediately travel along $\vec{\mathbf{B}}$ from its tail to its head. Your net trip, from where you started to where you stopped, will be $\vec{\mathbf{C}}$.

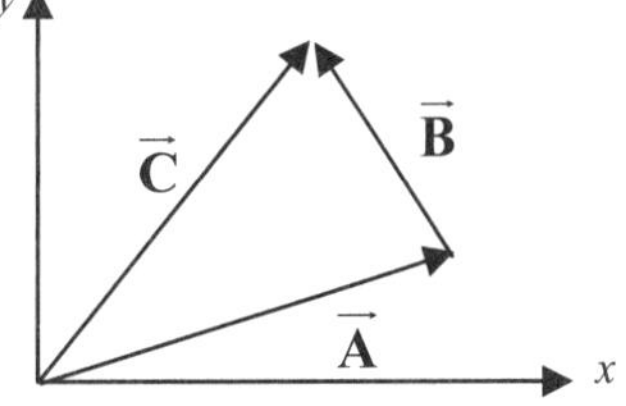

We can summarize this procedure by the following rule:

> *To add two vectors, place the tail of the second vector at the head of the first. The sum of the two is the vector extending from the tail of the first vector to the head of the second.*

The act of moving vector $\vec{\mathbf{B}}$ and placing it at the head of $\vec{\mathbf{A}}$ is acceptable because vectors are characterized only by their length (magnitude) and orientation (direction), not by their location in the coordinate system.

The order in which two vectors are added does not matter. To see this graphically, compare the following figure, which represents the addition $\vec{\mathbf{C}} = \vec{\mathbf{B}} + \vec{\mathbf{A}}$, with the diagram above for $\vec{\mathbf{C}} = \vec{\mathbf{A}} + \vec{\mathbf{B}}$.

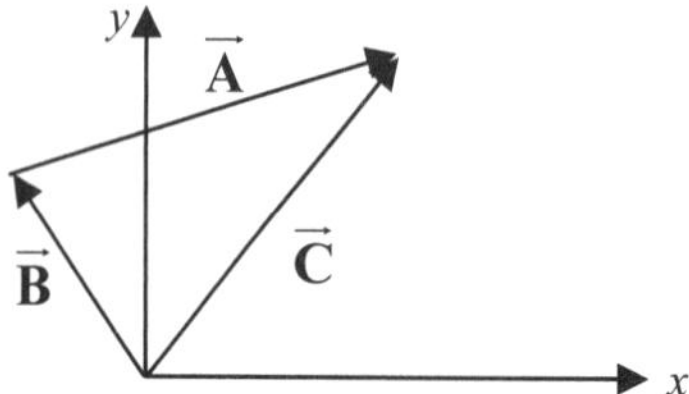


Observe that these two additions produce the same result.

We can approach vector subtraction the same way as addition by recalling that subtraction is the addition of the negative of a quantity; that is, the expression $\vec{\mathbf{D}} = \vec{\mathbf{A}} - \vec{\mathbf{B}}$ is equivalent to the expression $\vec{\mathbf{D}} = \vec{\mathbf{A}} + \left(-\vec{\mathbf{B}}\right)$. Therefore, once we form the vector $-\vec{\mathbf{B}}$ we can proceed with vector addition as already described. The negative of a vector is formed by rotating the vector 180°, so that you get a vector of equal length pointing in the opposite direction as shown below.

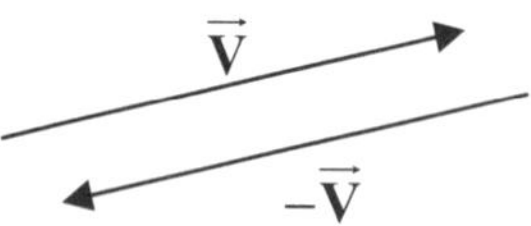


If we consider the same two vectors $\vec{\mathbf{A}}$ and $\vec{\mathbf{B}}$ from the addition example, the difference $\vec{\mathbf{A}} - \vec{\mathbf{B}}$ would be drawn as follows:

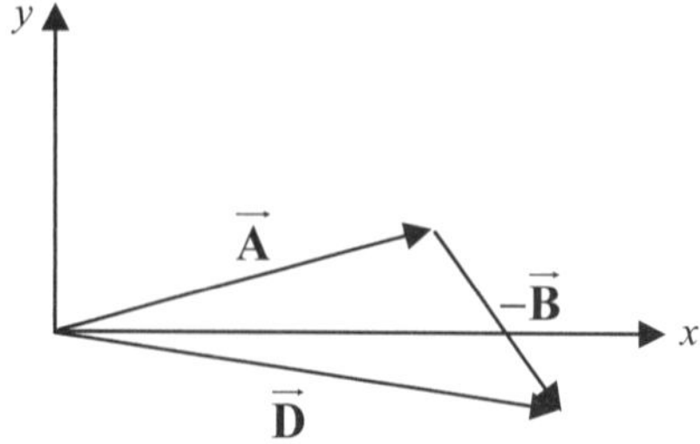

**Adding and Subtracting Vectors Using Components**

The graphical method of vector addition can also tell us how to add vectors using components if we resolve each of the vectors being added into its components. Consider the following figure showing the vector addition $\vec{\mathbf{E}} = \vec{\mathbf{F}} + \vec{\mathbf{G}}$.

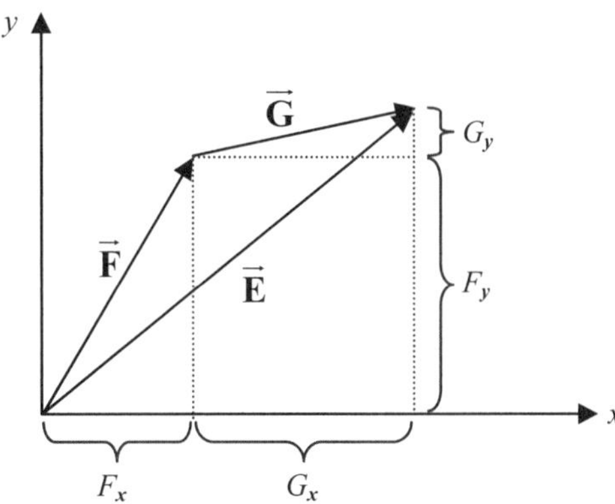


Inspection of this diagram shows that the extent of the vector $\vec{\mathbf{E}}$ along the $x$-axis, that is, $E_x$, is the sum of $F_x$ and $G_x$, and the extent of $\vec{\mathbf{E}}$ along the $y$-axis is the sum of $F_y$ and $G_y$. Thus,

$$E_x = F_x + G_x \text{ and } E_y = F_y + G_y$$

For vector subtraction, we follow the same procedure as we did with the graphical method by using the fact that subtraction is the addition of the negative of a vector. Algebraically, the negative of a vector is obtained by changing the sign of each scalar component. Therefore, if vector $\vec{\mathbf{A}}$ has components $A_x$ and $A_y$, then vector $-\vec{\mathbf{A}}$ will have components $-A_x$ and $-A_y$. Thus, for the vector difference $\vec{\mathbf{H}} = \vec{\mathbf{F}} - \vec{\mathbf{G}}$,

$$H_x = F_x - G_x \text{ and } H_y = F_y - G_y$$

Once the components are determined, the magnitude and direction can be calculated using trigonometry.

---

**Example 3–3 Vector Addition** A vector $\vec{\mathbf{r}}$ has components $r_x = -12.0$ and $r_y = 15.0$. Another vector $\vec{\mathbf{s}}$ has components $s_x = 9.00$ and $s_y = -4.00$. Determine the magnitude and direction of the sum $\vec{\mathbf{r}} + \vec{\mathbf{s}}$.

**Picture the Problem** The sketch shows the graphical addition of $\vec{\mathbf{r}}$ and $\vec{\mathbf{s}}$; the dashed vector is the sum $\vec{\mathbf{r}}+\vec{\mathbf{s}}$, which has been called $\vec{\mathbf{d}}$. The angle $\phi$ is used for the direction of $\vec{\mathbf{d}}$.

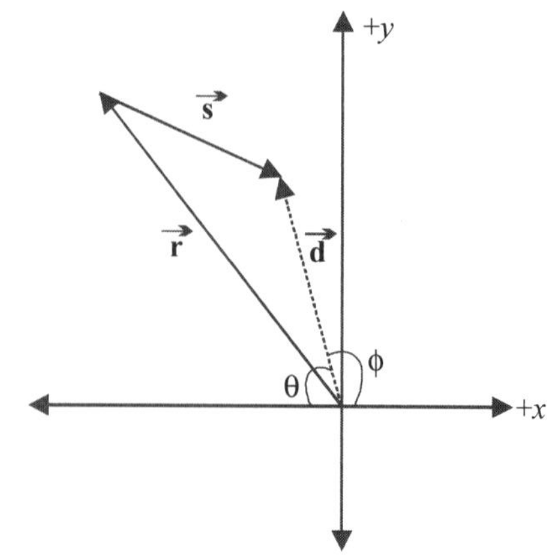


**Strategy** We know how to get the magnitude and direction of $\vec{\mathbf{d}}$ from its components, so we'll first find the components by using vector addition.

**Solution**

1. Obtain the $x$ component of $\vec{\mathbf{d}}$: $d_x = r_x + s_x = -12.0 + 9.00 = -3.00$

2. Obtain the $y$ component of $\vec{\mathbf{d}}$: $d_y = r_y + s_y = 15.0 + (-4.00) = 11.0$

3. Get the magnitude of $\vec{\mathbf{d}}$ from its components: $d = \left(d_x^2 + d_y^2\right)^{1/2} = \left[(-3.00)^2 + (11.0)^2\right]^{1/2} = 11.4$

4. Obtain the reference angle $\theta$ for the direction. This is the angle that $\vec{\mathbf{d}}$ makes with the negative $x$-axis: $\theta = \tan^{-1}\left(\dfrac{d_y}{|d_x|}\right) = \tan^{-1}\left(\dfrac{11.0}{3.00}\right) = 74.7^\circ$

5. Use $\theta$ to determine $\phi$: $\phi = 180^\circ - \theta = 180^\circ - 74.7^\circ = 105^\circ$

**Insight** Notice that the sketch alone tells us that $\vec{\mathbf{d}}$ should have a negative $x$ component and a positive $y$ component before any calculations are done. Always try to check for these kinds of consistencies.

---

**Example 3–4 Vector Subtraction** A vector $\vec{\mathbf{v}}_1$ has a magnitude of 25.0 and makes an angle of $-37.0^\circ$ with the positive $x$ direction. Another vector, $\vec{\mathbf{v}}_2$, has a magnitude of 15.0 and makes an angle of $70.0^\circ$ with the positive $x$ direction. Determine the magnitude and direction of the vector $\vec{\mathbf{v}}_3$ if $\vec{\mathbf{v}}_3 = \vec{\mathbf{v}}_1 - \vec{\mathbf{v}}_2$.

**Picture the Problem** The sketch shows the graphical subtraction of $\vec{\mathbf{v}}_2$ from $\vec{\mathbf{v}}_1$; the dashed vector is $\vec{\mathbf{v}}_3$, and $\phi$ is used for its direction.

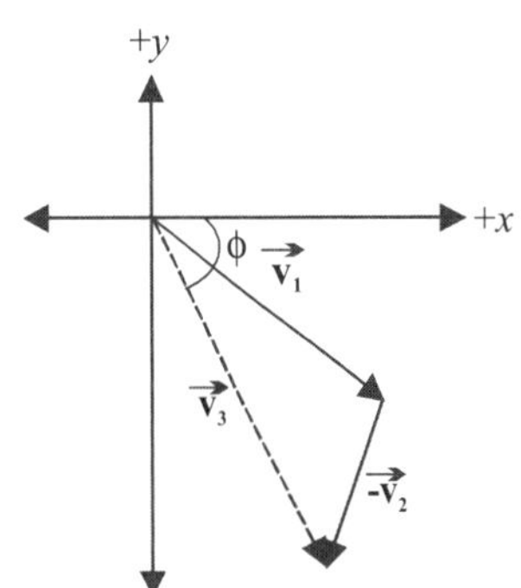


**Strategy** In this case, we know the magnitude and direction of each vector. From these quantities, we can determine the components of $\vec{\mathbf{v}}_1$ and $\vec{\mathbf{v}}_2$ and use these components to get the components of $\vec{\mathbf{v}}_3$. From the components of $\vec{\mathbf{v}}_3$, we can determine its magnitude and direction.

**Solution**

1. Obtain the x component of $\vec{\mathbf{v}}_1$: $V_{1x} = V_1 \cos\theta_1 = (25.0)\cos(-37.0^\circ) = 19.97$

2. Obtain the y component of $\vec{\mathbf{v}}_1$: $V_{1y} = V_1 \sin\theta_1 = (25.0)\sin(-37.0^\circ) = -15.05$

3. Obtain the x component of $\vec{\mathbf{v}}_2$: $V_{2x} = V_2 \cos\theta_2 = (15.0)\cos(70.0^\circ) = 5.13$

4. Obtain the y component of $\vec{\mathbf{v}}_2$: $V_{2y} = V_2 \sin\theta_2 = (15.0)\sin(70.0^\circ) = 14.10$

5. Obtain the x component of $\vec{\mathbf{v}}_3$: $V_{3x} = V_{1x} - V_{2x} = 19.97 - 5.13 = 14.84$

6. Obtain the y component of $\vec{\mathbf{v}}_3$: $V_{3y} = V_{1y} - V_{2y} = -15.05 - 14.10 = -29.15$

7. Get the magnitude of $\vec{\mathbf{v}}_3$ from its components: $V_3 = \left(V_{3x}^2 + V_{3y}^2\right)^{1/2} = \left[(14.84)^2 + (-29.15)^2\right]^{1/2} = 32.7$

8. Obtain the reference angle for the direction of $\vec{\mathbf{v}}_3$. Note that this is the angle $\phi$ that $\vec{\mathbf{v}}_3$ makes with the positive $x$-axis: $\phi = \tan^{-1}\left(\dfrac{V_{3y}}{V_{3x}}\right) = \tan^{-1}\left(\dfrac{-29.15}{14.84}\right) = -63.0^\circ$

**Insight** Notice that the solution to this problem followed a straightforward procedure with which you should become very familiar. In the final calculation of $\phi$, we don't need to use absolute values because this angle gives the final direction of the vector. We then rely on our sketch, or knowledge of the signs of the components, to accurately place the vector in the coordinate system.

## Practice Quiz

3. Which of the following sketches correctly represents the vector addition $\vec{\mathbf{R}}_3 = \vec{\mathbf{R}}_1 + \vec{\mathbf{R}}_2$?

(a) 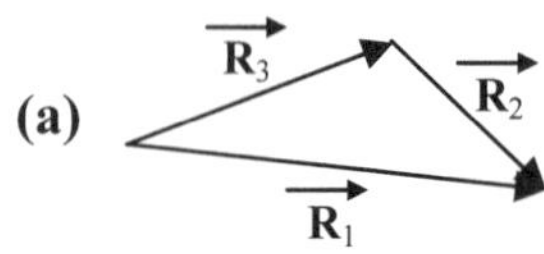


(b) 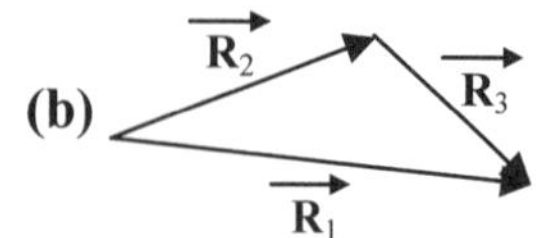


(c) 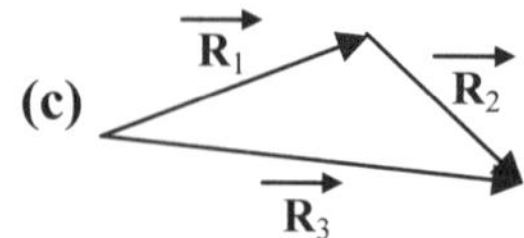


(d) 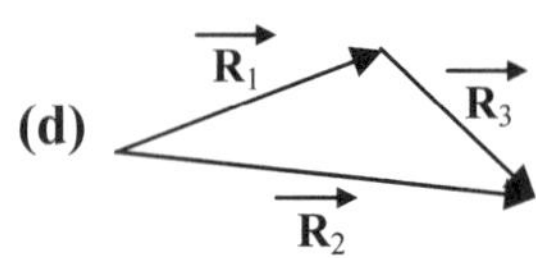


(e) 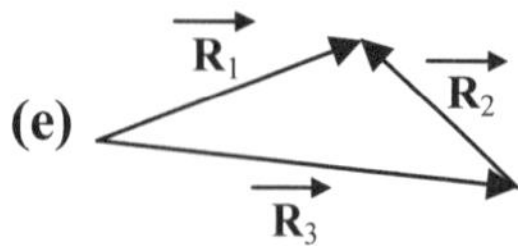


4. Which of the following sketches correctly represents the vector subtraction $\vec{\mathbf{R}}_3 = \vec{\mathbf{R}}_1 - \vec{\mathbf{R}}_2$?

(a) 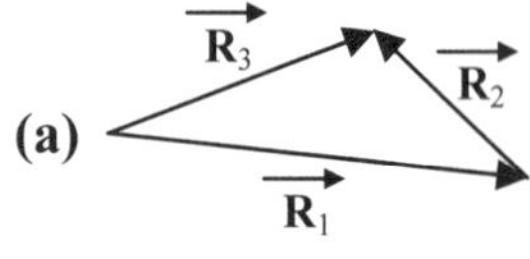


(b) 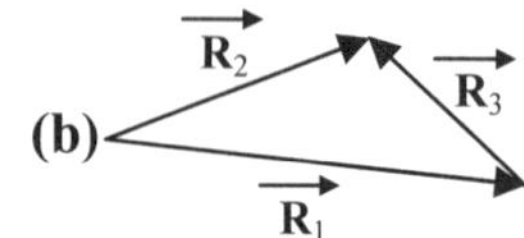


(c) 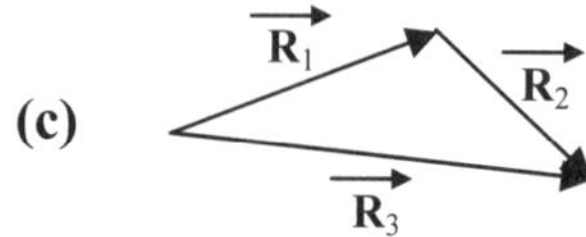


(d) 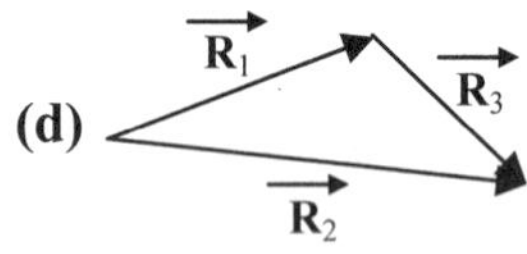


(e) 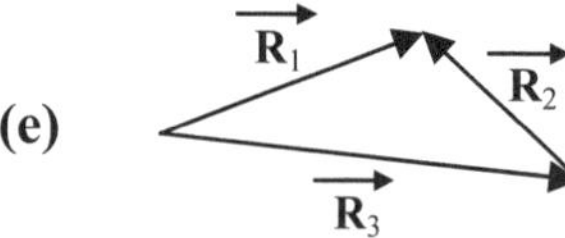


5. If a vector $\vec{\mathbf{Q}}$ has $x$ and $y$ components of –3.0 and 5.5 respectively, and vector $\vec{\mathbf{R}}$ has $x$ and $y$ components of 9.2 and 4.4 respectively, what are the $x$ and $y$ components $(x, y)$ of the sum $\vec{\mathbf{Q}} + \vec{\mathbf{R}}$?

**(a)** (14.7, 1.4) **(b)** (6.2, 9.9) **(c)** (3.7, 7.4) **(d)** (2.5, –4.8) **(e)** (–3.0, 4.4)

6. Given vectors $\vec{\mathbf{Q}}$ and $\vec{\mathbf{R}}$ from question 5, what are the $x$ and $y$ components $(x, y)$ of $\vec{\mathbf{R}} - \vec{\mathbf{Q}}$?

**(a)** (–7.4, –3.7) **(b)** (6.2, 9.9) **(c)** (6.2, 1.1) **(d)** (12.2, –1.1) **(e)** (9.2, –4.4)

7. Suppose a vector $\vec{\mathbf{A}}$ has $x$ and $y$ components of 41 and 28, respectively and vector $\vec{\mathbf{B}}$ has $x$ and $y$ components of 12 and –18, respectively. Determine the magnitude of the vector $\vec{\mathbf{C}}$ if $\vec{\mathbf{C}} = \vec{\mathbf{B}} + \vec{\mathbf{A}}$.

**(a)** 50 **(b)** 22 **(c)** 54 **(d)** 28 **(e)** 72

8. Given the vectors $\vec{\mathbf{A}}$, $\vec{\mathbf{B}}$, and $\vec{\mathbf{C}}$ in question 7, what is the direction of $\vec{\mathbf{C}}$ in a coordinate system in which the $+x$-axis points horizontally to the right and the $+y$-axis points vertically upward?

**(a)** 11° **(b)** 34° **(c)** –22° **(d)** 56° **(e)** 91°

## 3–4 Unit Vectors

A convenient way to completely specify a vector is by using **unit vectors.** A unit vector is a dimensionless vector whose magnitude equals 1. Generally, unit vectors are used to indicate specific directions. Our most common application of unit vectors will be for specifying the *x*- and *y*-directions. A unit vector will be distinguished from other vectors by having a ^ over it. Therefore, the unit vector for the *x* direction is $\hat{\mathbf{x}}$, and the unit vector for the *y* direction is $\hat{\mathbf{y}}$.

When a unit vector is multiplied by a scalar, the result is a vector whose magnitude equals the absolute value of the scalar and whose direction is the same as that of the unit vector if the scalar is positive, or the opposite of the unit vector if the scalar is negative. For example, the vector $\vec{\mathbf{A}} = 5\hat{\mathbf{x}}$ has a magnitude of $A = 5$ and points in the positive *x* direction; and the vector $\vec{\mathbf{B}} = -8\hat{\mathbf{y}}$ has a magnitude of $B = 8$ and points in the negative *y* direction. Using this fact, we can write any arbitrary vector in terms of unit vectors by multiplying its scalar components by the corresponding unit vectors and summing them:

$$\vec{\mathbf{A}} = A_x\hat{\mathbf{x}} + A_y\hat{\mathbf{y}}$$

The quantities $A_x\hat{\mathbf{x}}$ and $A_y\hat{\mathbf{y}}$ are called the *vector components* of $\vec{\mathbf{A}}$.

---

**Exercise 3–5 Unit Vectors** Express the following vectors in unit-vector notation: **(a)** the position vector $\vec{\mathbf{R}}$ from Example 3–1, **(b)** the position vector $\vec{\mathbf{R}}$ from Example 3–2, **(c)** the vector $\vec{\mathbf{d}}$ from Example 3–3, and **(d)** the vector $\vec{\mathbf{v}}_3$ from Example 3–4.

**Solution** Following the preceding discussion we construct the vectors by multiplying the scalar components determined in the relevant examples by the appropriate unit vectors and summing them.

**(a)** Since in Example 3–1 we determined that $R_x = -250$ m, and $R_y = 310$ m, we can write for vector $\vec{\mathbf{R}}$

$$\vec{\mathbf{R}} = -(250\text{ m})\hat{\mathbf{x}} + (310\text{ m})\hat{\mathbf{y}}$$

**(b)** Since in Example 3–2 we determined that $R_x = 0.38$ m and $R_y = 0.65$ m, we can write for vector $\vec{\mathbf{R}}$

$$\vec{\mathbf{R}} = (0.38\text{ m})\hat{\mathbf{x}} + (0.65\text{ m})\hat{\mathbf{y}}$$

**(c)** In Example 3–3, we found $d_x = -3.00$ and $d_y = 11.0$, so that

$$\vec{\mathbf{d}} = -3.00\hat{\mathbf{x}} + 11.0\hat{\mathbf{y}}$$

**(d)** In Example 3–4, we calculated $v_{3x} = 14.84$ and $v_{3y} = -29.15$, which gives

$$\vec{\mathbf{v}}_3 = 14.84\,\hat{\mathbf{x}} - 29.15\,\hat{\mathbf{y}}$$

---

## 3–5 Position, Displacement, Velocity, and Acceleration Vectors

If you recall the discussions of position, displacement, velocity, and acceleration from Chapter 2, you may remember that associated with each of these quantities was a size (magnitude) and a direction (positive or negative). Therefore, these are all vector quantities and in more than one dimension we wish to represent these quantities in the more general vector notation discussed previously. The SI units of all these quantities are given in Chapter 2.

The position vector $\vec{\mathbf{r}}$ for an object in a coordinate system is the arrow from the origin of the coordinate system to the location of the object. The magnitude of this vector is the distance of the object from the origin. The $x$ and $y$ scalar components of $\vec{\mathbf{r}}$ are the $x$ and $y$ coordinates of the object. Therefore, in unit vector notation, the two-dimensional position vector is written as

$$\vec{\mathbf{r}} = x\hat{\mathbf{x}} + y\hat{\mathbf{y}}$$

Examples 3–1 and 3–2 involved two-dimensional position vectors.

The displacement of an object is its change in position, so the displacement vector $\Delta\vec{\mathbf{r}}$ is the difference between its final and initial positions, $\Delta\vec{\mathbf{r}} = \vec{\mathbf{r}}_f - \vec{\mathbf{r}}_i$. Given what we learned in Section 3–3 about vector addition and subtraction, the unit vector notation for displacement in two dimensions is

$$\Delta\vec{\mathbf{r}} = (x_f - x_i)\hat{\mathbf{x}} + (y_f - y_i)\hat{\mathbf{y}}$$

The average velocity is the displacement divided by the elapsed time $\Delta t$. Dividing the vector $\Delta\vec{\mathbf{r}}$ by the scalar $\Delta t$, we divide each component of $\Delta\vec{\mathbf{r}}$ by $\Delta t$; hence,

$$\vec{\mathbf{v}}_{\text{av}} = \frac{\Delta\vec{\mathbf{r}}}{\Delta t} = \frac{\Delta x}{\Delta t}\hat{\mathbf{x}} + \frac{\Delta y}{\Delta t}\hat{\mathbf{y}}$$

where $\Delta x = x_f - x_i$ and $\Delta y = y_f - y_i$. The instantaneous velocity, once again, is given by the limit of the average velocity as the time interval approaches zero:

$$\vec{\mathbf{v}} = \lim_{\Delta t \to 0} \frac{\Delta\vec{\mathbf{r}}}{\Delta t}$$

The average acceleration is the change in velocity, $\Delta\vec{\mathbf{v}}$, divided by the elapsed time, $\Delta t$,

$$\vec{\mathbf{a}}_{\text{av}} = \frac{\Delta\vec{\mathbf{v}}}{\Delta t}$$

The change in velocity is the difference between the final and initial velocities, $\Delta\vec{\mathbf{v}} = \vec{\mathbf{v}}_f - \vec{\mathbf{v}}_i$. The instantaneous acceleration is the limit of the average acceleration as the time interval approaches zero:

$$\vec{\mathbf{a}} = \lim_{\Delta t \to 0} \frac{\Delta\vec{\mathbf{v}}}{\Delta t}$$

**Example 3–6 Out of Gas** Suppose you are driving northwest at 13.5 m/s for exactly half an hour when you run out of gas. Frustrated, you walk for 40.0 min to the nearest gas station, which is 2.40 km away at 70° north of west from where your car stopped. Determine your average velocity for this trip. State the answer both as a magnitude and a direction and using unit vectors.

**Picture the Problem** The diagram shows the three relevant displacement vectors for the trip. The vector $\vec{\mathbf{d}}_c$ is your displacement while driving the car, $\vec{\mathbf{d}}_w$ is your displacement while walking, and $\vec{\mathbf{r}}$ is your total displacement.

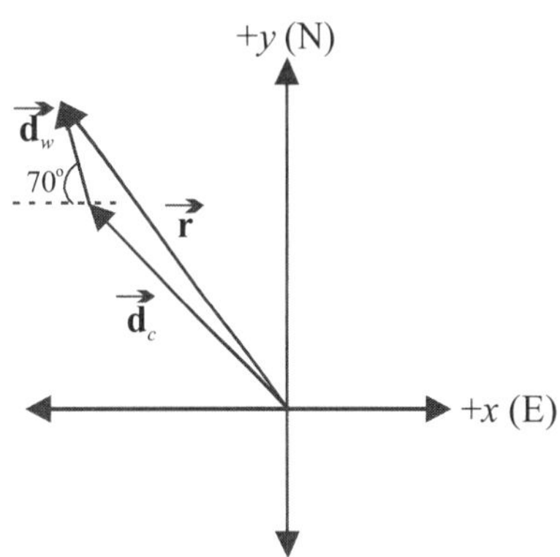


**Strategy** Because average velocity is displacement divided by elapsed time, we will first calculate the displacement vectors and then determine the average velocity.

**Solution**

**Step 1:** Determine the displacement $\vec{\mathbf{d}}_c$.

1. Calculate the magnitude of $\vec{\mathbf{d}}_c$ from the speed and time while in the car:

$$d_c = v_c t_c = (13.5 \text{ m/s})(30 \text{ min})\left(\frac{60 \text{ s}}{\text{min}}\right) = 24{,}300 \text{ m}$$

2. Going northwest implies a direction for $\vec{\mathbf{d}}_c$ of:

$$\phi_c = 90^\circ + 45^\circ = 135^\circ$$

3. Determine the $x$ and $y$ components of $\vec{\mathbf{d}}_c$:

$$d_{cx} = d_c \cos\phi_c = (24{,}300 \text{ m})\cos(135^\circ) = -17{,}183 \text{ m}$$
$$d_{cy} = d_c \sin\phi_c = (24{,}300 \text{ m})\sin(135^\circ) = 17{,}183 \text{ m}$$

**Step 2:** Determine the displacement $\vec{\mathbf{d}}_w$.

4. 30° north of west implies a direction for $\vec{\mathbf{d}}_w$ of:

$$\phi_w = 180^\circ - 70^\circ = 110^\circ$$

5. Determine the $x$ and $y$ components of $\vec{\mathbf{d}}_w$:

$$d_{wx} = d_w \cos\phi_w = (2400 \text{ m})\cos(110^\circ) = -820.8 \text{ m}$$
$$d_{wy} = d_w \sin\phi_w = (2400 \text{ m})\sin(110^\circ) = 2{,}255 \text{ m}$$

**Step 3:** The total displacement $\vec{\mathbf{r}}$ is $\vec{\mathbf{d}}_c + \vec{\mathbf{d}}_w$

6. Determine the $x$ and $y$ components of $\vec{\mathbf{r}}$:

$$r_x = d_{cx} + d_{wx} = -17{,}183 \text{ m} - 820.8 \text{ m} = -18{,}004 \text{ m}$$
$$r_y = d_{cy} + d_{wy} = 17{,}183 \text{ m} + 2{,}255 \text{ m} = 19{,}438 \text{ m}$$

**7**. Determine the total elapsed time, *t*: $t = t_c + t_w = 30\text{ min} + 40\text{ min} = 70\text{ min}\left(\dfrac{60\text{ s}}{\text{min}}\right)$

$= 4200\text{ s}$

**8**. Determine the *x* and *y* components of $\vec{\mathbf{v}}_{\text{av}}$:

$$v_{\text{av},x} = \frac{r_x}{t} = \frac{-18{,}004\text{ m}}{4200\text{ s}} = -4.287\text{ m/s}$$

$$v_{\text{av},y} = \frac{r_y}{t} = \frac{19{,}438\text{ m}}{4200\text{ s}} = 4.628\text{ m/s}$$

**9**. State the average velocity using unit vectors: $\vec{\mathbf{v}}_{\text{av}} = -(4.29\text{ m/s})\hat{\mathbf{x}} + (4.63\text{ m/s})\hat{\mathbf{y}}$

**10**. Find the magnitude of $\vec{\mathbf{v}}_{\text{av}}$:

$$v_{\text{av}} = \left(v_{\text{av},x}^2 + v_{\text{av},y}^2\right)^{1/2}$$

$$= \left[(-4.287\text{ m/s})^2 + (4.628\text{ m/s})^2\right]^{1/2} = 6.31\text{ m/s}$$

**11**. Calculate a reference angle $\theta$ for $\vec{\mathbf{v}}_{\text{av}}$:

$$\theta = \tan^{-1}\left(\frac{v_{av,y}}{v_{av,x}}\right) = \tan^{-1}\left(\frac{4.628\text{ m/s}}{-4.287\text{ m/s}}\right) = -47.2^\circ$$

**12**. From the signs of the components we can see that $\theta$ is measured above the $-x$-axis; therefore, the magnitude and direction of $\vec{\mathbf{v}}_{\text{av}}$ is: 6.31 m/s at 47.2° north of west.

**Insight** In the final calculation of $\theta$, the negative signs come strictly as a result of the negative *x* component. Without any further statement it would be assumed that the angle is measured clockwise from the $+x$-axis. This potential mistake is why the interpretation in the final step is needed.

---

**Exercise 3–7 Turn Left at the Light** Suppose that you are driving at 35 mi/h down Main Street while approaching a green light at the intersection with State Street. In order to turn left onto State Street you slow down, make the turn, then speed up to 35 mi/h on State Street. If it takes you a total time of 1.75 minutes from the time you begin to slow down on Main Street until you reach 35 mi/h on State Street, what is the magnitude and direction of your average acceleration?

**Solution** Try to sketch the velocity and average acceleration vectors for this problem. Let us choose the direction of the initial velocity on Main Street to be the $+x$ direction, and the direction of the final velocity on State Street to be the $+y$ direction.

**Given:** $v_i = 35$ mi/h, $\theta_i = 0.0°$, $v_f = 35$ mi/h, $\theta_f = 90°$, $t = 1.75$ min **Find:** $a_{\text{av}}$, $\theta_{\text{av}}$

Since the average acceleration is determined by the change in velocity, $\Delta\vec{\mathbf{v}}$, let us first calculate $\Delta v$ by getting the components of the initial and final velocities. For the initial velocity we have

$$v_{ix} = v_i \cos\theta_i = 35 \text{ mi/h} \cos(0.0^\circ) = 35 \text{ mi/h}\left(\frac{0.447 \text{ m/s}}{1 \text{ mi/h}}\right) = 15.6 \text{ m/s}$$

$$v_{iy} = v_i \sin\theta_i = 35 \text{ mi/h} \sin(0.0^\circ) = 0 \text{ m/s}$$

For the final velocity we have

$$v_{fx} = v_f \cos\theta_f = 35 \text{ mi/h} \cos(90^\circ) = 0 \text{ mi/h} = 0 \text{ m/s}$$

$$v_{fy} = v_f \sin\theta_f = 35 \text{ mi/h} \sin(90^\circ) = 35 \text{ mi/h} = 15.6 \text{ m/s}$$

Now that we have the components, we can determine the components of $\Delta\vec{\mathbf{v}}$.

$$\Delta v_x = v_{fx} - v_{ix} = 0 \text{ m/s} - 15.6 \text{ m/s} = -15.6 \text{ m/s}$$

$$\Delta v_y = v_{fy} - v_{iy} = 15.6 \text{ m/s} - 0 \text{ m/s} = 15.6 \text{ m/s}$$

From these components we can determine the magnitude:

$$\Delta v = \left[\Delta v_x^2 + \Delta v_y^2\right]^{1/2} = \left[(-15.6 \text{ m/s})^2 + (15.6 \text{ m/s})^2\right]^{1/2} = 22.06 \text{ m/s}$$

The total elapsed time is $1.75 \times 60$ s = 105 s, so the magnitude of the average acceleration is

$$a_{\text{av}} = \frac{\Delta v}{t} = \frac{22.06 \text{ m/s}}{105 \text{ s}} = 0.21 \text{ m/s}^2$$

To determine the direction, notice that the $a_{\text{av},x}$ will be negative, because $\Delta v_x$ is negative, and $a_{\text{av},y}$ will be positive for a similar reason. These facts mean that $\vec{\mathbf{a}}_{\text{av}}$ lies in the second quadrant of the coordinate system. (If you couldn't sketch $\vec{\mathbf{a}}_{\text{av}}$ before, then do so now.) Also notice that because the $x$ and $y$ components of $\Delta\vec{\mathbf{v}}$ have the same size, 15.6 m/s, then $\Delta\vec{\mathbf{v}}$, and therefore $\vec{\mathbf{a}}_{\text{av}}$ must make a 45 degree angle in that quadrant. Finally, we can say that the average acceleration is 0.21 m/s$^2$ at 135 degrees from the positive $x$-axis.

Notice that even though the magnitudes of the initial and final velocities are the same, the acceleration is not zero. This is because the directions of the velocities are different, and this difference is just as important for determining acceleration as a difference in speed.

## Practice Quiz

**9.** In unit-vector notation, determine the displacement of an object that goes from $(x, y)$ coordinates of $(-3.6, 2.1)$ to $(4.2, -9.8)$ in arbitrary units.

**(a)** $0.6\,\hat{\mathbf{x}} - 7.7\,\hat{\mathbf{y}}$ **(b)** $-0.6\,\hat{\mathbf{x}} + 7.7\,\hat{\mathbf{y}}$ **(c)** $-7.8\,\hat{\mathbf{x}} + 11.9\,\hat{\mathbf{y}}$ **(d)** $7.8\,\hat{\mathbf{x}} - 11.9\,\hat{\mathbf{y}}$ **(e)** $-0.6\,\hat{\mathbf{x}} - 7.7\,\hat{\mathbf{y}}$

**10**. Which of the following (in m/s) gives the correct average velocity of a particle that moves from position $\vec{\mathbf{r}}_1 = 2.0\hat{\mathbf{x}} + 5.0\hat{\mathbf{y}}$ to position $\vec{\mathbf{r}}_2 = 8.0\hat{\mathbf{x}} - 3.0\hat{\mathbf{y}}$, both in meters, in 5.0 s?

**(a)** $1.2\,\hat{\mathbf{x}} - 1.6\,\hat{\mathbf{y}}$ **(b)** $6.0\,\hat{\mathbf{x}} - 8.0\,\hat{\mathbf{y}}$ **(c)** $2.0\,\hat{\mathbf{x}} + 0.4\,\hat{\mathbf{y}}$ **(d)** $30\,\hat{\mathbf{x}} - 40\,\hat{\mathbf{y}}$ **(e)** $1.6\,\hat{\mathbf{x}} - 0.6\,\hat{\mathbf{y}}$

**11**. An object has an average acceleration of $\vec{\mathbf{a}}_{\text{av}} = 8.21\hat{\mathbf{x}} + 1.71\hat{\mathbf{y}}$ in m/s$^2$ after accelerating for 6.75 s. What was its change in velocity? (Answers are in m/s.)

**(a)** $187\,\hat{\mathbf{x}} + 39.0\,\hat{\mathbf{y}}$ **(b)** $1.22\,\hat{\mathbf{x}} + 0.253\,\hat{\mathbf{y}}$ **(c)** $55.4\,\hat{\mathbf{x}} + 11.5\,\hat{\mathbf{y}}$ **(d)** $4.11\,\hat{\mathbf{x}} - 0.855\,\hat{\mathbf{y}}$ **(e)** 0

## 3–6 Relative Motion

A relevant application of vector addition and subtraction is relative motion. Specifically, we will focus on the velocity of an object as measured by two observers who are moving with respect to each other at a constant relative velocity. To identify who is measuring what, we use a system of subscripts in which the first subscript refers to the object whose velocity is being measured, and the second subscript refers to the observer (or coordinate system) making the measurement. [Note that the observer can be fictitious; we really only need the coordinate system with respect to which the velocity has the value in question.] For example, a velocity labeled $\vec{\mathbf{v}}_{12}$ refers to the velocity *of* object 1 *relative to* (or as measured by) object 2. If you reverse the subscripts to get the velocity of 2 relative to 1, this velocity $\vec{\mathbf{v}}_{21}$ relates to $\vec{\mathbf{v}}_{12}$ by a minus sign: $\vec{\mathbf{v}}_{21} = -\,\vec{\mathbf{v}}_{12}$.

To state how relative velocity motion is handled in more general terms, consider the accompanying diagram, which shows two coordinate systems, *A* and *B*, and a point *P* that is in motion relative to each coordinate system. Observers in system *A* identify the velocity of *P* as $\vec{\mathbf{v}}_{PA}$, whereas observers in system *B* identify *P*'s velocity as $\vec{\mathbf{v}}_{PB}$. Let us assume that system *B* moves relative to system *A* with a velocity $\vec{\mathbf{v}}_{BA}$. The relationship between these velocities can be written as

$$\vec{\mathbf{v}}_{PA} = \vec{\mathbf{v}}_{PB} + \vec{\mathbf{v}}_{BA}$$

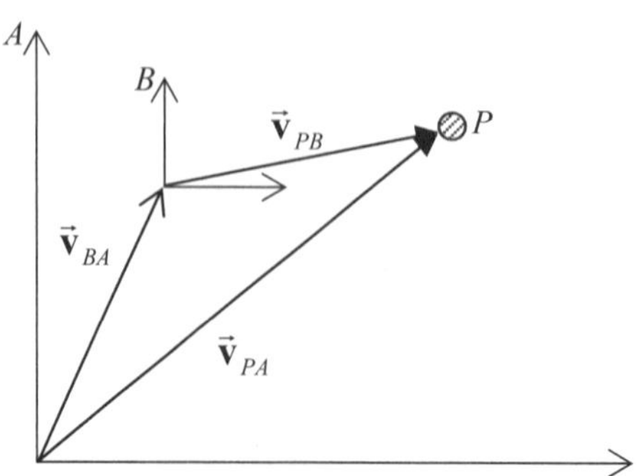

It is very helpful to notice how the subscript convention works in this velocity addition equation. The resultant velocity $\vec{\mathbf{v}}_{PA}$ has $P$ as the leftmost subscript and $A$ as the rightmost subscript. Now, the subscripts on the two velocities whose addition produces the resultant are ordered $PBBA$. In this ordering, $P$ is the leftmost subscript and $A$ is the rightmost, just as with the resultant. The subscript for the system not referenced by the resultant ($B$ in this case) is repeated in the middle. This mnemonic can be used to help analyze almost any relative velocity situation as long as you stay consistent with how you label the velocities. You should be careful, however, not to fall into the trap of using this mnemonic as a substitute for understanding the physical situation being described.

---

**Example 3–8 Crossing the Street** Tom and Jan are standing at the side of the street waiting to cross. A car is coming at 10.0 m/s. Jan decides to go ahead and run directly across the street at 2.80 m/s, whereas Tom chooses to wait. While Jan is running across, what is her velocity as measured by the driver of the car?

**Picture the Problem** The diagram indicates the motion of both Jan and the car. The velocity vectors are labeled such that 'g' means ground.

car $\vec{\mathbf{v}}_{\text{cg}}$ $\vec{\mathbf{v}}_{\text{Jg}}$ Jan Tom $y$ $x$

**Strategy** We need to determine Jan's velocity with respect to the car. In our system of labels this velocity is $\vec{\mathbf{v}}_{\text{Jc}}$.

**Solution**

1. The vector addition we need is: $\vec{\mathbf{v}}_{\text{Jc}} = \vec{\mathbf{v}}_{\text{Jg}} + \vec{\mathbf{v}}_{\text{gc}}$

2. Write out $\vec{\mathbf{v}}_{\text{Jg}}$ based on the given information: $\vec{\mathbf{v}}_{\text{Jg}} = 2.80 \text{ m/s } \hat{\mathbf{y}}$

3. Write out $\vec{\mathbf{v}}_{\text{gc}}$ based on the given information: $\vec{\mathbf{v}}_{\text{gc}} = -\vec{\mathbf{v}}_{\text{cg}} = -10.0 \text{ m/s } \hat{\mathbf{x}}$

4. Perform the vector addition: $\vec{\mathbf{v}}_{\text{Jc}} = -(10.0 \text{ m/s})\,\hat{\mathbf{x}} + (2.80 \text{ m/s})\,\hat{\mathbf{y}}$

**Insight** The mnemonic outlined previously made it straightforward to identify the correct vector addition to solve this problem. However, make sure that the final answer for $\vec{\mathbf{v}}_{\text{Jc}}$ makes intuitive sense to you as well. Why should its $x$ component be negative?

---

**Example 3–9 Holding an Umbrella** You are walking on campus at 1.3 m/s directly against a 20 mi/h, horizontal wind in a light snowstorm. If the snowflakes fall vertically downward with a speed of 3.9 m/s with respect to still air, at what angle, with respect to the vertical, should you hold your umbrella to best protect yourself from the snow?

**Picture the Problem** The sketch shows you, relative to the ground, walking in the storm holding the umbrella at an angle. The motion of the snow relative to the ground is also indicated by the lines about to strike the umbrella.

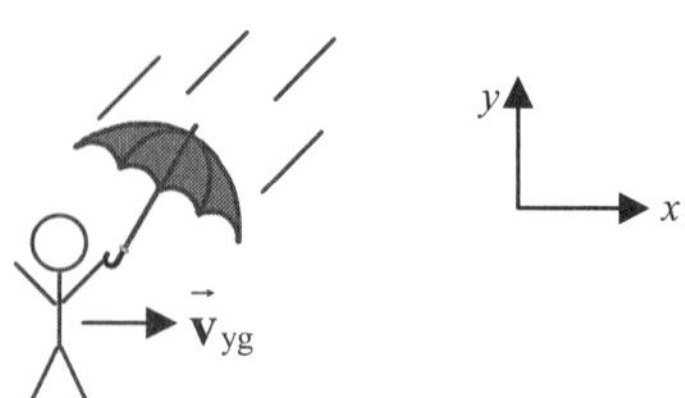


**Strategy** The umbrella protects best if it is oriented so that the snow strikes the greatest amount of its surface. For this to happen, the shaft of the umbrella should be oriented along the line of motion of the snow (as viewed by you). Therefore, the question will be answered if we determine the angle that the velocity of the snow makes with the vertical as measured by you, so we need $\vec{\mathbf{v}}_{sy}$.

**Solution**

The various reference frames here are *you* (y), the *ground* (g), the *snow* (s), and the *air* (a).

1. The vector addition that we need is: $\vec{\mathbf{v}}_{sy} = \vec{\mathbf{v}}_{sg} + \vec{\mathbf{v}}_{gy}$

2. We can immediately write down $\vec{\mathbf{v}}_{gy}$: $\vec{\mathbf{v}}_{gy} = -\vec{\mathbf{v}}_{yg} = -1.3 \text{ m/s } \hat{\mathbf{x}}$

3. The vector addition to get $\vec{\mathbf{v}}_{sg}$ is: $\vec{\mathbf{v}}_{sg} = \vec{\mathbf{v}}_{sa} + \vec{\mathbf{v}}_{ag}$

4. $\vec{\mathbf{v}}_{sa}$ is given in the problem: $\vec{\mathbf{v}}_{sa} = -3.9 \text{ m/s } \hat{\mathbf{y}}$

5. From the given information we also have $\vec{\mathbf{v}}_{ag}$: $\vec{\mathbf{v}}_{ag} = -20 \text{ mi/h} \left(\dfrac{0.447 \text{ m/s}}{1 \text{ mi/h}}\right) \hat{\mathbf{x}} = -8.94 \text{ m/s } \hat{\mathbf{x}}$

6. The velocity of the snow relative to the ground: $\vec{\mathbf{v}}_{sg} = -8.94 \text{ m/s } \hat{\mathbf{x}} - 3.9 \text{ m/s } \hat{\mathbf{y}}$

7. From step 1, the velocity of the snow relative to you is:

$$\vec{\mathbf{v}}_{\text{sy}} = (-8.94\text{ m/s} - 1.3\text{ m/s})\,\hat{\mathbf{x}} - 3.9\text{ m/s}\,\hat{\mathbf{y}}$$
$$= -10.24\text{ m/s}\,\hat{\mathbf{x}} - 3.9\text{ m/s}\,\hat{\mathbf{y}}$$

8. As shown in the following sketch, the angle $\theta$ this velocity makes with the vertical is:

$$\theta = \tan^{-1}\left(\frac{v_{\text{sy},x}}{v_{\text{sy},y}}\right) = \tan^{-1}\left(\frac{-10.24\text{ m/s}}{-3.9\text{ m/s}}\right) = 69^\circ$$

9. As discussed in the strategy section, this must also be the angle at which you should hold the umbrella.

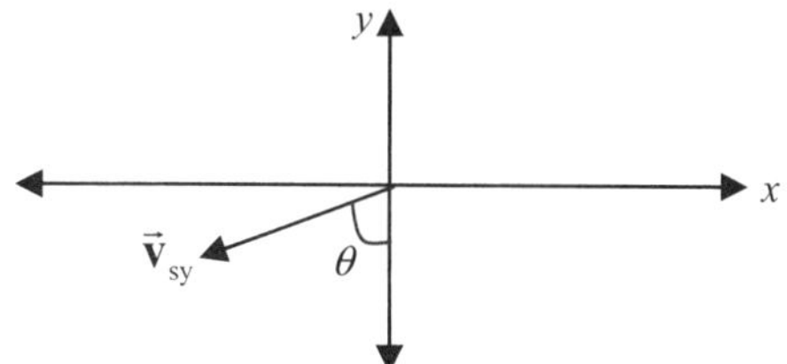


**Insight** It is easy to forget that the angle of the umbrella is determined relative to you. A common mistake here is to calculate only the angle of the snow relative to the ground. When dealing with relative motion, always remember to ask yourself *with respect to whom* is this question being asked.

---

## Practice Quiz

12. If $\vec{\mathbf{v}}_{PA}$ is the velocity of object *P* with respect to frame *A*, and $\vec{\mathbf{v}}_{PB}$ is the velocity of object *P* with respect to frame *B*, which of the following equals the velocity of frame *B* with respect to frame *A*?

**(a)** $\vec{\mathbf{v}}_{AP} + \vec{\mathbf{v}}_{BP}$ **(b)** $\vec{\mathbf{v}}_{PA} + \vec{\mathbf{v}}_{PB}$ **(c)** $\vec{\mathbf{v}}_{AP} - \vec{\mathbf{v}}_{BP}$ **(d)** $\vec{\mathbf{v}}_{BP} - \vec{\mathbf{v}}_{PA}$ **(e)** $\vec{\mathbf{v}}_{BP} + \vec{\mathbf{v}}_{PA}$

13. If $\vec{\mathbf{v}}_{\text{BA}} = 12\,\hat{\mathbf{x}} + 22\,\hat{\mathbf{y}}$ in m/s and $\vec{\mathbf{v}}_{\text{PA}} = -8\,\hat{\mathbf{x}} - 8\,\hat{\mathbf{y}}$ in m/s, then what is $\vec{\mathbf{v}}_{\text{PB}}$ in SI units?

**(a)** $4\,\hat{\mathbf{x}} + 14\,\hat{\mathbf{y}}$ **(b)** $-20\,\hat{\mathbf{x}} - 30\,\hat{\mathbf{y}}$ **(c)** $-20\,\hat{\mathbf{x}} + 14\,\hat{\mathbf{y}}$ **(d)** $20\,\hat{\mathbf{x}} + 30\,\hat{\mathbf{y}}$ **(e)** $-4\,\hat{\mathbf{x}} - 14\,\hat{\mathbf{y}}$

# Reference Tools and Resources

## I. Key Terms and Phrases

**vector** a mathematical quantity having both magnitude and direction (with appropriate units)

**scalar** a numerical value with appropriate units

**magnitude of a vector** the full numerical value of the quantity being represented

**direction of a vector** the orientation within a coordinate system of the quantity being represented

**component of a vector** the part of a vector associated with a specific direction

**unit vector** a dimensionless vector of unit magnitude

## II. Important Equations

| Name/Topic | Equation | Explanation |
|---|---|---|
| scalar components of a vector $\vec{\mathbf{V}}$ | $V_x = V\cos\theta,\ V_y = V\sin\theta$ | Scalar components of a vector from its magnitude and direction. The angle $\theta$ is measured from the $+x$ direction |
| reference angle | $\theta = \tan^{-1}\left(\frac{V_y}{V_x}\right)$ | The reference angle gives the direction of a vector with respect to the coordinate axes |
| magnitude of a vector | $V = \sqrt{V_x^2 + V_y^2}$ | The magnitude of a vector from its $x$ and $y$ scalar components |
| unit vectors | $\vec{\mathbf{V}} = V_x\hat{\mathbf{x}} + V_y\hat{\mathbf{y}}$ | A two-dimensional vector written in terms of unit vectors |
| vector addition/subtraction | $\vec{\mathbf{A}} \pm \vec{\mathbf{B}} = (A_x \pm B_x)\hat{\mathbf{x}} + (A_y \pm B_y)\hat{\mathbf{y}}$ | Vector addition and subtraction with components |
| relative motion | $\vec{\mathbf{v}}_{\mathrm{PA}} = \vec{\mathbf{v}}_{\mathrm{PB}} + \vec{\mathbf{v}}_{\mathrm{BA}}$ | Velocity addition for relative motion using the subscripting mnemonic |

## III. Know Your Units

| Quantity | Dimension | SI Unit |
|---|---|---|
| unit vector | — | dimensionless |

## IV. Tips

The component method of vector addition described in this chapter is very powerful and always works. However, the method can sometimes be cumbersome, and increasing the number of calculations provides greater opportunities for some sort of mistake to slip in. For cases when you know the magnitudes and direction of the two vectors you need to add, the law of cosines can provide a more direct route to the result than the component method. The law of cosines is illustrated below.

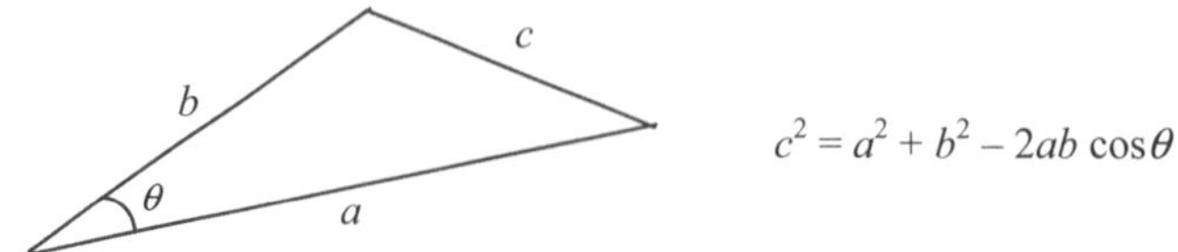

Consider the many calculations in the solution of Example 3–4. If we only needed the magnitude of $\vec{\mathbf{v}}_3$, having been given the magnitude and direction of both $\vec{\mathbf{v}}_1$ and $\vec{\mathbf{v}}_2$, this would have allowed us to immediately draw the following triangle:

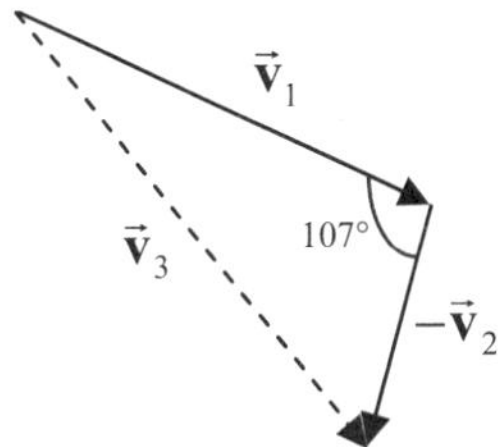


The law of cosines would then yield the result with just one additional step

$$v_3 = \left[25.0^2 + 15.0^2 - 2(25.0)(15.0)\cos(107^\circ)\right]^{1/2} = 32.7$$

## Puzzle

**SHARING THE BURDEN**

The helpful twin brothers are swinging their little 30-lb sister on a rope. Is it possible to arrange the rope so that each of the twins holds up 30 lbs?

## Answers to Selected Conceptual Questions

**2.** Vectors $\vec{\mathbf{A}}$, $\vec{\mathbf{G}}$, and $\vec{\mathbf{J}}$ are equal to one another because they have the same magnitude (length) and orientation with the axes. In addition, vector $\vec{\mathbf{I}}$ is the same as vector $\vec{\mathbf{L}}$.

**4.** No. The component and the magnitude can be equal if the vector has only one component. However, if it has more than one nonzero component, its magnitude will be greater than either of its components.

**6.** No. If a vector has a nonzero component, the smallest magnitude it can have is the magnitude of the component.

## Solutions to Selected End-of-Chapter Problems and Conceptual Exercises

21. **Picture the Problem**: The vectors involved in the problem are depicted at right. The control tower (CT) is at the origin and north is up in the diagram.

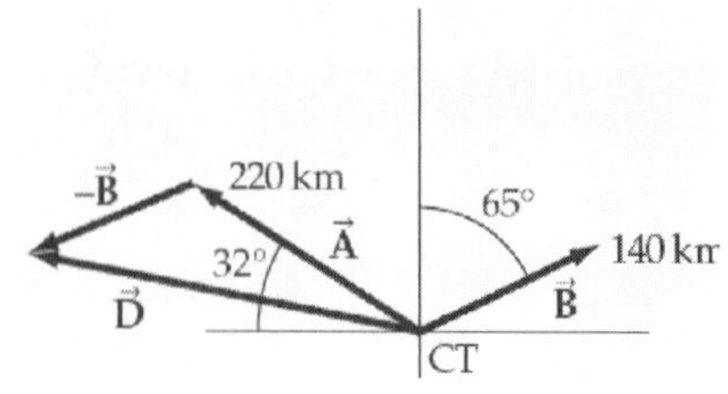


**Strategy:** Subtract vector $\vec{\mathbf{B}}$ from $\vec{\mathbf{A}}$ using the vector component method.

**Solution: 1. (a)** A sketch of the vectors and their difference is shown at right.

**2. (b)** Subtract the $x$ components: $D_x = A_x - B_x = (220\text{ km})\cos(180-32°)-(140\text{ km})\cos(90-65°) = \underline{\underline{-310\text{ km}}}$

**3.** Subtract the $y$ components: $D_y = A_y - B_y = (220\text{ km})\sin(180-32°)-(140\text{ km})\sin(90-65°) = \underline{\underline{57\text{ km}}}$

**4.** Find the magnitude of $D$: $D=\sqrt{D_x^2+D_y^2}=\sqrt{(310\text{ km})^2+(57\text{ km})^2}=\boxed{320\text{ km}}=3.2\times10^5\text{ m}$

**5.** Find the direction of $D$: $\theta_D=\tan^{-1}\left(\dfrac{D_y}{D_x}\right)=\tan^{-1}\left(\dfrac{57\text{ km}}{-310\text{ km}}\right)=-10°+180°=170°$ or $\boxed{10°\text{ north of west}}$

**Insight:** Resolving vectors into components takes a little bit of extra effort, but you can get much more accurate answers using this approach than by adding the vectors graphically. Notice, however, that when your calculator returns −10° as the angle in step 5, you must have a picture of the vectors in your head (or on paper) to correctly determine the direction.

31. **Picture the Problem**: The vectors involved in the problem are depicted at right.

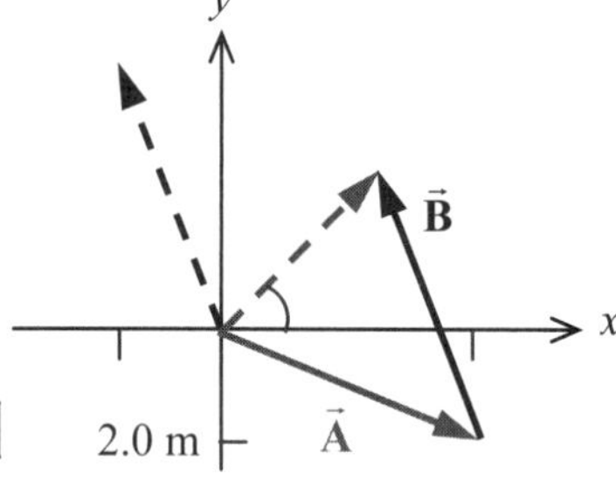


**Strategy:** Determine the lengths and directions of the various vectors by using their $x$ and $y$ components.

**Solution: 1. (a)** Find the direction of $\vec{\mathbf{A}}$ from its components: $\theta_{\vec{A}}=\tan^{-1}\left(\dfrac{-2.0\text{ m}}{5.0\text{ m}}\right)=\boxed{-22°}$

**2.** Find the magnitude of $\vec{\mathbf{A}}$: $A=\sqrt{(5.0\text{ m})^2+(-2.0\text{ m})^2}=\boxed{5.4\text{ m}}$

**3. (b)** Find the direction of $\vec{\mathbf{B}}$ from its components: $\theta_{\vec{B}}=\tan^{-1}\left(\dfrac{5.0\text{ m}}{-2.0\text{ m}}\right)=-68°+180°=\boxed{110°}$

**4.** Find the magnitude of $\vec{\mathbf{B}}$: $B=\sqrt{(-2.0\text{ m})^2+(5.0\text{ m})^2}=\boxed{5.4\text{ m}}$

**5. (c)** Find the components of $\vec{\mathbf{A}}+\vec{\mathbf{B}}$: $\vec{\mathbf{A}}+\vec{\mathbf{B}}=(5.0-2.0\text{ m})\hat{\mathbf{x}}+(-2.0+5.0\text{ m})\hat{\mathbf{y}}=(3.0\text{ m})\hat{\mathbf{x}}+(3.0\text{ m})\hat{\mathbf{y}}$

**6.** Find the direction of $\vec{\mathbf{A}}+\vec{\mathbf{B}}$ from its components: $\theta_{\vec{A}+\vec{B}}=\tan^{-1}\left(\dfrac{3.0\text{ m}}{3.0\text{ m}}\right)=\boxed{45°}$

**7.** Find the magnitude of $\vec{\mathbf{A}}+\vec{\mathbf{B}}$: $\left|\vec{\mathbf{A}}+\vec{\mathbf{B}}\right|=\sqrt{(3.0\text{ m})^2+(3.0\text{ m})^2}=\boxed{4.2\text{ m}}$

**Insight:** In the world of vectors 5.4 m + 5.4 m can be anything between 0 m and 10.8 m, depending upon the directions that the vectors point. In this case their sum is 4.2 m.

40. **Picture the Problem**: You travel due west for 125 s at 27 m/s, then due south at 14 m/s for 66 s.

**Strategy:** Find the components of the displacement vector. Once the components are known the magnitude and direction can be easily found. Let north be the positive $y$ direction and east be the positive $x$ direction.

**Solution: 1.** Find the westward displacement: $r_x = v_x t = (-27\text{ m/s})(125\text{ s}) = -3375\text{ m}$

**2.** Find the southward displacement: $r_y = v_y t = (-14\text{ m/s})(66\text{ s}) = -924\text{ m}$

**3.** Find the direction of the displacement: $\theta = \tan^{-1}\left(\dfrac{r_y}{r_x}\right) = \tan^{-1}\left(\dfrac{-924\text{ m}}{-3375\text{ m}}\right) = 15° + 180° = 195°$

or $\boxed{15° \text{ south of west}}$

**4.** Find the magnitude of the displacement: $r = \sqrt{(-3375\text{ m})^2 + (-924\text{ m})^2} = 3500\text{ m} = \boxed{3.5\text{ km}}$

**Insight:** The 15° refers to the angle below the negative $x$ axis (west) because the argument of the inverse tangent function is $r_y/r_x$, or south divided by west.

53. **Picture the Problem**: The vectors involved in this problem are depicted at right.

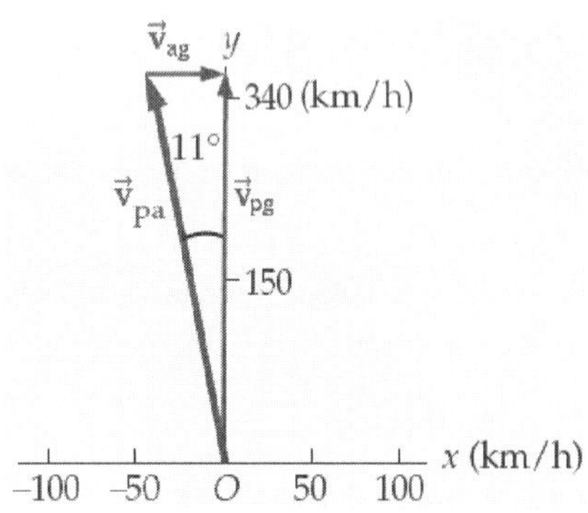


**Strategy:** Let $\vec{\mathbf{v}}_{pg}$ = velocity of the plane relative to the ground, $\vec{\mathbf{v}}_{pa}$ = velocity of the plane relative to the air, and $\vec{\mathbf{v}}_{ag}$ = velocity of the air relative to the ground. The drawing at right depicts the vectors added according to equation 3-8, $\vec{\mathbf{v}}_{pg} = \vec{\mathbf{v}}_{pa} + \vec{\mathbf{v}}_{ag}$. Determine the angle of the triangle from the inverse sine function.

**Solution: 1. (a)** Use the inverse sine function to find $\theta$: $\theta = \sin^{-1}\left(\dfrac{v_{ag}}{v_{pa}}\right) = \sin^{-1}\left(\dfrac{65\text{ km/h}}{340\text{ km/h}}\right) = \boxed{11° \text{ west of north}}$

**2. (b)** The drawing above depicts the vectors.

**3. (c)** If the plane reduces its speed but the wind velocity remains the same, the angle found in part (a) should be $\boxed{\text{increased}}$ in order for the plane to continue flying due north.

**Insight:** If the plane's speed were to be reduced to 240 km/h, the required angle would become 16°.

67. **Picture the Problem**: The three-dimensional vector is depicted at right.

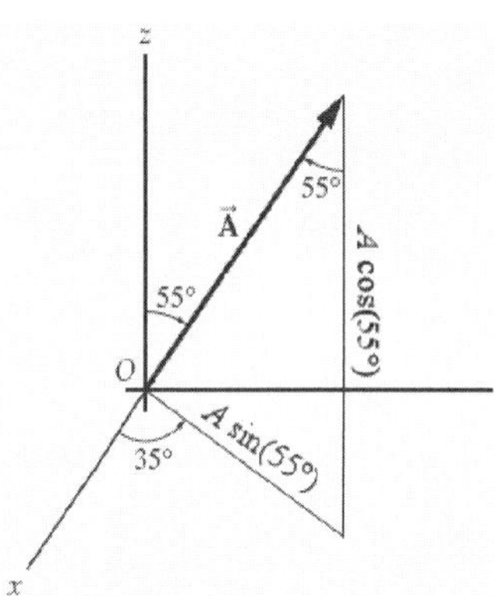


**Strategy:** Determine the $z$ component of $\vec{\mathbf{A}}$ by applying the cosine function to the right triangle formed in the $z$ direction. Then find the projection of $\vec{\mathbf{A}}$ onto the $xy$ plane ($A \sin 55°$) in order to find the $x$ and $y$ components of $\vec{\mathbf{A}}$.

**Solution: 1.** Find the $z$ component of $\vec{\mathbf{A}}$: $A_z = (65\text{ m})\cos 55° = \boxed{37\text{ m}}$

**2.** Find the projection onto the $xy$ plane: $A_{xy} = A\sin 55° = (65\text{ m})\sin 55°$

**3.** Find the $x$ component of $\vec{\mathbf{A}}$: $A_x = \left[(65\text{ m})\sin 55°\right]\cos 35° = \boxed{44\text{ m}}$

**4.** Find the $y$ component of $\vec{\mathbf{A}}$ : $A_y = \left[(65\text{ m})\sin 55°\right]\sin 35° = \boxed{31\text{ m}}$

**Insight:** A knowledge of right triangles can help you find the components of even a three-dimensional vector. Once the components are known, then addition and subtraction of vectors become straightforward procedures.

71. **Picture the Problem**: The vectors involved in this problem are illustrated at right.

**Strategy:** To use the graphical method you must make a scale drawing of the vectors and then measure the vector sum with a ruler. To use the component method you must independently add the $x$ and $y$ components of each vector.

**Solution: 1. (a)** Using the scale drawing above you can measure the length of the vector sum: $|\Delta\vec{\mathbf{r}}| \approx \boxed{38\text{ ft}}$

**2. (b)** Independently add the $x$ and $y$ components of the vector sum:

$$\Delta\vec{\mathbf{r}} = \vec{\mathbf{A}} + \vec{\mathbf{B}} + \vec{\mathbf{C}} + \vec{\mathbf{D}}$$
$$= (0 + 45\text{ ft} - 35\text{ ft} + 0)\hat{\mathbf{x}} + (51\text{ft} + 0 + 0 - 13\text{ ft})\hat{\mathbf{y}}$$
$$\Delta\vec{\mathbf{r}} = (10\text{ ft})\hat{\mathbf{x}} + (38\text{ ft})\hat{\mathbf{y}}$$

**3.** Find the magnitude of the sum: $|\Delta\vec{\mathbf{r}}| = \sqrt{(10\text{ ft})^2 + (38\text{ ft})^2} = \boxed{39\text{ ft}}$

**4.** Find the direction of the vector sum: $\theta = \tan^{-1}\left(\dfrac{\Delta r_x}{\Delta r_y}\right) = \tan^{-1}\left(\dfrac{10\text{ ft}}{38\text{ ft}}\right) = \boxed{15°\text{ clockwise from }\vec{\mathbf{A}}}$

**Insight:** When adding vectors graphically you must always ensure you are adding them head-to-tail. The vector sum is a vector that starts at the beginning of the first vector ( $\vec{\mathbf{A}}$ ) and ends at the end of the last vector ( $\vec{\mathbf{D}}$ ).

## Answers to Practice Quiz

**1.** (e) **2.** (b) **3.** (c) **4.** (e) **5.** (b) **6.** (d) **7.** (c) **8.** (a) **9.** (d) **10.** (a) **11.** (c) **12.** (e) **13.** (b)

# CHAPTER 4
# TWO-DIMENSIONAL KINEMATICS

## Chapter Objectives

After studying this chapter, you should

1. know how to treat motion with constant velocity in two dimensions.
2. know how to treat motion with constant acceleration in two dimensions.
3. be able to apply the equations for two-dimensional motion to a projectile.
4. be able to calculate positions, velocities, and times for various types of projectile motion.

## Warm-Ups

1. To attain the maximum range, a projectile has to be launched at 45 degrees if the landing spot and the launch spot are at the same height (neglecting air resistance effects). Explain in a few sentences how the relation between the vertical and the horizontal components of the initial velocity affects the projectile range.
2. On the Moon the acceleration due to gravity is about one-sixth that on Earth. If a golfer on the Moon imparts the same initial velocity to the ball as she does on Earth, how much farther will the ball go?
3. Three swimmers start across the river at the same time. They all swim with the same speed relative to the water. Peter swims in a direction perpendicular to the current. Albert swims slightly upstream along a line that makes an 80° angle with the shore. Dawn swims slightly downstream along a line that makes an 80° angle with the shore. Which of the three swimmers will reach the opposite shore first?
4. The pilot of a small plane is trying to maintain a course due north. His air speed is 120 mi/h. There is a 10 mi/h wind from the East. Estimate the direction in which the plane should be pointed to maintain the correct course.

## Chapter Review

In Chapter 2, you studied motion in one dimension. Everything you learned there applies to two-dimensional motion. In fact, two-dimensional motion is really just two completely independent cases of motion in one dimension. We call these two dimensions the horizontal ($x$) and vertical ($y$) directions. The running theme of this chapter is the following: **horizontal and vertical motions are independent.**

## 4–1 Motion in Two Dimensions

### Constant Velocity in Two Dimensions

Using the fact that two-dimensional motion is really two separate cases of one-dimensional motion makes it easy to find the equations for describing constant velocity in two dimensions: we simply take the one-dimensional equation, $\Delta x = vt$, and apply it to both the $x$ and $y$ directions. The key difference is that (a) *only the horizontal component of velocity,* $v_{0x}$*, applies to the horizontal motion*, and (b*) only the vertical component of velocity,* $v_{0y}$*, applies to the vertical motion*. Therefore, the equations are

$$x = x_0 + v_{0x}t$$
$$y = y_0 + v_{0y}t$$

Notice that because the velocity is constant, $v = v_0$ at all times during the motion.

---

**Example 4–1 Pocket Billiards** A billiard ball is struck so that it rolls across the pool table with a velocity of 25.0 cm/s at an angle of 52.0° above the horizontal. The horizontal direction is the short side of a standard 122-cm × 244-cm table. If the ball was initially located at a point $(x_0, y_0)$ = (11.0 cm, 17.0 cm) from the left corner on the short side, **(a)** how much time will pass before it strikes a side of the table? **(b)** Will it go into a pocket?

**Picture the Problem** The rough sketch shows the table and the ball. The lower left corner of the pool table is the origin of the coordinate system.

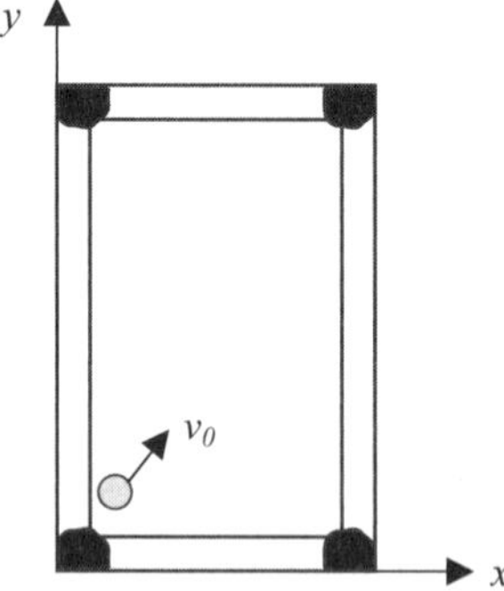


**Strategy** Treating the vertical and horizontal motions as independent, we need the $x$ and $y$ components of the velocity. We can then solve this problem by finding out where the ball will be at certain times.

**Solution**

Determine the components of the velocity:

$$v_{0x} = v_0 \cos\theta = (25.0 \text{ cm/s})\cos(52.0^\circ) = 15.39 \text{ cm/s}$$
$$v_{0y} = v_0 \sin\theta = (25.0 \text{ cm/s})\sin(52.0^\circ) = 19.70 \text{ cm/s}$$

**Part (a)**

1. Solve for the time it takes the ball to reach the right edge along the $x$-direction:

$$t = \frac{x - x_0}{v_{0x}} = \frac{122 \text{ cm} - 11.0 \text{ cm}}{15.39 \text{ cm/s}} = 7.212 \text{ s}$$

2. Determine the vertical position of the ball: $y = y_0 + v_{0y}t = 17.0\text{ cm} + \left(19.70\ \tfrac{\text{cm}}{\text{s}}\right)\left(7.212\text{ s}\right) = 159\text{ cm}$

Since the upper edge is 244 cm from the lower edge, the ball has not reached the upper edge by the time it reaches the right edge. It strikes the right edge first.

**Part (b)**

3. Determine the ball's distance from the corner when it hits the right edge: $\Delta y = 244\text{ cm} - 159\text{ cm} = 85\text{ cm}$

The ball is 85 cm from the corner, which is too far for it to fall into the pocket.

**Insight** Clearly, we were able to treat the horizontal and vertical motions as simultaneous yet independent. Notice also that an additional digit was retained for the values of $t$, $v_{0x}$, and $v_{0y}$ because they were used in intermediate steps.

---

## Practice Quiz

1. If you walk in a direction 50.0° north of east at a pace of 2.30 m/s for one hour, how far north of your starting position will you be?

   **(a)** 8.28 km **(b)** 138 m **(c)** 6.34 km **(d)** 5.32 km

2. A bird flies with a westerly speed of 20 m/s and a northerly speed of 15 m/s. How far will it fly over a time period of 32 seconds?

   **(a)** 800 m **(b)** 640 m **(c)** 480 m **(d)** 160 m **(e)** 1100 m

### Constant Acceleration in Two Dimensions

The same strategy used with constant velocity applies to motion with constant acceleration: we just apply the one-dimensional equations from Chapter 2 to both directions using only $x$ components of velocity and acceleration for the horizontal motion and only $y$ components for the vertical motion. Thus, the four equations become eight:

| Horizontal Motion | Vertical Motion |
|---|---|
| $v_x = v_{0x} + a_x t$ | $v_y = v_{0y} + a_y t$ |
| $x = x_0 + \frac{1}{2}(v_{0x} + v_x)t$ | $y = y_0 + \frac{1}{2}(v_{0y} + v_y)t$ |
| $v_x^2 = v_{0x}^2 + 2a_x(x - x_0)$ | $v_y^2 = v_{0y}^2 + 2a_y(y - y_0)$ |
| $x = x_0 + v_{0x}t + \frac{1}{2}a_x t^2$ | $y = y_0 + v_{0y}t + \frac{1}{2}a_y t^2$ |

These equations can be applied precisely as in Chapter 2. Here, they represent two independent yet simultaneous cases of one-dimensional motion.

---

**Example 4–2 Horizontal Rocket Launch** A toy rocket is launched horizontally, from rest, off the edge of a 0.850-m-high table. While in flight the rocket manages to remain pointed horizontally, and its thrust causes it to accelerate forward at 7.48 m/s$^2$ as gravity pulls it down. **(a)** How far from the edge of the table will it be when it hits the floor? **(b)** With what speed will it hit the floor?

**Picture the Problem** The diagram shows the rocket leaving the edge of the table and the path it takes. The positive coordinate directions are also indicated.

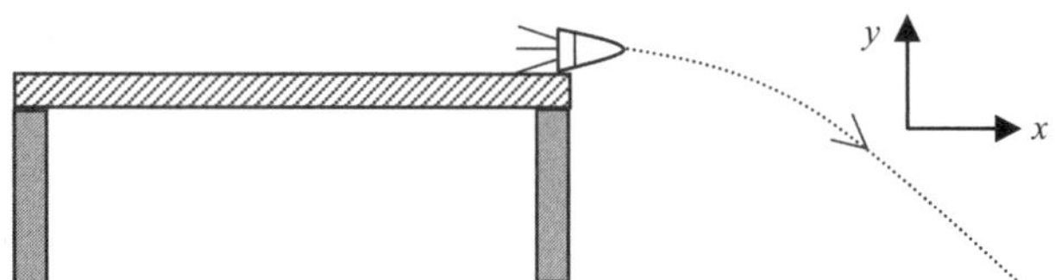


**Strategy** The rocket has constant acceleration in both directions, so the full set of eight equations applies.

**Solution**

**Part (a):** We need to determine when it hits the floor and get its horizontal position. Let's choose $(x_0, y_0) = (0, 0)$. Also note that the initial velocities are zero.

1. Choose a vertical equation that contains $t$ with all other quantities known: $y = \frac{1}{2}a_y t^2$

2. Solve this equation for the time to hit the floor: $t = \sqrt{\dfrac{2y}{a_y}} = \sqrt{\dfrac{2(-0.850\text{ m})}{-9.81\text{ m/s}^2}} = 0.4163\text{ s}$

**3.** Choose a horizontal equation involving $x$ with all other quantities known: $x = \frac{1}{2}a_x t^2$

**4.** Solve to get the horizontal position: $x = \frac{1}{2}(7.48 \text{ m/s}^2)(0.4163 \text{ s})^2 = 0.648 \text{ m}$

**Part (b):**

**5.** Get the horizontal velocity when it hits the floor: $v_x = a_x t = (7.48 \text{ m/s}^2)(0.4163 \text{ s}) = 3.114 \text{ m/s}$

**6.** Get the vertical velocity when it hits the floor: $v_y = a_y t = (-9.81 \text{ m/s}^2)(0.4163 \text{ s}) = -4.084 \text{ m/s}$

**7.** Determine the magnitude of $v$ to get the speed: $v = \sqrt{v_x^2 + v_y^2} = \sqrt{(3.114 \text{ m/s})^2 + (-4.084 \text{ m/s})^2}$
$= 5.14 \text{ m/s}$

**Insight** Be sure you understand the minus signs in part (a)-2.

## 4–2 – 4–4 Projectile Motion

The main application that is considered in this chapter is **projectile motion**. This motion is that of an object projected with some initial velocity and then allowed to fall freely. Effects such as air resistance and Earth's rotation that sometimes cause an object's motion to differ significantly from that of pure free fall are ignored. The key point for understanding projectile motion is that the acceleration due to gravity acts only vertically, and since gravity is the only influence, there is no acceleration in the horizontal direction. This fact means that **although there is gravitational acceleration vertically, the horizontal motion is that of constant velocity**.

It is convenient to adopt a standard coordinate system for projectile motion. This standard takes the positive direction for vertical motion to be upward, making downward the negative direction. This means that the acceleration of gravity is given a negative value. For horizontal motion, the positive direction is to the right, and the negative direction to the left. Given this coordinate system, we can rewrite the equations for two-dimensional motion in a form specific to projectile motion. To accomplish this rewriting, we make use of several facts:

$$a_x = 0, \quad v_x = v_{0x} = v_0 \cos\theta$$
$$a_y = -g, \quad v_{0y} = v_0 \sin\theta$$

where $\theta$ is the **launch angle**, which is the angle of the initial velocity as measured from the horizontal direction. With these substitutions, the equations of projectile motion are as follows:

| Horizontal Motion | Vertical Motion |
|---|---|
| $v_x = v_0 \cos\theta$ | $v_y = v_0 \sin\theta - gt$ |
| $x = x_0 + (v_0 \cos\theta)t$ | $y = y_0 + \frac{1}{2}(v_0 \sin\theta + v_y)t$ |
| | $v_y^2 = v_0^2 \sin^2\theta - 2g(y - y_0)$ |
| | $y = y_0 + (v_0 \sin\theta)t - \frac{1}{2}gt^2$ |

This set of equations describes projectile motion for both zero and nonzero launch angles.

---

**Example 4–3 Over the Edge** Wagging his tail, a dog knocks an object horizontally off the edge of a table that is 0.750 m high. If the object hits the floor 0.995 m away from the edge of the table, with what speed did the object leave the table?

**Picture the Problem** The sketch shows the object leaving the edge of the table and the path it takes. The positive coordinate directions are also indicated.

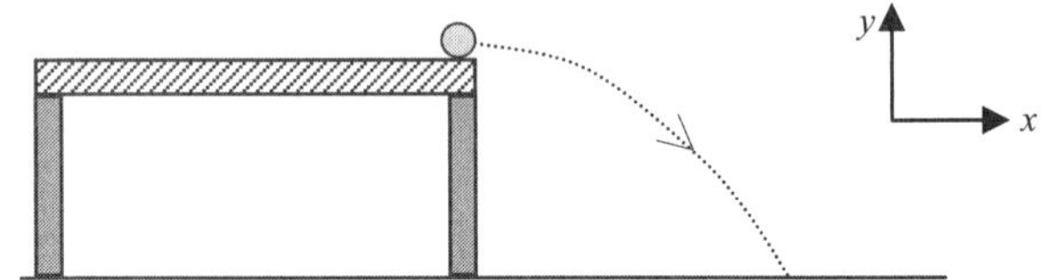


**Strategy** Let's choose $(x_0, y_0) = (0, 0)$. Also notice that $v_{0y} = 0$, and $\cos\theta = 1$. The only equation involving $v_0$ is the one giving $x$ as a function of time.

**Solution**

1. Solve the horizontal equation for $v_0$: $v_0 = \dfrac{x}{t}$

2. Use the vertical equation with all known quantities to get an expression for $t$: $y = -\frac{1}{2}gt^2 \Rightarrow t = \sqrt{\dfrac{-2y}{g}}$

3. Put these results together to get an expression for $v_0$ in terms of known quantities and solve: $v_0 = x\sqrt{\dfrac{-g}{2y}} = (0.995\text{ m})\sqrt{\dfrac{-9.81\text{ m/s}^2}{2(-0.750\text{ m})}} = 2.54\text{ m/s}$

**Insight** Here, no values were calculated at an intermediate step. Instead, the formula was carried through to the end. This approach is better for handling round-off error, but problems can sometimes get messy this way. Compare this horizontal launch problem to that of Example 4–2.

**Example 4–4 Toss Me That Wrench** A person stands 12.5 m from a building that is 40.0 m tall. He wants to toss a wrench to his coworker who is working on the roof. If he releases the tool at a height of 1.00 m above the ground, what velocity must he give the tool if it is to just make it onto the roof?

**Picture the Problem** Our sketch shows the wrench being tossed so that it just barely makes it onto the roof.

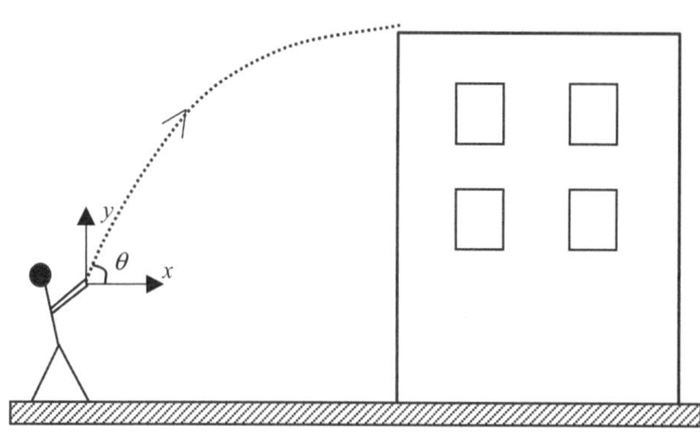


**Strategy** Because the wrench just barely makes it onto the roof, the top of the roof is the maximum height of the wrench, where $v_y = 0$. Treating the motions separately, we need to find the components of $v_0$. From these values we can determine the magnitude and direction.

**Solution**

Since we choose the origin at the point of launch, we take $x_0 = 0$, $y_0 = 0$, and $y = 39.0$ m.

1. Choose an equation containing $v_{0y}$ with all other quantities known and solve:

$$v_y^2 = v_{0y}^2 - 2g(y - y_0) \Rightarrow v_{0y} = \sqrt{2g(y - y_0)}$$
$$\therefore v_{0y} = \sqrt{2(9.81 \text{ m/s}^2)(39.0 \text{ m})} = 27.66 \text{ m/s}$$

2. The only equation involving $v_{0x}$ requires the time. Choose a vertical equation involving time with all other quantities known:

$$v_y = v_{0y} - gt \quad \Rightarrow \quad t = \frac{v_{0y}}{g} = \frac{27.66 \text{ m/s}}{9.81 \text{ m/s}^2} = 2.820 \text{ s}$$

3. Solve the horizontal equation for $v_{0x}$:

$$v_{0x} = \frac{x - x_0}{t} = \frac{12.5 \text{ m}}{2.820 \text{ s}} = 4.433 \text{ m/s}$$

4. Calculate the magnitude of the initial velocity:

$$v_0 = \sqrt{v_{0x}^2 + v_{0y}^2} = \sqrt{(4.433 \text{ m/s})^2 + (27.66 \text{ m/s})^2} = 28.0 \text{ m/s}$$

5. Calculate the direction of the initial velocity:

$$\theta = \tan^{-1}\left(\frac{v_{0y}}{v_{0x}}\right) = \tan^{-1}\left(\frac{27.66 \text{ m/s}}{4.433 \text{ m/s}}\right) = 80.9^\circ$$

**Insight** In both this example and in the previous one, the origin was taken to be the initial position. Although this is not necessary, it is sometimes useful to adopt a convention such as this and not have to worry about what choices to make for every problem.

## Practice Quiz

3. A stone is thrown horizontally from the top of a 50.0-m building with a speed of 12.3 m/s. How far from the building will it land?
   **(a)** 30.8 m **(b)** 39.3 m **(c)** 3.19 m **(d)** 615 m

4. A projectile is launched from the ground with a speed of 23.7 m/s at 33.0° above the horizontal. What is its speed when it is at its maximum height above the ground?
   **(a)** 19.9 m/s **(b)** 0.00 m/s **(c)** 12.9 m/s **(d)** 23.7 m/s

5. A projectile is launched from the ground with a speed of 53.4 m/s at 68.0° above the horizontal. Take its initial position as the origin, what are its $(x, y)$ coordinates after 8.24 seconds?
   **(a)** (440 m, 440 m) **(b)** (440 m, 107 m) **(c)** (408 m, 333 m) **(d)** (165 m, 74.9 m)

## 4–5 Projectile Motion: Key Characteristics

The equations for projectile motion can be used to derive several important properties of the motion. The key characteristics and symmetries are as follows:

* The path, or trajectory, that a projectile follows is a parabola.

* If the initial and final elevations are the same, the **range** ($R$) of a projectile, which is the horizontal distance it travels before landing, is given by

$$R = \frac{v_0^2}{g}\sin 2\theta$$

* If the initial and final elevations are the same, the launch angle that produces the maximum range is $\theta = 45°$.

* If the initial and final elevations are the same, the amount of time the projectile spends in the air, sometimes called the *time-of-flight*, is given by

$$t = \frac{2v_0}{g}\sin\theta$$

* If the initial and final elevations are the same, the time it takes a projectile launched at some upward angle to reach its maximum height equals the time it takes to fall back down from its maximum height.

* The maximum height that a projectile will reach above its initial height is given by

$$y_{\text{max}} = \frac{(v_0 \sin\theta)^2}{2g}$$

* The speed that a projectile has at a given height on its way up is equal to the speed it will have at that same height on its way back down.

* At a given height, the angle of the velocity of a projectile above the horizontal on its way up equals the angle of the velocity below the horizontal on its way down.

Notice that all these characteristics are determined by the initial velocity given to the projectile.

---

**Exercise 4–5 At the Driving Range** A golf ball sitting on level ground is struck and given an initial velocity of 41.2 m/s at an angle of 58.0°. **(a)** How high does the ball go into the air? **(b)** How far does it travel? **(c)** How long is the ball in the air?

**Solution** Try to sketch a picture for this problem; the ball moves in a parabolic path starting and ending on the ground. The following information is supplied in the problem:

**Given:** $v_0 = 41.2$ m/s, $\theta = 58.0°$; **Find: (a)** $y_{\text{max}}$, **(b)** $R$, **(c)** $t$

We are given the initial velocity and we know that it completely specifies the motion. Making use of the known results, we can directly solve for each of these quantities.

**(a)** $$y_{\text{max}} = \frac{(v_0 \sin\theta)^2}{2g} = \frac{\left[(41.2\text{ m/s})\sin(58.0^\circ)\right]^2}{2(9.81\text{ m/s}^2)} = 62.2\text{ m}$$

**(b)** $$R = \frac{2v_0^2}{g}\sin 2\theta = \frac{(41.2\text{ m/s})^2}{9.81\text{ m/s}^2}\sin(116^\circ) = 156\text{ m}$$

**(c)** $$t = \frac{2v_0}{g}\sin\theta = \frac{2(41.2\text{ m/s})}{9.81\text{ m/s}^2}\sin(58.0^\circ) = 7.12\text{ s}$$

The questions asked in this problem are some of the basic things you might want to know about a projectile. This exercise illustrates the utility of working out equations for certain quantities.

---

## Practice Quiz

6. In general, what is the shape of the path of a projectile?

   **(a)** a hyperbola **(b)** a parabola **(c)** a straight line **(d)** a circle

7. A projectile is launched from the ground with an initial velocity of 44.0 m/s at a launch angle of 26.0°. How long will this projectile be in the air?

   **(a)** 8.97 s **(b)** 8.06 s **(c)** 1.97 s **(d)** 3.93 s

8. A projectile reaches a height $h$ when launched at an angle $\theta$ with speed $v_0$. If this projectile is launched from the same level at the same angle with speed $2v_0$, how high will it go?

   **(a)** $h$ **(b)** $2h$ **(c)** $4h$ **(d)** $h/2$

9. A projectile remains airborne for a time $t$ when launched at an angle $\theta$ with speed $v_0$. If this projectile is launched from the same level at the same angle with speed $2v_0$, how long will it be airborne?

   **(a)** $t$ **(b)** $2t$ **(c)** $4t$ **(d)** $t/2$

10. A projectile travels a distance $R$ when launched at an angle $\theta$ with speed $v_0$. If this projectile is launched from the same level at the same angle with speed $2v_0$, how far will it go?

    **(a)** $R$ **(b)** $2R$ **(c)** $4R$ **(d)** $R/2$

## Reference Tools and Resources

### I. Key Terms and Phrases

**projectile motion** the motion of an object that is projected with an initial velocity and then moves under the influence of gravity only

**launch angle** the angle of the initial velocity of a projectile measured relative to the horizontal

**range** the horizontal distance traveled by a projectile before it lands

### II. Important Equations

| Name/Topic | Equation | Explanation |
|---|---|---|
| constant velocity | $x = x_0 + v_{0x}t$<br>$y = y_0 + v_{0y}t$ | Each direction obeys the constant velocity equation independently |

| | | |
|---|---|---|
| constant acceleration: horizontal component | $v_x = v_{0x} + a_x t$<br>$x = x_0 + \frac{1}{2}(v_{0x} + v_x)t$<br>$v_x^2 = v_{0x}^2 + 2a_x(x - x_0)$<br>$x = x_0 + v_{0x}t + \frac{1}{2}a_x t^2$ | The constant acceleration equations for horizontal components |
| constant acceleration: vertical component | $v_y = v_{0y} + a_y t$<br>$y = y_0 + \frac{1}{2}(v_{0y} + v_y)t$<br>$v_y^2 = v_{0y}^2 + 2a_y(y - y_0)$<br>$y = y_0 + v_{0y}t + \frac{1}{2}a_y t^2$ | The constant acceleration equations for vertical components |
| range of a projectile | $R = \frac{v_0^2}{g}\sin 2\theta$ | The horizontal distance traveled if the initial height equals the final height |
| time-of-flight of a projectile | $t = \frac{2v_0}{g}\sin\theta$ | The total time in the air if the initial height equals the final height |
| maximum height of a projectile | $y_{\max} = \frac{(v_0 \sin\theta)^2}{2g}$ | The maximum height above the initial height of launch |

## III. Tips

As discussed in Chapter 2, you may find it useful to add a fifth equation to the list of equations describing motion with constant acceleration. This equation does not contain the initial velocity. In the context of two-dimensional motion, the additional equations are

$$x = x_0 + v_x t - \tfrac{1}{2}a_x t^2$$
$$y = y_0 + v_y t - \tfrac{1}{2}a_y t^2$$

or, in the context of projectile motion,

$$y = y_0 + v_y t + \tfrac{1}{2}gt^2$$

In addition to the equations derived in your text and listed above for projectile motion, it may also be useful to know the equation for the path of the projectile, that is, $y$ as a function of $x$. This equation can be determined by solving for $t$ from the horizontal equation $x - x_0 = (v_0 \cos\theta)t$ and substituting this result into the vertical equation $y - y_0 = (v_0 \sin\theta)t - \frac{1}{2}gt^2$. After some manipulations the result is

$$y - y_0 = (\tan\theta)(x - x_0) - \frac{g}{2v_0^2 \cos^2\theta}(x - x_0)^2$$

Try to rework Example 4–4 using this equation.

## Puzzle

### THE LONG SHOT

You should know by now, from reading the text and from working the projectile motion problems, that the maximum range for a projectile is achieved when the projectile is fired at 45°. This is true if the launch starts and ends at the same altitude. What happens if you fire at a target at a lower elevation? Is the optimal angle still 45°? Is it more? Is it less? Answer these questions in words, not equations, briefly explaining how you obtained your answers.

## Answers to Selected Conceptual Questions

**8.** Maximum height depends on the square of the initial speed. Therefore, to reach twice the height, projectile 1 must have an initial speed that is the square root of 2 times greater than the initial speed of projectile 2. It follows that the ratio of the speeds is the square root of 2.

**10.** The tomato lands on the road in front of you. This follows from the fact that its horizontal speed is the same as yours during the entire time of its fall.

## Solutions to Selected End-of-Chapter Problems and Conceptual Exercises

14. **Picture the Problem**: The baseball is launched horizontally and follows a parabolic trajectory like that pictured at right.

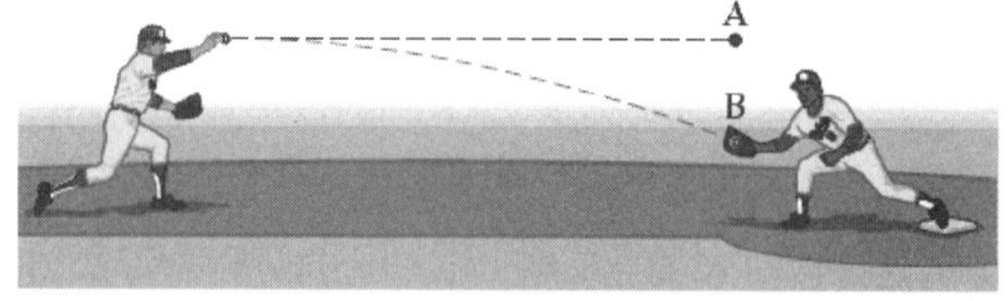


**Strategy:** Using equation 4-7 find the time it takes the baseball to travel the 18 m to the plate, and then calculate the vertical distance it will fall during that time.

**Solution: 1. (a)** Find the time it takes the ball to reach the plate knowing that its horizontal velocity $v_0$ remains unchanged throughout the flight:

$$t = \frac{x}{v_0} = \frac{18\text{ m}}{32\text{ m/s}} = \underline{\underline{0.563\text{ s}}}$$

**2.** Find the vertical drop during the time of flight: $h - y = \frac{1}{2}gt^2 = \frac{1}{2}(9.81\text{ m/s}^2)(0.563\text{ s})^2 = \boxed{1.6\text{ m}}$

**3. (b)** If the pitch speed is increased the time the ball travels is less, therefore the drop distance $\boxed{\text{decreases}}$.

**4. (c)** Since the moon's gravity is less the drop distance $\boxed{\text{decreases.}}$

**Insight:** The 1.6 m drop corresponds to about 5 ft, nearly the height of a player. The 32 m/s (72 mi/h) speed is a bit slow; a sizzling fastball at 44 m/s (98 mi/h) would drop about 0.82 m, half as much.

35. **Picture the Problem**: The trajectories of the snowballs are depicted at right.

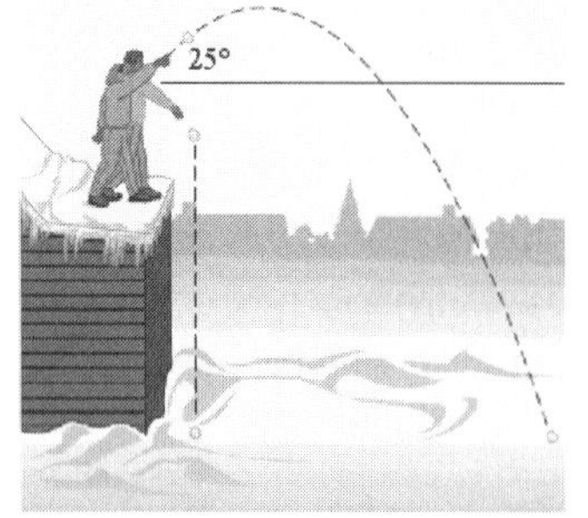


**Strategy:** Use equation 4-10 to find the vertical component of the final velocity of each snowball, as well as the horizontal component of each velocity. Use the known components to determine the landing speed.

**Solution: 1. (a)** The landing speed of snowball A is the same as that of snowball B, because the landing speed is independent of launch angle.

**2. (b)** Use equation 4-10 to find the vertical component of the final velocity for snowball A:

$$v_y = \pm\sqrt{v_0^2 \sin^2\theta - 2g\Delta y}$$
$$= \pm\sqrt{(13\text{ m/s})^2 \sin^2(-90°) - 2(9.81\text{ m/s}^2)(-7.0\text{ m})}$$
$$= \underline{\underline{-18\text{ m/s}}}\ \text{(The snowball is traveling downward.)}$$

**3.** Find the horizontal component of the velocity for snowball A:

$$v_x = v_{0x} = v_0 \cos(-90°) = 0\text{ m/s}$$

**4.** Find the landing speed of snowball A:

$$v = \sqrt{v_y^2 + v_x^2} = \sqrt{(-18\text{ m/s})^2 + 0^2} = \boxed{18\text{ m/s}}$$

**5.** Use equation 4-10 to find the vertical component of the final velocity for snowball B:

$$v_y = \pm\sqrt{v_0^2 \sin^2\theta - 2g\Delta y}$$
$$= \pm\sqrt{(13\text{ m/s})^2 \sin^2(25°) - 2(9.81\text{ m/s}^2)(-7.0\text{ m})}$$
$$= \underline{\underline{-13\text{ m/s}}}\ \text{(The snowball is traveling downward.)}$$

**6.** Find the horizontal component of the velocity for snowball B:

$$v_x = v_{0x} = (13\text{ m/s})\cos(25°) = \underline{\underline{12\text{ m/s}}}$$

**7.** Find the landing speed of snowball B:

$$v = \sqrt{v_y^2 + v_x^2} = \sqrt{(-13\text{ m/s})^2 + (12\text{ m/s})^2} = \boxed{18\text{ m/s}}$$

**Insight:** Conservation of mechanical energy provides an even clearer explanation of why the landing speeds of the snowballs are the same regardless of launch angle. We will explore the concept in Chapter 8.

56. **Picture the Problem**: The hay bale travels along a parabolic arc, maintaining its horizontal velocity but changing its vertical speed due to the constant downward acceleration of gravity.

**Strategy:** The initial velocity of the bale is given as $\vec{\mathbf{v}}_0 = (1.12\text{ m/s})\hat{\mathbf{x}} + (8.85\text{ m/s})\hat{\mathbf{y}}$ . Use the fact that the horizontal component of the bale's velocity never changes throughout the flight in order to find the vertical component of the velocity when the total speed is 5.00 m/s. Then find the time elapsed between the initial throw and the instant the bale has that vertical speed. For part (b) set the vertical speed equal to the (constant) horizontal speed in magnitude but negative in direction (pointing downward). Find the time elapsed between the initial throw and the instant the bale has that new vertical speed.

**Solution: 1. (a)** Determine the $y$ component of the velocity when the total speed is 5.00 m/s:

$$v^2 = v_{0x}^2 + v_y^2$$
$$v_y = \pm\sqrt{v^2 - v_{0x}^2} = \pm\sqrt{(5.00\text{ m/s})^2 - (1.12\text{ m/s})^2} = \underline{\underline{\pm 4.873\text{ m/s}}}$$

**2.** Use the positive value of $v_y$ because the bale is rising when the speed first equals 5.00 m/s. Find the time elapsed to this point from equation 4-6:

$$t = \frac{v_y - v_{0y}}{-g} = \frac{4.87\text{ m/s} - 8.85\text{ m/s}}{-9.81\text{ m/s}^2} = \boxed{0.405\text{ s}}$$

**3. (b)** If the bale's velocity points 45.0° below the horizontal then we know the vertical velocity:

$$v_y = -v_{0x} = -1.12\text{ m/s}$$

**4.** Find the time elapsed from equation 4-6: $t = \dfrac{v_y - v_{0y}}{-g} = \dfrac{-1.12\text{ m/s} - 8.85\text{ m/s}}{-9.81\text{ m/s}^2} = \boxed{1.02\text{ s}}$

**5. (c)** If $\vec{\mathbf{v}}_0$ is pointed straight upward then the initial vertical velocity component will be larger, so it will rise higher and its time in the air will $\boxed{\text{increase}}$.

**Insight:** The bale will have a speed of 5.00 m/s again after 1.40 s has elapsed. If it were thrown straight upward with the same initial speed (8.92 m/s), it would rise to a height of 4.06 m, as opposed to 3.99 m as in the original case.

64. **Picture the Problem**: The trajectory of the cork and its initial velocity $\vec{\mathbf{v}}_{cg}$ as seen by an observer on the ground are depicted at right.

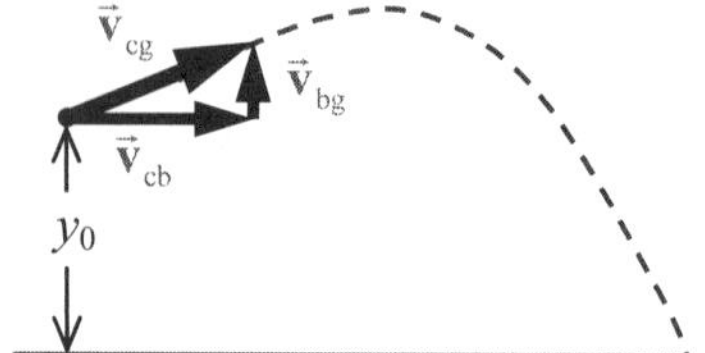


**Strategy:** Add the velocity of the balloon to the velocity of the cork as seen by an observer in the balloon to find the velocity of the cork as seen by an observer on the ground. Use the components of the initial velocity to find its magnitude and direction. Then use equation 4-10 to determine the maximum height of the cork above ground and the time of flight.

**Solution: 1. (a)** Add $\vec{\mathbf{v}}_{bg}$ (the velocity of the balloon relative to ground) to $\vec{\mathbf{v}}_{cb}$ (the velocity of the cork relative to the balloon):

$$\vec{\mathbf{v}}_{cg} = \vec{\mathbf{v}}_{cb} + \vec{\mathbf{v}}_{bg} = \left[(5.00\text{ m/s})\hat{\mathbf{x}}\right] + \left[(2.00\text{ m/s})\hat{\mathbf{y}}\right]$$

$$\vec{\mathbf{v}}_{cg} = \boxed{(5.00\text{ m/s})\hat{\mathbf{x}} + (2.00\text{ m/s})\hat{\mathbf{y}}}$$

**2. (b)** Use the components of $\vec{\mathbf{v}}_{cg}$ to find the speed:

$$v_{cg} = \sqrt{(5.00\text{ m/s})^2 + (2.00\text{ m/s}^2)} = \boxed{5.39\text{ m/s}}$$

**3.** Use the components of $\vec{\mathbf{v}}_{cg}$ to find the direction:

$$\theta = \tan^{-1}\left(\frac{v_y}{v_x}\right) = \tan^{-1}\left(\frac{2.00\text{ m/s}}{5.00\text{ m/s}}\right) = \boxed{21.8^\circ\text{ above horizontal}}$$

**4. (c)** Use equation 4-6 to find the maximum height of the cork, noting its initial height is 6.00 m:

$$v_y^2 = 0 = v_{0y}^2 - 2g\Delta y$$

$$\Delta y = \frac{v_{0y}^2}{2g} = \frac{(2.00\text{ m/s})^2}{2(9.81\text{ m/s}^2)} = 0.204\text{ m}$$

$$y_{max} = y_0 + \Delta y = 6.00 + 0.204\text{ m} = \boxed{6.20\text{ m}}$$

**5. (d)** Use equation 4-6 to find the time to reach the peak of flight:

$$t_{up} = \frac{v_y - v_{0y}}{-g} = \frac{0 - 2.00\text{ m/s}}{-9.81\text{ m/s}^2} = \underline{\underline{0.204\text{ s}}}$$

**6.** Use equation 4-6 to find the time to fall from the peak of flight:

$$y = y_0 + v_{0y}t - \tfrac{1}{2}gt^2 \Rightarrow 0 = y_0 + 0 - \tfrac{1}{2}gt_{down}^2$$

$$t_{down} = \sqrt{\frac{2y_0}{g}} = \sqrt{\frac{2(6.20\text{ m})}{9.81\text{ m/s}^2}} = \underline{\underline{1.124\text{ s}}}$$

**7.** Add the times to find the total time of flight: $t_{flight} = t_{up} + t_{down} = 0.204 + 1.124\text{ s} = \boxed{1.33\text{ s}}$

**Insight:** Another way to solve part (d) is to use $v_y^2 = v_{0y}^2 - 2g\Delta y$ to find $v_y$, then use $v_y$, $v_{0y}$, and the acceleration of gravity to find the total time of flight. It saves a step or two but doesn't reveal $t_{up}$ or $t_{down}$.

81. **Picture the Problem**: The archerfish hits the bug while the water droplet is still moving upward toward the peak of its flight.

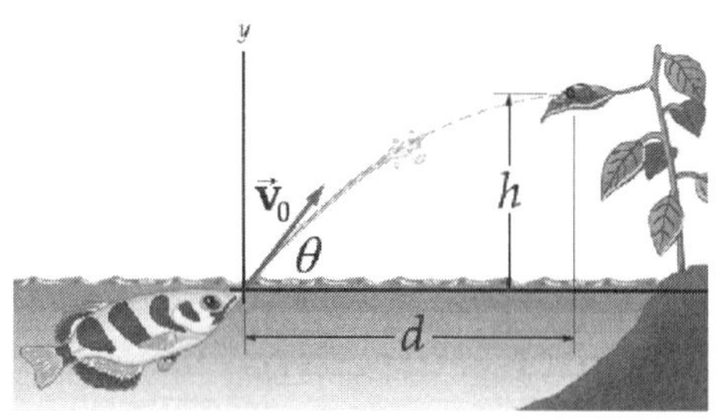


**Strategy:** Use equation 4-6 to find the time of flight of the water droplet that would place it at 3.00 cm above the water's surface after being launched with a speed of 2.15 m/s at 52.0° above the horizontal. Use the time of flight and the horizontal speed of the droplet to determine the horizontal distance at which the fish should be located.

**Solution: 1. (a)** Find the initial vertical speed $v_{0y}$:

$$v_{0y} = v_0 \sin\theta = (2.15\text{ m/s})\sin 52.0° = \underline{\underline{1.694\text{ m/s}}}$$

**2.** Solve $y = v_{0y}t - \frac{1}{2}gt^2$ for $t$ using the quadratic formula:

$$t = \frac{-b \pm \sqrt{b^2 - 4ac}}{2a}$$

$$= \frac{-v_{0y} \pm \sqrt{v_{0y}^2 - 4\left(-\frac{1}{2}g\right)(-y)}}{2\left(-\frac{1}{2}g\right)}$$

$$= \frac{1.694\text{ m/s} \mp \sqrt{(1.694\text{ m/s})^2 - 2(9.81\text{ m/s}^2)(0.0300\text{ m})}}{9.81\text{ m/s}^2}$$

$$t = \underline{\underline{0.0187,\ 0.327\text{ s}}}$$

**3.** Choose the smaller of the two times (the larger time corresponds to when the drop is moving downward) and find the horizontal distance:

$$x = (v_0\cos\theta)t = (2.15\text{ m/s})\cos 52.0°(0.0187\text{ s})$$

$$x = 0.0248\text{ m} = \boxed{2.48\text{ cm}}$$

**4. (b)** The bug has only $\boxed{0.0187\text{ s}}$ or 18.7 ms to react to the incoming water droplet.

**Insight:** The 18.7 ms is about one-tenth of the human reaction time. The bug is most likely going to be dinner!

## Answers to Practice Quiz

**1.** (c) **2.** (a) **3.** (b) **4.** (a) **5.** (d) **6.** (b) **7.** (d) **8.** (c) **9.** (b) **10.** (c)

# CHAPTER 5
# NEWTON'S LAWS OF MOTION

## Chapter Objectives

After studying this chapter, you should

1. be able to state, and understand the meaning of, Newton's three laws of motion.
2. be able to apply Newton's laws to simple situations in one and two dimensions.
3. be able to draw free-body diagrams.
4. know the difference between weight and mass.
5. be able to apply Newton's laws on inclined surfaces.

## Warm-Ups

1. The word *push* is a reasonably good synonym for the word force.
   Is *hold* a synonym for force?
   How about *support*?
   How many others can you think of?
2. The engine on a fighter airplane can exert a force of 105,840 N (24,000 lb). The take-off mass of the plane is 16,875 kg. (It weighs 37,500 lb.) If you mounted this aircraft's engine on your car, what acceleration would you get? (Use metric units. The data in pounds are given for comparison. Use a reasonable estimate for the mass of your car. One kilogram of mass weighs 2.2 lb.)
3. A 72-kg skydiver is descending with a parachute. His speed is still increasing at 1.2 m/s$^2$. What are the magnitude and direction of the force of the parachute harness on the skydiver? What are the magnitude and direction of the net force on the skydiver?
4. Suppose you run into a wall at 4.5 m/s (about 10 mi/h). Let's say the wall brings you to a complete stop in 0.5 s. Find your average deceleration, and estimate the force (in newtons) that the wall exerted on you while you were stopping. Compare that force with your weight.

## Chapter Review

In the previous chapters, we studied kinematics, which you may recall is the description of motion. In this chapter, we begin our study of **dynamics**. To begin this study we start with the concept of **force**. Force relates not only to a description of motion but also to the causes of various types of motion. The concept of force is governed by three laws known as Newton's laws of motion; these three laws, and how to apply them, are the principal focus of this chapter.

## 5–1 Force and Mass

A force is a push or a pull. When a nonzero net force is applied to an object, the object's response can be detected by its subsequent motion. Precisely how an object moves in response to a force depends on the object's **mass**. Mass can be thought of as a measure of the matter content of an object. More importantly for motion, though, mass is a measure of an object's *natural* tendency to move with constant velocity, referred to as its **inertia**, that is, **mass is a measure of inertia**. As discussed in Chapter 1, the SI unit of mass is the kilogram (kg).

## 5–2 – 5–5 Newton's Laws of Motion

Newton's first law of motion is also known as the *law of inertia*. This law essentially defines the concept of inertia in the context of motion.

> **Newton's First Law**
>
> *An object moving with constant velocity continues to do so unless acted upon by a nonzero net force.*

This law says that the natural state of motion is that of constant velocity. Since the law does not distinguish the constant velocity of zero (at rest) from any other constant velocity, it says that all constant velocities are equivalent.

Often, we loosely refer to anything that can detect an object as an observer. Any observer can be treated as the origin of a coordinate system in which it makes measurements. Moving observers carry their coordinate systems with them. We refer to an observer's coordinate system as its *frame of reference*. An **inertial frame of reference** (or just *inertial frame*, for short) is one in which the law of inertia holds. Inertial frames move with constant velocity relative to other inertial frames. Newton's three laws of motion are true with respect to these constant-velocity reference frames, and it is in this sense that all constant velocities are equivalent.

The law of inertia says that an object's velocity will change (i.e., it will accelerate) if a nonzero force is applied to it; therefore, **force causes acceleration**. Newton's second law picks up on this theme and tells us, quantitatively, how the acceleration relates to the net force.

**Newton's Second Law**

*The acceleration of an object equals the ratio of the net force on the object to its mass.*

As an equation this law is most commonly written as

$$\vec{\mathbf{a}} = \frac{\sum \vec{\mathbf{F}}}{m} \quad \text{or} \quad \sum \vec{\mathbf{F}} = m\vec{\mathbf{a}}$$

In these expressions, $\Sigma\vec{\mathbf{F}}$ represents the net force as the vector sum of all forces acting on the object. From this form of Newton's second law, we can also see, from the fact that the acceleration is inversely proportional to the mass, why mass is interpreted as a measure of inertia. Saying that inertia is the tendency to maintain a constant velocity is the same as saying that it is the resistance to acceleration. The greater an object's mass, the smaller its acceleration will be for a given force, and vice versa. Thus, mass is what determines how strongly an object "wants" to keep its velocity constant.

The SI unit of force is the newton (N). The mathematical statement of Newton's second law makes it clear that the unit of force is the product of the units of mass and acceleration: $1\text{ N} = 1\text{ kg}\cdot\text{m/s}^2$. In the application of Newton's second law, it will frequently be convenient to resolve the vector equation into scalar components

$$\sum F_x = ma_x \quad \sum F_y = ma_y \quad \sum F_z = ma_z$$

In all situations involving force, each of these equations must be satisfied independently. Another very important aspect of force analysis is the use of a **free-body diagram**. In such a diagram, we isolate the pertinent body, making it a "free body," then we draw all the force vectors acting on this body. In cases of nonrotational motion, we will usually (but not always) idealize the body as a point particle without loss of accuracy. As detailed in your text, a good general strategy for doing force analysis is as follows:

1. Sketch the physical picture with forces drawn.
2. Draw a free-body diagram of the object(s) of interest.
3. Choose a convenient coordinate system.
4. Resolve the forces on your free-body diagram into components.
5. Apply Newton's second law to each coordinate direction.

---

**Example 5–1 Skydiving** Skydiving equipment, including the parachute, typically has a mass of about 9.1 kg. At one point during a jump, an 80.0-kg skydiver is accelerating downward at 1.50 m/s$^2$. Determine the net force on the skydiver in vector notation.

**Picture the Problem** The leftmost sketch shows the skydiver in the air and the two main forces involved. The rightmost sketch is a free-body diagram of an idealized skydiver. The choice of the coordinate system is indicated between them.

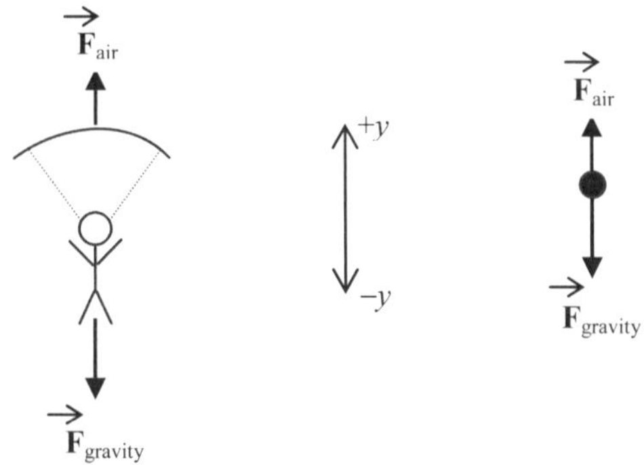


**Strategy** We need only the net force on the skydiver, so we should draw a free-body diagram and go straight to Newton's second law for the vertical direction, with "up" as $+y$.

**Solution**

1. Determine the total mass, $M$: $M = 80.0 \text{ kg} + 9.1 \text{ kg} = 89.1 \text{ kg}$

2. By the choice of coordinate system, the acceleration is in the negative $y$ direction: $\vec{\mathbf{a}} = -(1.50 \text{ m/s}^2)\hat{\mathbf{y}}$

3. Use Newton's second law to get the net force: $\vec{\mathbf{F}}_{net} = M\vec{\mathbf{a}} = -(89.1 \text{ kg})(1.50 \text{ m/s}^2)\hat{\mathbf{y}} = -(134 \text{ N})\hat{\mathbf{y}}$

**Insight** The acceleration and net force are written as negative only by choice. We could easily have chosen downward to be the positive direction, and no negative signs would have been needed.

---

**Example 5–2 Vehicle Performance** With a full tank of gas the Pontiac Grand AM SE has a mass of about 1500 kg; its 0 → 60 mi/h time is listed as 6.7 seconds. What net forward force must have been exerted on the vehicle during this performance test?

**Picture the Problem** The sketch shows the car and the net forward force on it.

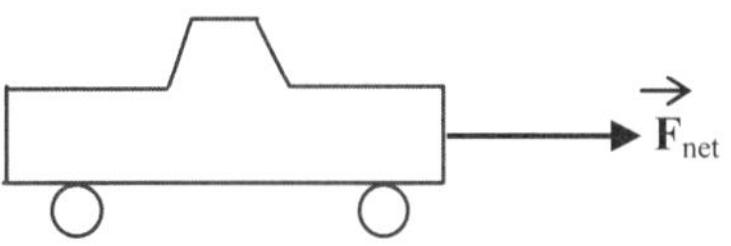


**Strategy** Since we don't know for certain that the acceleration is uniform, we can determine only the average net force on the vehicle from the average acceleration.

**Solution**

1. Convert the speed to SI units: $v = 60 \text{ mi/h} \times \dfrac{1 \text{ m/s}}{2.24 \text{ mi/h}} = 26.8 \text{ m/s}$

**2**. Use one of the equations for constant acceleration to determine $a$: $a_{av} = \dfrac{v - v_0}{\Delta t} = \dfrac{26.8 \text{ m/s}}{6.7 \text{ s}} = 4.0 \text{ m/s}^2$

**3**. Use Newton's laws to get the average net force: $F_{av} = ma_{av} = (1500 \text{ kg})(4.0 \text{ m/s}^2) = 6000 \text{ N}$

**Insight** We calculated only the magnitude because we already stipulated that the direction is "forward."

---

Newton's third law of motion is commonly known as the law of action and reaction. The words *action* and *reaction* refer to forces, and so *force* is the word we'll use:

> **Newton's Third Law**
>
> *For every force that an agent applies to an object, there is a reaction force equal in magnitude and opposite in direction applied by the object to the original agent.*

An important point to remember here is that a force and its reaction *always act on different objects*; therefore, these forces *never* cancel each other. Basically, this law says that a single object cannot act on others without being acted on; two objects always interact, applying equal and opposite forces to each other.

## Practice Quiz

1. If the same force is applied to two different objects, the object with greater inertia will have

   **(a)** greater acceleration **(b)** smaller acceleration **(c)** the same acceleration as the other object
   **(d)** zero acceleration **(e)** none of the above

2. If two different objects have the same acceleration, the object with greater inertia has

   **(a)** greater force applied to it
   **(b)** less force applied to it
   **(c)** the same force applied to it as the other object has applied to it
   **(d)** zero force applied to it
   **(e)** none of the above

3. An object is observed to have an acceleration of 8.3 m/s$^2$ when a net force of 12.2 N is applied to it. What is its mass?

   **(a)** 100 kg **(b)** 0.68 kg **(c)** 1.5 kg **(d)** 21 kg **(e)** 3.9 kg

4. Under the action of a constant net force, a 3.1-kg object moves a distance of 15 m in 8.99 s starting from rest. What is the magnitude of this net force?
 **(a)** 0.37 N **(b)** 1.2 N **(c)** 420 N **(d)** 5.2 N **(e)** 0.58 N

## 5–6 – 5–7 Weight, Normal Force, and Inclined Surfaces

### Weight

Our everyday lives are partly governed by the fact that everything has **weight**. By definition, a body's weight on Earth is the downward gravitational force exerted on the body by the earth. You may recall that near Earth's surface all bodies fall with the gravitational acceleration, $g = 9.81\ \text{m/s}^2$. In accordance with Newton's second law, therefore, the gravitational force on a body, its weight, equals the product of its mass and this gravitational acceleration:

$$\vec{\mathbf{W}} = m\vec{\mathbf{g}}$$

The direction of the vectors $\vec{\mathbf{W}}$ and $\vec{\mathbf{g}}$ is toward the center of Earth.

The sensation of having weight is clear to us because of the force of contact between our feet and the ground beneath us. Earth's gravity pulls us into the floor, and the floor reacts by pressing back against us. This reaction is what feels like our weight. However, if the object on which we are standing is accelerating, this reaction force can trick us into feeling either heavier or lighter depending on the direction of the acceleration. This reaction force is called our **apparent weight** because it is the weight we perceive ourselves to have even when it differs from the actual force of gravity on us.

### Normal Force

Another everyday force occurs when two surfaces come into direct physical contact. The force of contact between the surfaces can be resolved into components that are parallel and perpendicular to the surfaces. The perpendicular component of this force is called the **normal force** (*normal* means "perpendicular"). The normal force that the floor exerts on your feet when you are standing is what we just called the apparent weight. As you'll see in the next chapter, the normal force between two surfaces is an important factor in determining the friction between these surfaces.

---

**Example 5–3 Going Up in an Elevator** Low-acceleration elevators that might be found in a hospital, typically accelerate at about $3.00\ \text{ft/s}^2$ when first starting upward. If a person knows her weight to be 125 lb, what would a scale read if she is standing on it when the elevator starts upward?

**Picture the Problem** The left-hand sketch shows a person standing on a scale in an upward-accelerating elevator. The right-hand sketch is the free-body diagram.

**Strategy** The reading on the scale is determined by the force that the person exerts on the scale. By Newton's third law we know that the force on the scale equals the normal force $\vec{\mathbf{N}}$ of the scale on the person (her apparent weight, $\vec{\mathbf{N}} = \vec{\mathbf{W}}_a$). So, instead, we draw a free-body diagram of the person because we know the person's true weight. Let's choose "up" to be +y.

**Solution**

1. Apply Newton's second law to the free-body diagram:

$$\sum F_y = N_y + W_y = N - W = ma$$
$$\Rightarrow N = W + ma$$

2. Solving for $\vec{\mathbf{N}}$ gives us the apparent weight:

$$N = 125 \text{ lb} + \left(\frac{125 \text{ lb}}{32 \text{ ft/s}^2}\right)(3.00 \text{ ft/s}^2) = 137 \text{ lb}$$

**Insight** Notice that in the sum of forces in step 1, $W_y = -W = -125$ lb.

---

**Exercise 5–4 Rearranging** While rearranging the living room, a student pushes a 27.5-kg sofa across the room at constant speed by applying a force that makes an angle of 35.0° below the horizontal. If the floor opposes the sliding sofa (by friction) with a backward force of 160 N, then **(a)** what is the normal force of the floor on the sofa, and **(b)** what magnitude of force does the student apply to the sofa?

**Solution** Try to sketch a physical picture for this problem; the free-body diagram for the sofa is shown.

**Given**: $m = 27.5$ kg, $\theta = -35.0^\circ$, $f = 160$ N; **Find**: **(a)** $N$, **(b)** $F$

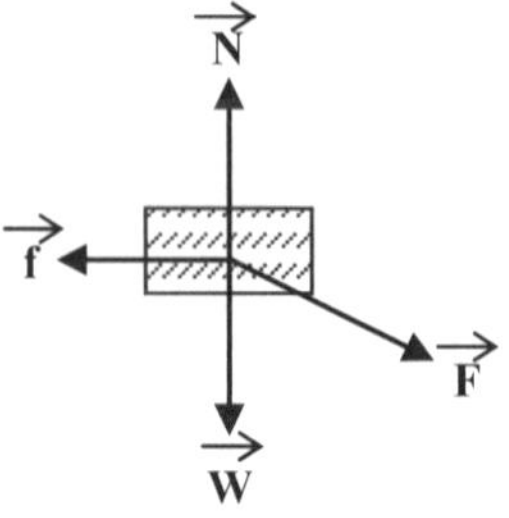

Taking the $+x$ direction to the right and the $+y$ direction up, we can apply Newton's second law to both the $x$ and $y$ directions. Because part (a) asks about the normal force, let's start with the $y$ direction. The sofa does not accelerate vertically up or down, so the net vertical force must be zero:

$$\sum F_y^{all} = N_y + W_y + F_y = 0 \Rightarrow N - W - F_y = 0$$
$$\therefore N = W - F_y$$

We cannot solve this equation immediately for $N$ because we don't yet know $F_y$. Notice, however, that since we know the direction of $\vec{\mathbf{F}}$, we can determine $F_y$, provided we can figure out $F_x$. Thus, we next apply Newton's second law to the $x$ direction. The sofa is being pushed at constant speed; therefore, the horizontal acceleration is zero, which means that the net horizontal force must also be zero,

$$\sum F_x^{all} = F_x + f_x = 0 \Rightarrow F_x - f = 0$$
$$\therefore F_x = f = 160 \text{ N}$$

Now that we know $F_x$ we can determine $F_y$ through the relation $\tan\theta = F_y / F_x$:

$$F_y = F_x \tan\theta = (160 \text{ N})\tan(-35.0°) = -112 \text{ N}$$

The minus sign just means that $F_y$ is along the $-y$ direction. Having $F_y$, we are now ready to calculate the normal force on the sofa,

$$N = W - F_y = mg - F_y = (27.5 \text{ kg})(9.81 \text{ m/s}^2) + 112 \text{ N} = 382 \text{ N}$$

which is the final result for part (a).

For part (b), we seek the magnitude of $\vec{\mathbf{F}}$. Because we have already calculated the two components of $\vec{\mathbf{F}}$, we can determine its magnitude from the Pythagorean theorem:

$$F = \sqrt{F_x^2 + F_y^2} = \sqrt{(160 \text{ N})^2 + (-112 \text{ N})^2} = 195 \text{ N}$$

An important point here is that Newton's second law must be satisfied in every direction, and that often a result from an equation for one direction is useful in solving an equation for a different direction.

## Practice Quiz

5. An object is at rest on a horizontal table. The normal force exerted by the table on the object
   **(a)** points vertically up **(b)** points vertically down **(c)** is parallel to the table
   **(d)** is zero **(e)** none of the above

6. An object sits undisturbed on a horizontal table. The normal force exerted by the table on the object
   **(a)** equals its weight **(b)** is greater than its weight **(c)** is less than its weight
   **(d)** is zero **(e)** none of the above

7. An object has a mass of 3.25 kg. Its weight is

**(a)** 3.25 kg **(b)** 3.25 N **(c)** 3.02 N **(d)** 7.15 N **(e)** none of the above

8. A man of mass 55.0 kg sits in a chair of mass 62.0 kg. The normal force of the chair on the man is

**(a)** 540 N **(b)** 608 N **(c)** 1150 N **(d)** 117 N **(e)** 7.00 N

9. A woman stands in an elevator that is moving upward at constant speed. Compared to her actual weight on Earth's surface, her apparent weight is

**(a)** greater **(b)** less **(c)** the same **(d)** zero **(e)** none of the above

10. By pushing on the floor, a person accelerates a 120-lb box across a room at 1.3 $m/s^2$. The floor opposes the motion of the box with a frictional force of 110 N. What horizontal force does the person apply to the box?

**(a)** 230 N **(b)** 71 N **(c)** 180 N **(d)** 110 N **(e)** 423 N

**Inclined Surfaces**

Life isn't conveniently arranged to take place only on smooth, flat, horizontal surfaces. Sometimes we just have to go uphill or downhill. In these situations, it is best to choose a coordinate system with axes that are parallel and perpendicular to the surface. Generally, we take the $x$-axis to be parallel to the surface and the $y$-axis to be perpendicular to the surface as shown below.

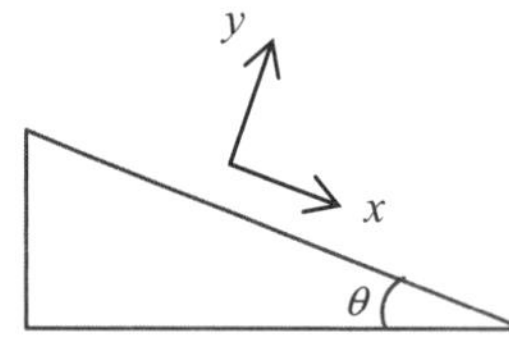


In this coordinate system, the weight, which always acts vertically downward, can be resolved into $x$ and $y$ components. If $\theta$ is the angle that the surface makes with the horizontal, then

$$W_x = W\sin\theta = mg\sin\theta \quad W_y = -W\cos\theta = -mg\cos\theta$$

Because the weight has components in both directions, the rest of the analysis on inclined surfaces follows precisely as it does on horizontal surfaces.

**Example 5–5 Finding a Better Way** A heavy suitcase that weighs 450 N and has wheels needs to be placed onto the back of a truck 0.65 m above the ground. Instead of lifting it straight up, you decide to get a thick piece of wood to use as a ramp and roll it up onto the truck. If the extended ramp makes an angle of 60° with the horizontal, **(a)** determine the minimum force needed to roll the suitcase up the ramp, and **(b)** compare this value with the minimum force needed to lift it up onto the truck.

**Picture the Problem** The upper sketch shows the truck with the wooden ramp and the suitcase. The lower sketch is a free-body diagram of the suitcase with $F_{min}$ as the magnitude of the minimum applied force needed to pull the suitcase up the ramp.

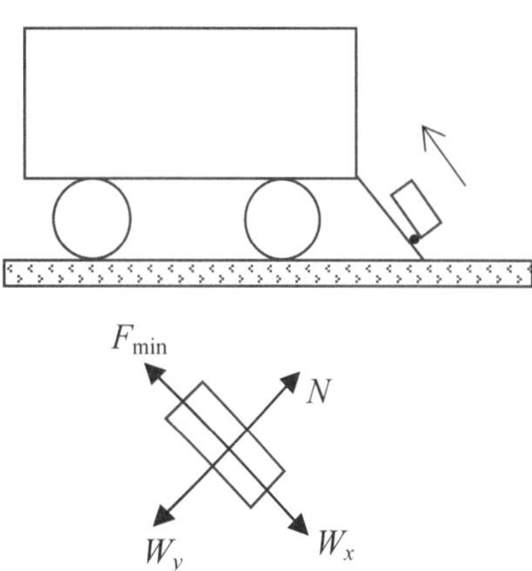


**Strategy** For part (a) we first recognize that the *minimum* force to roll the suitcase up must be parallel to the ramp (so all the force is used) and just enough to balance the suitcase's tendency to roll down the ramp (due to gravity). Therefore, the suitcase will roll up at constant speed. Likewise, for part (b), the minimum force just balances the weight of the suitcase.

**Solution**

**Part (a)**

1. Apply Newton's second law to the $x$ direction: $\sum F_x = F_{min,x} + W_x = 0 \;\Rightarrow\; -F_{min} + W\sin\theta = 0$
2. Solve this equation for $F_{min}$: $F_{min} = W\sin\theta = (450\text{ N})\sin(60^\circ) = 390\text{ N}$

**Part (b)**

3. Compare the result of (a) with the weight of the suitcase: $W - F_{min} = 60\text{ N}$
4. Calculate the percent difference: $\dfrac{60\text{ N}}{450\text{ N}} \times 100\% = 13\%$

**Insights** In the free-body diagram for this problem, the components of the weight were drawn directly on the diagram, rather than the vertically downward weight vector. This procedure is sometimes more convenient. Either way, the vectors are resolved into components, as part of the recommended strategy for force analysis, whether before or after the free-body diagram is drawn.

## Practice Quiz

11. An object is held at rest on a surface inclined at an angle θ with the horizontal. If the force holding it is parallel to the surface, the magnitude of the normal force exerted by the surface on the object

**(a)** equals its weight **(b)** is greater than its weight **(c)** is less than its weight

**(d)** is zero **(e)** none of the above

# Reference Tools and Resources

## I. Key Terms and Phrases

**dynamics** the branch of physics that studies force and the causes of various types of motion

**force** a push or a pull applied to an object

**mass** a measure of an object's inertia

**inertia** an object's natural tendency to move with constant velocity

**inertial reference frame** a frame of reference in which the law of inertia holds

**free-body diagram** a diagram of an isolated object showing all the force vectors acting on the object

**weight** the downward force due to gravity

**apparent weight** the perceived weight of an object as its force of contact with the ground or a scale

**normal force** the component of the contact force on a surface that is perpendicular to the surface

## II. Important Equations

| Name/Topic | Equation | Explanation |
|---|---|---|
| Newton's second law | $\vec{\mathbf{F}}_{net} = \sum \vec{\mathbf{F}} = m\vec{\mathbf{a}}$ | The vector sum of all forces acting on an object equals its mass times its acceleration |
| inclined surfaces | $W_x = mg\sin\theta$<br>$W_y = -mg\cos\theta$ | The components of an object's weight when on a surface inclined at an angle $\theta$ to the horizontal |

## III. Know Your Units

| Quantity | Dimension | SI Unit |
|---|---|---|
| force ( $F$ ) | $[\mathrm{M}]\cdot[\mathrm{L}]/[\mathrm{T}^2]$ | N |

### IV. Tips

You may have noticed that in the free-body diagrams in the preceding examples the forces were always drawn with their tails on the idealized body and their heads pointing away. This is the usual tradition and it is better to be consistent; so always draw the forces on your free-body diagrams with tails on the body and heads pointing away from the body.

## Puzzle

**STRETCHING**

A 10-kg mass (right) and a 5-kg mass (left) are suspended from pulleys as shown in the diagram. The lower mass is tied to the floor by another string. Two small metric scales (calibrated in newtons) are spliced into the rope. One of them is between the two pulley discs, the other just above the smaller mass. Assume that the masses of the scales are small enough to be neglected.

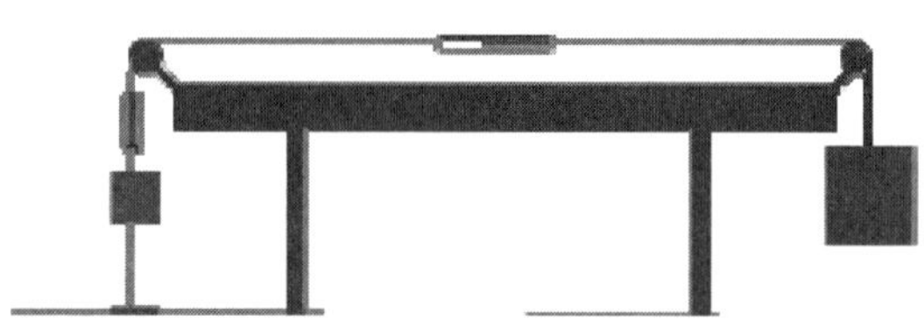

What do the scales read?

The string under the lighter mass is now cut. What do the scales read after the masses start to move.

## Answers to Selected Conceptual Questions

**2.** If the tablecloth is pulled rapidly, it can exert a force on the place settings for only a very short time. In this brief time, the objects on the table accelerate, but only slightly. Therefore, the objects may have barely moved by the time the tablecloth is completely removed.

**4.** As the dog shakes its body, water in its fur starts moving in one direction. Then it begins to shake its body in the opposite direction, much of the water continues in the same direction – due to the law of inertia (Newton's first law). As a result, water leaves the fur with each reversal in direction.

**14.** No. If only a single force acts on the object, it will not stay at rest; instead, it will accelerate in the direction of the force.

# Solutions to Selected End-of-Chapter Problems and Conceptual Exercises

11. **Picture the Problem**: The car is accelerated horizontally in the direction opposite its motion in order to slow it down from 16.0 m/s to 9.50 m/s.

**Strategy:** Use equation 5-1 and the definition of acceleration to determine the net force on the car as it slows down. Then use equation 2-10 to find the distance traveled by the car as it slows down.

**Solution: 1. (a)** Use equation 5-1 and the definition of acceleration to find the net force on the car:

$$\vec{\mathbf{F}} = m\vec{\mathbf{a}} = m\frac{\Delta\vec{\mathbf{v}}}{\Delta t} = (950 \text{ kg})\frac{(9.50-16.0 \text{ m/s})}{(1.20 \text{ s})}\hat{\mathbf{x}} = (-5100 \text{ N})\hat{\mathbf{x}}$$

$$= \boxed{5.1 \text{ kN opposite to the direction of motion}}$$

**2. (b)** Use equation 2-10 to find the distance traveled by the car as it slows down:

$$\Delta x = \tfrac{1}{2}(v_0 + v)\Delta t = \tfrac{1}{2}(16.0+9.50 \text{ m/s})(1.20 \text{ s}) = \boxed{15.3 \text{ m}}$$

**Insight:** We must consider "950 kg" as having only two significant figures because the zero is ambiguous. That limits the net force to two significant figures, even though the acceleration $(-5.42 \text{ m/s}^2)\hat{\mathbf{x}}$ has three significant figures.

21. **Picture the Problem**: The force pushes on box 3 in the manner indicated by the figure at right.

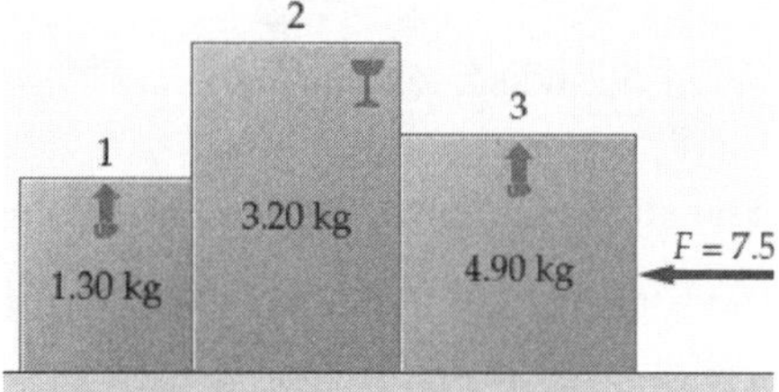


**Strategy:** The boxes must each have the same acceleration, but because they have different masses the net force on each must be different. These observations allow you to use Newton's Second Law for each individual box to determine the magnitudes of the contact forces. First find the acceleration of all the boxes and then apply equation 5-1 to find the contact forces.

**Solution: 1. (a)** The 7.50 N force accelerates all the boxes together:

$$F = (m_1 + m_2 + m_3)a \;\Rightarrow\; a = \frac{F}{m_1 + m_2 + m_3}$$

$$a = \frac{7.50\text{ N}}{1.30\text{ kg} + 3.20\text{ kg} + 4.90\text{ kg}} = \underline{\underline{0.798 \text{ m/s}^2}}$$

**2.** Write Newton's Second Law for the first box:

$$\sum\vec{\mathbf{F}} = F_{c12} = m_1\, a = (1.30 \text{ kg})(0.798 \text{ m/s}^2) = \boxed{1.04 \text{ N}}$$

**3. (b)** The contact force between boxes 2 and 3 is responsible for accelerating boxes 1 and 2:

$$\sum\vec{\mathbf{F}} = F_{c23} = (m_1 + m_2)a = (1.30+3.20 \text{ kg})(0.798 \text{ m/s}^2)$$

$$= \boxed{3.59 \text{ N}}$$

**Insight:** Another way to solve part (b) is to write Newton's Second Law for box 3 alone:
$\sum\vec{\mathbf{F}} = F - F_{c23} = m_3\, a \;\Rightarrow\; F_{c23} = F - m_3\, a = 7.50 \text{ N } -(4.90 \text{ kg})(0.798 \text{ m/s}^2) = 3.59 \text{ N}$ as we found in step 3.

30. **Picture the Problem**: The two teenagers pull on the sled in the directions indicated by the figure at right.

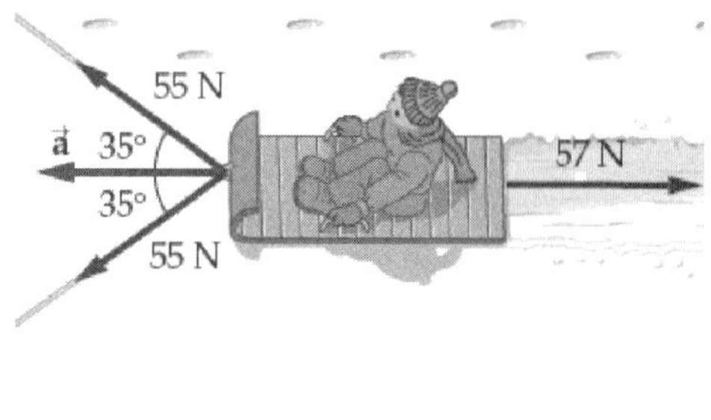


**Strategy:** Write Newton's Second Law in the $x$ direction (parallel to $\vec{a}$ ) in order to find the acceleration of the sled.

**Solution:** Write Newton's Second Law in the $x$ direction:

$$\sum F_x = 2F\cos 35° - 57\text{ N} = (m_{\text{sled}} + m_{\text{child}})a_x$$

$$a_x = \frac{2F\cos 35° - 57\text{ N}}{m_{\text{sled}} + m_{\text{child}}}$$

$$a_x = \frac{2(55\text{ N})\cos 35° - 57\text{ N}}{19 + 3.7\text{ kg}} = \boxed{1.5\text{ m/s}^2}$$

**Insight:** Some of the force exerted by the teenagers is exerted in the $y$ direction and cancels out; only the $x$ components of the forces move the sled.

40. **Picture the Problem**: The elevator accelerates up and down, changing your apparent weight $W_a$. A free body diagram of the situation is depicted at right.

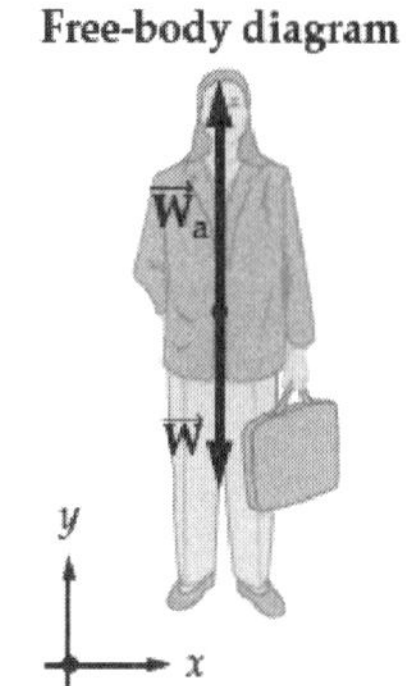


**Strategy:** There are two forces acting on you: the applied force $\vec{F} = \vec{W}_a$ of the scale acting upward and the force of gravity $\vec{W}$ acting downward. The force $W_a$ represents your apparent weight because it is both the force the scale exerts on you and the force you exert on the scale. Use Newton's Second Law together with the known force $W_a$ acceleration to determine the acceleration $a$.

**Solution: 1. (a)** The direction of acceleration is $\boxed{\text{upward}}$. An upward acceleration results in an apparent weight greater than the actual weight.

**2. (b)** Use Newton's Second Law together with the known forces to determine the acceleration $a$.

$$\sum F_y = W_a - W = ma$$

$$a = \frac{W_a - W}{m} = \frac{W_a - W}{W/g} = \frac{730 - 610\text{ N}}{610\text{ N}}(9.81\text{ m/s}^2) = \boxed{1.9\text{ m/s}^2}$$

**3. (c)** The only thing we can say about the velocity is that it is changing in the upward direction. That means $\boxed{\text{the elevator is either speeding up if it is traveling upward or slowing down if it is traveling downward}}$.

**Insight:** You feel the effects of apparent weight twice for each ride in an elevator, once as it accelerates from rest and again when it slows down and comes to rest.

50. **Picture the Problem**: The free body diagram of the lawn mower is shown at right.

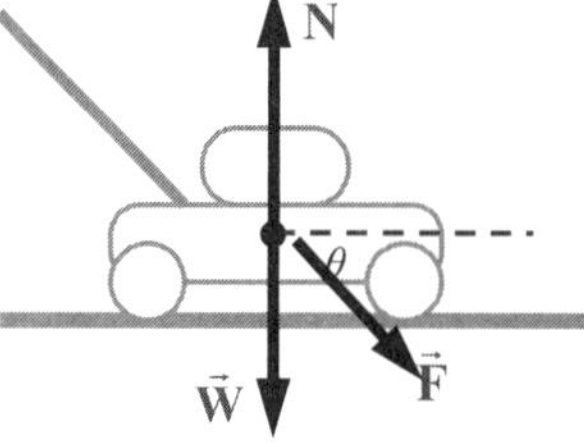


**Strategy:** Write Newton's Second Law in the vertical direction to determine the normal force.

**Solution: 1. (a)** Use Newton's Second Law to find $N$:

$$\sum F_y = N - F\sin\theta - mg = ma_y = 0$$

$$N = F\sin\theta + mg$$

$$= (219\text{ N})\sin 35° + (19\text{ kg})(9.81\text{ m/s}^2)$$

$$N = 310\text{ N} = \boxed{0.31\text{ kN}}$$

**2. (b)** If the angle between the handle and the horizontal is increased, the normal force exerted by the lawn will $\boxed{\text{increase}}$ because it must still balance the weight plus a larger downward force than before.

**Insight:** The vertical acceleration of the lawn mower will always remain zero because the ground prevents any vertical motion.

69. **Picture the Problem**: The spacecraft is accelerated in a straight line by the thrust from its engine.

**Strategy:** Convert the force of its engine from ounces to Newtons, and then find its acceleration in m/s$^2$. Finally, use Newton's Second Law to find the mass of the spacecraft.

**Solution: 1.** Convert the thrust to Newtons: $F = 0.064\text{ oz} \times \dfrac{1\text{ lb}}{16\text{ oz}} \times \dfrac{4.45\text{ N}}{\text{lb}} = \underline{\underline{0.0178\text{ N}}}$

**2.** Convert the acceleration to m/s$^2$: $a = \dfrac{\Delta v}{t} = \dfrac{7900\text{ mi/h}}{16000\text{ h}} \times \dfrac{1609\text{ m}}{\text{mi}} \times \left(\dfrac{1\text{ h}}{3600\text{ s}}\right)^2 = \underline{\underline{6.1\times10^{-5}\text{ m/s}^2}}$

**3.** Apply Newton's Second Law: $m = \dfrac{F}{a} = \dfrac{0.0178\text{ N}}{6.1\times10^{-5}\text{ m/s}^2} = \boxed{290\text{ kg}}$

**Insight:** We bent the rules of significant figures a bit in step 1 in order to avoid rounding error. A small amount of force exerted for a long time can make a big difference! We'll explore this more when we discuss impulse in Chapter 9.

## Answers to Practice Quiz

**1.** (b) **2.** (a) **3.** (c) **4.** (b) **5.** (a) **6.** (a) **7.** (e) **8.** (a) **9.** (c) **10.** (c) **11.** (c)

# CHAPTER 6

# APPLICATIONS OF NEWTON'S LAWS

## Chapter Objectives

After studying this chapter, you should

1. be able to perform force analysis in situations involving both static and kinetic friction.
2. be able to perform force analysis in situations involving string tensions and spring forces.
3. have a thorough understanding of translational equilibrium.
4. understand the roles of force and acceleration in circular motion.

## Warm-Ups

1. People could not walk without friction. Which type of friction are we referring to here, kinetic friction or static friction?
2. The coefficient of friction between a "safe" walking surface and your shoes is supposed to be about 0.6. Estimate the maximum acceleration you could attain under these conditions.
3. List all the forces acting on a car negotiating a turn at constant speed on a banked road. What should be the magnitude of the vector sum of all the forces acting on this car?
4. Estimate the coefficient of friction between skis and snow that would allow a skier to move down a 30-degree slope with constant speed if other factors were neglected. Is it reasonable to neglect other factors, such as air drag on the skier?

## Chapter Review

This chapter is a continuation of Chapter 5. Here we discuss further details in the application of Newton's laws by adding several specific forces, namely, friction, tension, and the Hooke's law force. We also introduce the concepts of centripetal force and the associated centripetal acceleration that relate to circular motion.

### 6–1 Frictional Forces

When two surfaces are in direct contact and one surface either slides or attempts to slide across the other, a force called *friction* that opposes the motion (or attempted motion) is generated between the surfaces.

When the surfaces are sliding across each other we call the force **kinetic friction**. When there is only attempted motion that is halted by friction we call it **static friction**. For example, if you apply a small horizontal force to a heavy object in an attempt to slide it, the object may not move at all because of the friction between it and the floor. In this case there is definitely friction but no motion, i.e., static friction.

The origin of friction is based, in complicated ways, on the microscopic structure of the surfaces involved. Often in physics, we handle these types of situations by identifying those factors on which the dependence is fairly simple and representing the rest of the complicated physics by one measured quantity. For kinetic friction we note that the magnitude of the frictional force, $f_k$, is directly proportional to the magnitude of the normal force, $N$, between the surfaces. The measured proportionality factor, $\mu_k$, is called the *coefficient of kinetic friction*. Thus, we have

$$f_k = \mu_k N$$

Note that this equation relates only the magnitudes of $\vec{\mathbf{f}}_k$ and $\vec{\mathbf{N}}$ because they are not in the same direction; they are perpendicular to each other.

For static friction, the force, $\vec{\mathbf{f}}_s$, takes on a range of values depending on the strength of the force it opposes. Static friction will cancel out any force trying to slide two surfaces across each other up to some maximum magnitude beyond which static friction is overcome. The magnitude of this maximum force of static friction is also directly proportional to the magnitude of the normal force. The measured proportionality factor, $\mu_s$, is called the *coefficient of static friction*. Thus, we have

$$f_{s,\max} = \mu_s N$$

which, for the more general case yields

$$f_s \leq \mu_s N$$

Typically, for a given pair of surfaces we have $\mu_k < \mu_s$.

---

**Exercise 6–1 Getting It Going** A 105-lb crate sits on a floor. If the minimum horizontal force required to start it sliding is 33 lb, what is the coefficient of static friction between the crate and the floor?

**Solution:** The free-body diagram is shown below. The following information is given in the problem:

**Given:** $W = 105$ lb, $F_{\min} = 33$ lb; **Find:** $\mu_s$

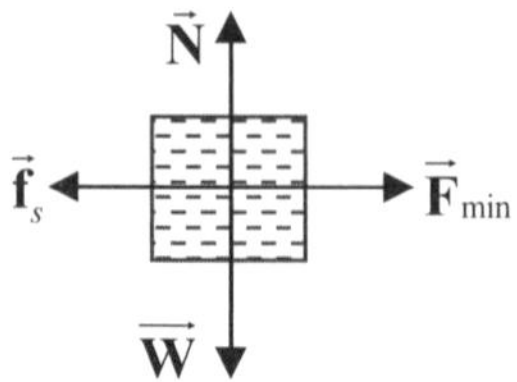

The fact that we seek the minimum force needed to get the crate moving means that we want the force that is just barely large enough to overcome static friction. The force right at this critical condition equals the maximum force that static friction can apply

$$F_{\text{min}} = f_{s,\text{max}} = \mu_s N$$

To determine $N$, we apply Newton's second law in the $y$-direction (recognizing that the vertical acceleration is zero): $N - W = 0$. This tells us that $N = W$. Hence, we have

$$F_{\text{min}} = \mu_s W \therefore \mu_s = \frac{F_{\text{min}}}{W} = \frac{33\text{ lb}}{105\text{ lb}} = 0.31$$

## Practice Quiz

**1.** A person pushes an object across a room, on a horizontal floor, by applying a horizontal force. If, instead, he pushed it with a force that made an angle $\theta$ below the horizontal, the object would have experienced

**(a)** greater friction. **(b)** less friction. **(c)** the same frictional force.

**(d)** zero friction.

**2.** A person pulls an object across a room, on a horizontal floor, by applying a horizontal force. If, instead, she pulled it with a force that made an angle $\theta$ above the horizontal, the object would have experienced

**(a)** greater friction. **(b)** less friction. **(c)** the same frictional force.

**(d)** zero friction.

**3.** The coefficient of kinetic friction represents

**(a)** the force that one surface applies to another when they slide across each other.

**(b)** the force that one surface applies to another when they do not slide across each other.

**(c)** the force that one surface applies to another that is perpendicular to the interface between them.

**(d)** the ratio of the normal force to the force of friction between two sliding surfaces.

**(e)** None of the above.

## 6–2 Strings and Springs

A common way in which forces are applied is by the action of a string (or any rope, cord, or cable wire). The force transmitted through the string is called the **tension** in the string. Unless otherwise stated, we shall, for now, consider strings of negligible mass for which the tension can be considered constant throughout. Therefore, the force applied on one end of a string gets transmitted through to objects attached to the other end.

Another important force in physics is governed by a principle known as **Hooke's law**. The prototype object that applies a Hooke's law force is an ideal spring. When a spring is deformed from its equilibrium length, that is, when it is stretched or compressed, it responds to this deformation by applying a force that tries to restore itself to equilibrium. For this reason, a Hooke's law force is also referred to as a *restoring force*. The precise strength of this restoring force is found to be directly proportional to the amount of deformation (provided the spring is not stretched beyond a certain limit). The proportionality factor between the force and the amount of deformation is called the **force constant** $k$. Taking the direction along which the string is deformed as the $x$ direction, we write Hooke's law as

$$F_x = -kx$$

where the minus sign indicates that the force is exerted by the spring in the direction opposite the deformation $x$. If $x$ is in the positive direction, then $F_x$ points in the negative direction, and vice versa.

---

**Example 6–2 Rearranging (Again)** Suppose the student in Exercise 5–4 doesn't like the way he rearranged his room and decides to try again. This time he attaches a cord to the sofa and pulls the 27.5-kg sofa across the room, accelerating it at 0.150 m/s$^2$. If the cord makes an angle of 55.0° above the horizontal, and the coefficient of kinetic friction between the sofa and the floor is 0.240, what is the tension in the cord?

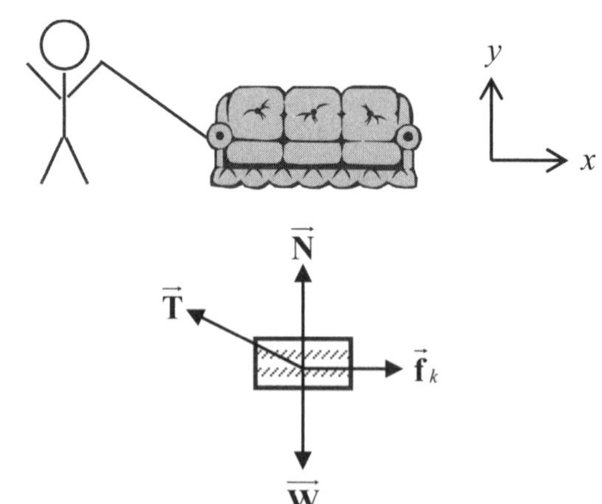


**Picture the Problem** The top sketch shows the sofa with the cord being pulled by the student. Below it is the free-body diagram of the sofa.

**Strategy** In this example, we follow the same steps for force analysis as in the previous chapter. First we resolve the forces into their $x$ and $y$ components and then apply Newton's second law to each direction.

**Solution**

1. Resolve the components of the tension: $T_x = -T\cos\theta,\ T_y = T\sin\theta$

2. Apply Newton's law to the $x$ direction: $\sum F_x = f_{k,x} + T_x = ma \Rightarrow \mu_k N - T\cos\theta = ma$

3. Apply Newton's law to the $y$ direction (there is no $y$ component of acceleration): $\sum F_y = N_y + W_y + T_y = 0 \Rightarrow N - mg + T\sin\theta = 0$

4. Solve for $N$ using the equation for the $y$ direction: $N = mg - T\sin\theta$

5. Place $N$ into the equation for the $x$ direction: $\mu_k(mg - T\sin\theta) - T\cos\theta = ma$

6. Solve for $T$ and evaluate (note that the acceleration is in the negative direction):

$$\mu_k mg - T(\mu_k \sin\theta + \cos\theta) = ma \therefore$$

$$T = \frac{m(\mu_k g - a)}{\mu_k \sin\theta + \cos\theta}$$

$$= \frac{27.5\text{ kg}\left[0.240(9.81\text{ m/s}^2) + 0.150\text{ m/s}^2\right]}{0.240\sin(55.0^\circ) + \cos(55.0^\circ)} = 89.4\text{ N}$$

**Insights** Notice here that we had to be careful with signs (as always). Both $T_x$ and $a$ were explicitly given minus signs because of the choice of coordinate system. If the problem asked for the force applied by the student instead, we would still work the problem exactly the same way using the approximation that the student's force is transmitted unchanged throughout the length of the cord.

---

**Example 6–3 Industrial Strength** Certain industrial springs, used to absorb shock and minimize damage, have force constants of about 225 N/cm. If such a spring breaks when stretched more than 5.5 cm, what is the maximum mass that can be hung from this spring without breaking it?

**Picture the Problem** The left-hand sketch shows a mass hanging from a spring. The right-hand sketch is the free-body diagram of the mass.

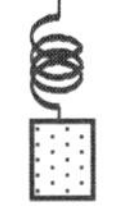

**Strategy** The maximum mass that can be hung from the spring without breaking it is a mass whose weight stretches it 5.5 cm. When such an object is hung from the spring, it sits motionless with an acceleration of zero.

**Solution**

1. Apply Newton's second law to the forces in the free-body diagram: $\vec{\mathbf{F}}_s + \vec{\mathbf{W}} = 0 \Rightarrow kx - mg = 0$

**2.** Rearrange and solve for mass:

$$mg = kx \Rightarrow m = \frac{kx}{g}$$

$$\therefore m = \frac{(225 \text{ N/cm})(5.5 \text{ cm})}{9.81 \text{ m/s}^2} = 130 \text{ kg}$$

**Insight** This mass of 130 kg weighs approximately 285 lb.

---

## Practice Quiz

4. Two people pull on opposite ends of a rope. If each one applies 30 N, what is the tension in the rope?

   **(a)** 60 N **(b)** 30 N **(c)** 15 N **(d)** 7.5 N **(e)** 0 N

5. If a spring stretches 3.5 cm from equilibrium when a 2.7-kg mass is hung from its end, what force is required to compress the spring 1.8 cm from its equilibrium length?

   **(a)** 26 N **(b)** 760 N **(c)** 17 N **(d)** 14 N **(e)** 130 N

## 6–3 Translational Equilibrium

A particularly important special case arises when the net force on an object is zero. Objects for which this is true are said to be in **translational equilibrium**. Since the net force equals zero, the acceleration is also zero. Why is this case important? Look around you; objects in translational equilibrium are everywhere. Notice that being in equilibrium (for short) does not mean being motionless. Exercise 5–4 and Example 5–5 were cases of translational equilibrium in which the object moved with constant velocity, whereas Example 6–3 is one in which the object of interest was motionless.

---

**Example 6–4 Held in Place** A 5.32-kg box is held stationary on a ramp inclined at 40.0 degrees to the horizontal by a cord that is attached to a vertical wall. The length of the cord is parallel to the ramp, and it provides just enough force to hold the box in place. If the coefficient of static friction between the box and the ramp is 0.101, what is the tension in the cord?

**Picture the Problem** The left-hand diagram shows the box on the ramp being held in place by the cord. The right-hand sketch is the free-body diagram of the box.

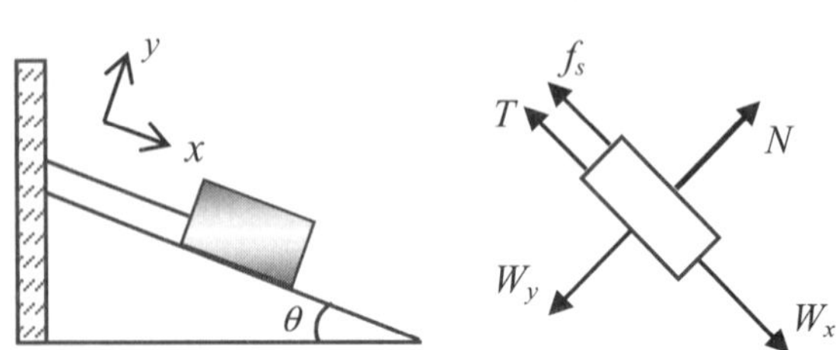


**Strategy** The case of translational equilibrium has only one condition to apply, $\sum \vec{\mathbf{F}} = 0$, so we

apply this condition for each direction if necessary. Because the tension is just enough to balance the forces, static friction must have its maximum value.

**Solution**

1. Apply the equilibrium condition to the $x$ direction: $W_x + T_x + f_{s,\max} = 0 \Rightarrow W_x - T - \mu_s N = 0$

2. Solve this equation for $T$: $T = mg\sin\theta - \mu_s N$

3. Apply the equilibrium condition to the $y$-direction: $N_y + W_y = 0 \Rightarrow N - mg\cos\theta = 0$
   $\therefore\ N = mg\cos\theta$

4. Substitute $N$ into the equation for $T$: $T = mg\sin\theta - \mu_s mg\cos\theta = mg(\sin\theta - \mu_s\cos\theta)$

5. Solve for the tension:
   $$T = (5.32\text{ kg})(9.81\tfrac{\text{m}}{\text{s}^2})\left[\sin(40^\circ) - 0.101\cos(40^\circ)\right] = 29.5\text{ N}$$

**Insights** Notice that for translational equilibrium in any direction the condition is always that the sum of the forces equals zero.

---

## Practice Quiz

6. If an object is in translational equilibrium,
   **(a)** it must be at rest.
   **(b)** there are no forces acting on it.
   **(c)** it is not rotating.
   **(d)** it moves with constant acceleration.
   **(e)** None of the above.

7. Can an object be in translational equilibrium if only one force acts on it?
   **(a)** Yes, if it is a gravitational force.
   **(b)** No, because even one force can have more than one component.
   **(c)** Yes, trying to lift an object that is too heavy for you to lift is an example.
   **(d)** No, an object with only one force on it must accelerate.
   **(e)** None of the above.

## 6–4 Connected Objects

In many applications, two or more objects are connected to each other. Typically, objects are connected by strings, springs, or direct contact. With regard to using Newton's laws to analyze connected objects there are two issues that we address in this section. The first issue concerns how to define the system on which you base your analysis. Do you consider each object in isolation, a subset of the objects, or all of the objects collectively as a single system? The answer is that you may do any or all of these things depending on the details of the problem and what you need to determine. The thing to keep in mind when working with connected objects is that no matter how you define your system, Newton's second law only involves the net *external* force acting *on* your system.

The second issue concerns the fact that in many cases, different connected objects may move in different directions. This raises the issue of how to define the coordinate system(s) for your analysis so as to properly represent the motions of interest. This too is usually a case-by-case decision depending on what you find most convenient.

**Exercise 6–5 Connected** Two objects are connected by a cord of negligible mass. One of mass 1.20 kg is pulled along a horizontal surface by a horizontal force $F$ = 8.11 N. The cord attached to this object extends over a light pulley and pulls the second object, of mass 0.832 kg, up a ramp inclined at 30.0° from the horizontal. If the coefficient of kinetic friction between the 1.20-kg object and the horizontal surface is 0.221 and that between the 0.832-kg object and the ramp is 0.147, determine **(a)** the acceleration of the objects and **(b)** the tension in the cord.

**Solution:** A diagram is shown below. The following information is given in the problem:

**Given:** $m_1 = 1.20$ kg, $F = 8.11$ N, $m_2 = 0.832$ kg, $\theta = 30.0°$, $\mu_1 = 0.221$, $\mu_2 = 0.147$

**Find:** (a) $a$, (b) $T$

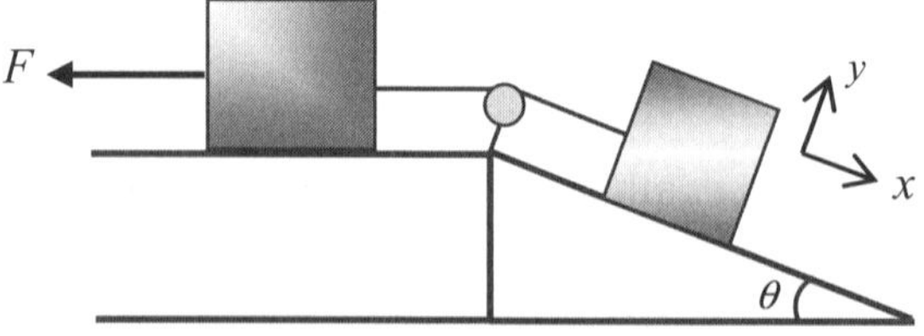


**(a)** Because we seek the acceleration of both objects, we really seek the acceleration of the entire system. The fact that the masses of the string and pulley are negligible means that the effect of the pulley is to change the direction of the tension in the cord undiminished. Taking both blocks as the system, the external forces are the applied force $F$, gravity, the normal forces, and the frictional forces; the tension in the cord is internal to our chosen system.

Given that the two objects move along different lines of motion, can we treat them as a single system? The answer is "yes." Furthermore, because the forces on each object are either parallel or perpendicular to the motion of that object, we are even able to treat the entire system as if they moved along the same line. An *effective* free-body diagram for the system could be drawn as

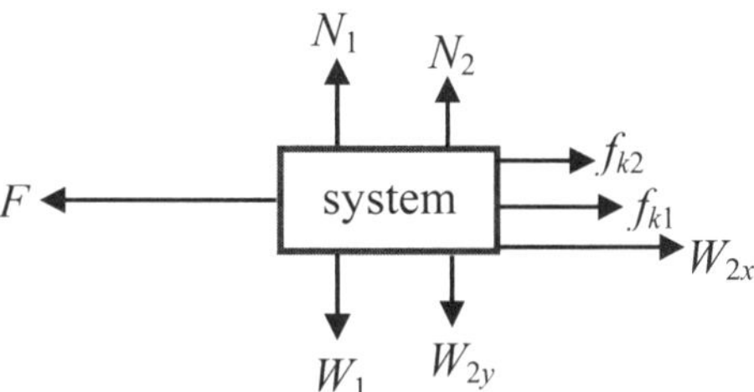


Of course, a free-body diagram is just a representation; when determining the values of these forces, their actual directions must be taken into account.

The acceleration of the system can be determined by Newton's second law

$$F - f_{k1} - f_{k2} - W_{2y} = M_{\text{sys}}a \quad \Rightarrow \quad a = \frac{1}{m_1 + m_2}\left(F - \mu_1 N_1 - \mu_2 N_2 - W_{2y}\right)$$

The fact that neither object accelerates perpendicular to the surface it is on requires that $N_1 = m_1 g$, $N_2 = m_2 g\cos\theta$, and $W_{2y} = m_2 g\sin\theta$ (verify these). So we can now write

$$a = \frac{F - \mu_1 m_1 g - \mu_2 m_2 g\cos\theta - m_2 g\sin\theta}{m_1 + m_2}$$

$$= \frac{8.11\text{ N} - \left[0.221(1.20\text{ kg}) + 0.147(0.832\text{ kg})\cos(30.0^\circ) + (0.832\text{ kg})\sin(30.0^\circ)\right](9.81\tfrac{\text{m}}{\text{s}^2})}{1.20\text{ kg} + 0.832\text{ kg}} = 0.191\tfrac{\text{m}}{\text{s}^2}$$

**(b)** With the acceleration known, we can now take a system consisting of either object alone to determine the tension in the cord. Taking the top one gives the free-body diagram

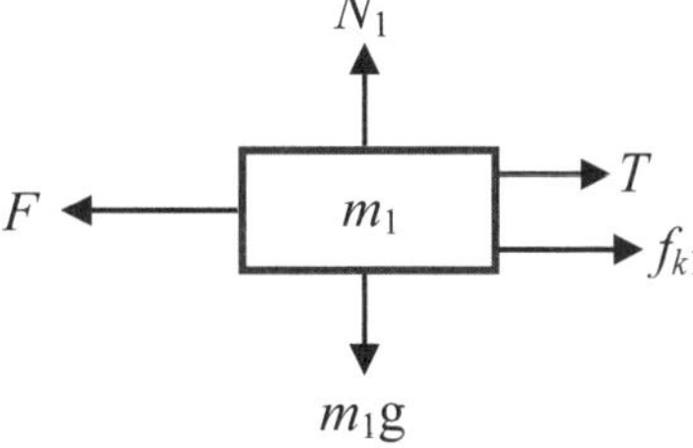


This free-body diagram gives

$$F - f_{k1} - T = m_1 a \quad \Rightarrow \quad T = F - f_{k1} - m_1 a$$

The final result is then

$$T = F - \mu_1 m_1 g - m_1 a = 8.11\text{ N} - 1.20\text{ kg}\left[0.221(9.81\tfrac{\text{m}}{\text{s}^2}) + 0.1911\tfrac{\text{m}}{\text{s}^2}\right] = 5.28\text{ N}$$

Just to show you that it can be done, this example was intentionally solved treating the forces that act along the directions of motion as if they acted along the same line even though they do not. Another approach would be to apply Newton's second law to each object individually. Then you would get a system of two equations with $a$ and $T$ as the two unknowns. Try solving it this way. Getting the acceleration by looking at both objects simultaneously, as done in this exercise, saves you the trouble of having to solve a system of equations.

## Practice Quiz

8. A horizontal force of 30 N is applied directly to a 3.0-kg block. A cord connects this block to a 2.0-kg block that is pulled along by the cord. If the blocks are on a frictionless horizontal surface, what is the acceleration of the blocks?

   **(a)** 6.0 m/s$^2$ **(b)** 10 m/s$^2$ **(c)** 15 m/s$^2$ **(d)** 1.0 m/s$^2$ **(e)** 30 m/s$^2$

9. What is the tension in the cord in question 8?

   **(a)** 30 N **(b)** 0 N **(c)** 12 N **(d)** 49 N **(e)** 18 N

## 6–5 Circular Motion

In the previous chapter we noted that life doesn't take place only on smooth, flat, horizontal surfaces, which is why we needed to study inclined surfaces. Likewise, motion doesn't always take place in straight-line paths. Therefore, we need to study how to handle motion along curves. Specifically, this section focuses on circular motion, or at least motion along a circular section. If the object that moves along this circular path does so at constant speed, then the acceleration must be perpendicular to the velocity and always point toward the center of the path. For this reason, the acceleration is called **centripetal acceleration**; its magnitude is given by

$$a_{cp} = \frac{v^2}{r}$$

where $v$ is its speed, and $r$ is the radius of the circular path.

According to Newton's second law, where there is an acceleration there must be a force that causes it. In the case of circular motion, the force must also point toward the center of the circular path and is therefore called a **centripetal force**. Also, in accordance with Newton's law, the magnitude of this centripetal force equals the product of the mass and the centripetal acceleration:

$$f_{cp} = m\frac{v^2}{r}$$

It is important to recognize that centripetal force is not a new kind of force. The above expression is merely a condition that must be met by whatever force holds the object in this *uniform circular motion*, as it is often called. In your studies, and in everyday life, you will come across circular motion caused by many different sources such as a normal force, tension, static friction, gravity, and in later chapters, electric and magnetic forces as well.

---

**Example 6–6 Making a Turn** Suppose that the combined mass of you and your bike is 75 kg. You are riding down the street and have to make a turn in a circular section of road whose radius is approximately 6.5 m. The speedometer on your bike reads 20 km/h. If the coefficient of static friction between your wheels and the road is 0.79, are you going too fast to make the turn safely?

**Picture the Problem** The sketch shows a side view for the free-body diagram of you and your bike. The center of the circular path is to the right.

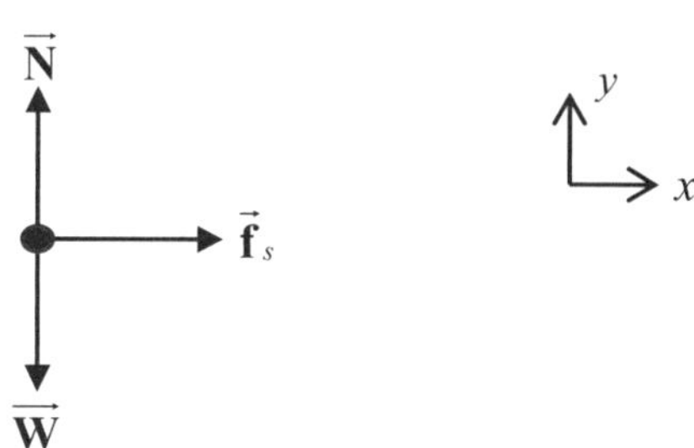


**Strategy** The first thing to recognize is that static friction must supply the centripetal force to sustain the circular motion, so the problem is really asking for a comparison between the required centripetal force and $f_{s,\max}$.

**Solution**

1. Convert the speed to m/s: $$v = 20\ \text{km/h}\left(\frac{1000\ \text{m}}{\text{km}}\right)\left(\frac{1\ \text{h}}{3600\ \text{s}}\right) = 5.56\ \text{m/s}$$

2. Evaluate the centripetal force: $$f_{cp} = m\frac{v^2}{r} = (75\ \text{kg})\frac{(5.56\ \text{m/s})^2}{6.5\ \text{m}} = 360\ \text{N}$$

3. Determine the maximum force of static friction: $$f_{s,\max} = \mu_s N = \mu_s mg = (0.79)(75\ \text{kg})(9.81\ \text{m/s}^2) = 580\ \text{N}$$

**Insights** The results show that the needed centripetal force is well within the range of static friction, so you're safe. Notice that in step 3, we set $N = mg$. We did this, without having to write out a separate Newton's second law equation for the $y$ direction, by inspecting the free-body diagram and recognizing that $\vec{\mathbf{N}}$ and $\vec{\mathbf{W}}$ must balance. As you do more and more problems you should begin to gain a similar feel for the free-body diagrams.

---

## Practice Quiz

**10.** For an object undergoing circular motion at constant speed,

**(a)** the velocity is constant.

**(b)** the acceleration is constant.

**(c)** the direction of the velocity is constant.

**(d)** the direction of the acceleration is constant.

**(e)** None of the above.

**11.** Which of the following correctly identifies the relationship between the directions of velocity and acceleration for objects in circular motion at constant speed?

**(a)** They are in the same direction.

**(b)** They are in opposite directions.

**(c)** They are perpendicular to each other.

**(d)** Their directions may differ by any angle depending on the curve of the arc.

**(e)** None of the above.

**12.** With what constant speed must an object traverse a circular path of radius 1.34 m in order to have a centripetal acceleration equal to the acceleration of gravity?

**(a)** 3.13 m/s **(b)** 13.1 m/s **(c)** 1.34 m/s **(d)** 9.81 m/s **(e)** 3.63 m/s

# Reference Tools and Resources

## I. Key Terms and Phrases

**kinetic friction** the contact force between two sliding surfaces that opposes their motion

**static friction** the contact force between two non-sliding surfaces that opposes their attempt to slide

**tension** the force transmitted through a string or taut wire

**Hooke's law** the force law for an ideal spring

**force constant** the proportionality factor between the force and the deformation in Hooke's law

**translational equilibrium** the situation in which the net force on an object is zero

**centripetal acceleration** the center-pointing acceleration of objects in circular motion

**centripetal force** the center-pointing force on objects in circular motion

## II. Important Equations

| Name/Topic | Equation | Explanation |
|---|---|---|
| kinetic friction | $f_k = \mu_k N$ | The force of kinetic friction is directly proportional to the normal force |
| static friction | $0 \le f_s \le \mu_s N$<br>$f_{s,\max} = \mu_s N$ | The force of static friction takes on a range of values up to a maximum |
| Hooke's law | $F_x = -kx$ | The force law for an ideal spring; $x$ is the deformation from its equilibrium length |
| centripetal acceleration | $a_{cp} = \dfrac{v^2}{r}$ | The magnitude of the acceleration for circular motion at a constant speed $v$ |
| centripetal force | $f_{cp} = m\dfrac{v^2}{r}$ | The magnitude of force on a mass $m$ moving in circular motion of radius $r$ at constant speed $v$ |

## III. Know Your Units

| Quantity | Dimension | SI Unit |
|---|---|---|
| coefficient of friction | — | dimensionless |

# Puzzle

**DOUBLE THE FUN?**

Experimenting with his toy rockets, Danny fires his model rockets three different ways.

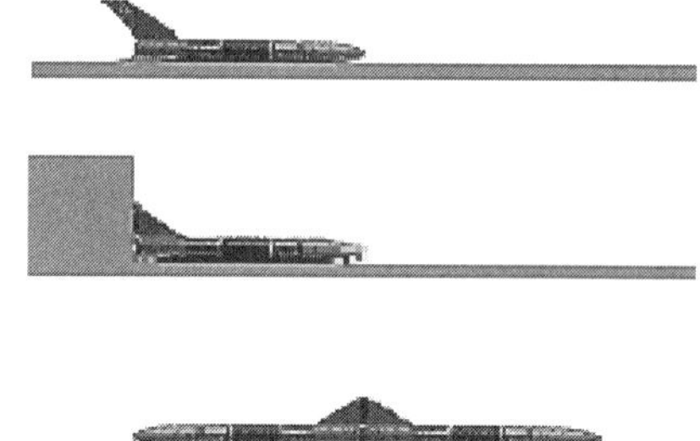

1. First, he places a model on the ground, free of obstructions and fires.
2. Next, he braces an identical model against a stiff wall and fires.
3. Third, he positions two identical models on the ground with their tails butting and fires them simultaneously.

   How do the rocket accelerations in the three experiments compare?

## Answers to Selected Conceptual Questions

**4.** The maximum acceleration is determined by the normal force exerted on the drive wheels. If the engine of the car is in the front and the drive wheels are in the rear, the normal force is less than it would be with front-wheel drive. During braking, however, all four wheels participate – including the wheels that sit under the engine.

**12.** Astronauts feel weightless because they are in constant free fall as they orbit, just as you would feel weightless inside an elevator that drops downward in free fall.

**20.** People on the outer rim of a rotating space station must experience a force directed toward the center of the station in order to follow a circular path. This force is applied by the "floor" of the station, which is really its outermost wall. Because people feel an upward force acting on them from the floor, just as they would on Earth, the sensation is like an "artificial gravity."

## Solutions to Selected End-of-Chapter Problems and Conceptual Exercises

10. **Picture the Problem**: The free-body diagram of this situation is depicted at right.

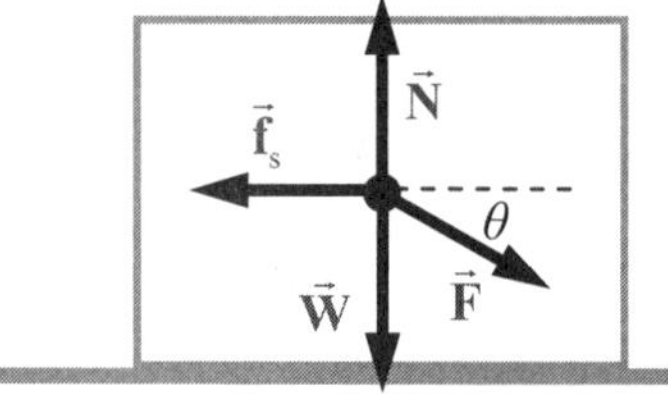


**Strategy:** Write Newton's Second Law in the vertical direction to determine the normal force of the floor on the crate. Then write Newton's Second Law in the horizontal direction to determine the minimum force necessary to start the crate moving.

**Solution: 1.** Write Newton's Second Law in the vertical direction:

$$\sum F_y = N - mg - F\sin\theta = ma_y = 0$$

$$N = mg + F\sin\theta$$

**2.** Write Newton's Second Law in the horizontal direction:

$$\sum F_x = F\cos\theta - \mu_s N = ma_x = 0$$

$$F\cos\theta = \mu_s N$$

**3.** Now substitute for $N$ and solve for the maximum force $F$ that would produce no acceleration:

$$F\cos\theta = \mu_s\left(mg + F\sin\theta\right)$$

$$F\left(\cos\theta - \mu_s\sin\theta\right) = \mu_s mg$$

$$F = \frac{\mu_s mg}{\left(\cos\theta - \mu_s\sin\theta\right)} = \frac{(0.57)(32\text{ kg})(9.81\text{ m/s}^2)}{\left[\cos 21° - (0.57)\sin 21°\right]}$$

$$F = 250\text{ N} = \boxed{0.25\text{ kN}}$$

**Insight:** The required pushing force is about 56 lb and represents about 78% of the crate's weight. If the person simply pushed horizontally they would need only 180 N or 40 lb of force.

32. **Picture the Problem**: The free-body diagram for each pulley is depicted at right.

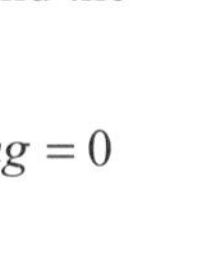

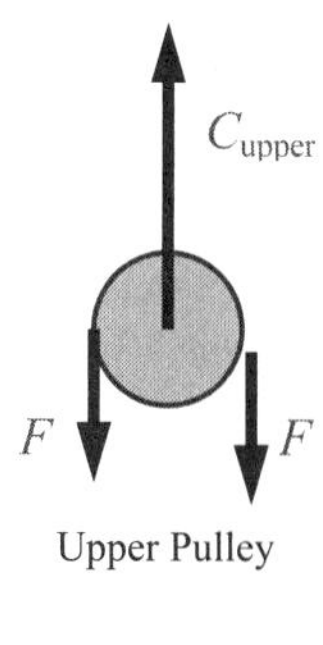


**Strategy:** The tension in the rope is the same everywhere and is equal to $F$ as long as the pulleys rotate without friction. Use the free-body diagrams of the pulleys to write Newton's Second Law in the vertical direction for each pulley, and use the resulting equations to find the tensions. The acceleration is zero everywhere in the system.

**Solution: 1. (a)** Write Newton's Second Law for the box:

$$\sum F_y = C_{\text{lower}} - mg = 0$$
$$C_{\text{lower}} = mg$$

**2.** Now write Newton's Second Law for the lower pulley and solve for $F$:

$$\sum F_y = F + F - C_{\text{lower}} = 0$$
$$F = \tfrac{1}{2}C_{\text{lower}} = \tfrac{1}{2}mg = \tfrac{1}{2}(52\text{ kg})(9.81\text{ m/s}^2)$$
$$= 255\text{ N} = \boxed{0.26\text{ kN}}\ \text{downwards}$$

**3. (b)** Write Newton's Second Law for the upper pulley and solve for $C_{\text{upper}}$:

$$\sum F_y = C_{\text{upper}} - F - F = 0$$
$$C_{\text{upper}} = 2F = 2(0.255\text{ kN}) = \boxed{0.51\text{ kN}}$$

**4. (c)** Use the result of step 1 to find $C_{\text{lower}}$: $C_{\text{lower}} = mg = (52\text{ kg})(9.81\text{ m/s}^2) = 510\text{ N} = \boxed{0.51\text{ kN}}$

**Insight:** Note that there are two rope tensions pulling upward on the lower pulley. Therefore each one supports half the weight of the crate and the tension in the rope is half that of the chains.

36. **Picture the Problem**: The forces acting on the small pulley are depicted at right.

**Strategy:** The rope tension is equal to the weight of the 2.50 kg mass and will everywhere be the same because the pulleys are assumed frictionless. Write Newton's Second Law in the $x$ direction and solve for the magnitude of $\vec{\mathbf{F}}$.

**Solution: 1.** Find the rope tension:

$$T_1 = T_2 = mg$$

**2.** Write Newton's Second Law in the $x$ direction and solve for $F$:

$$\sum F_x = -F + T_1\cos 30.0° + T_2\cos 30.0° = 0$$
$$F = 2mg\cos 30.0° = 2(2.50\text{ kg})(9.81\text{ m/s}^2)\cos 30.0° = \boxed{42.5\text{ N}} = 9.54\text{ lb}$$

**Insight:** The traction device is arranged to produce a force that is parallel to the leg bone so that it can heal straight. However, the force of gravity on the leg has been ignored here and in real life there would have to be upward component of the force exerted on the leg.

47. **Picture the Problem**: Refer to the figure at right.

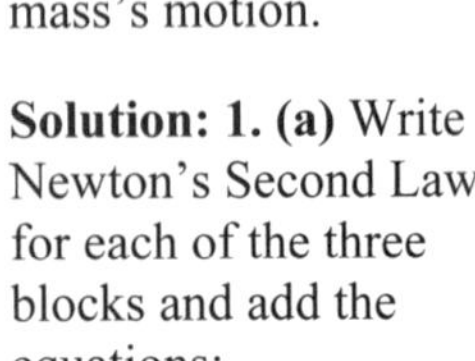

**Strategy:** Write Newton's Second Law for each of the three blocks and add the equations to eliminate the unknowns $T_1$ and $T_2$. Solve the resulting equation for the acceleration $a$. Use the acceleration to find the tensions in the strings. Let $x$ be positive in the direction of each mass's motion.

**Solution: 1. (a)** Write Newton's Second Law for each of the three blocks and add the equations:

$$\sum_{\text{block 1}} F_x = T_1 = m_1 a$$
$$\sum_{\text{block 2}} F_x = -T_1 + T_2 = m_2 a$$
$$\sum_{\text{block 3}} F_x = -T_2 + m_3 g = m_3 a$$
$$m_3 g = (m_1 + m_2 + m_3) a$$

**2.** Solve the resulting equation for $a$:

$$a = \left(\frac{m_3}{m_1 + m_2 + m_3}\right) g = \frac{3.0\ \text{kg}}{6.0\ \text{kg}}\left(9.81\ \text{m/s}^2\right) = \underline{\underline{4.9\ \text{m/s}^2}}$$

**3.** Use $a$ to find $T_1$:

$$T_1 = m_1 a = (1.0\ \text{kg})\left(4.9\ \text{m/s}^2\right) = \boxed{4.9\ \text{N}}$$

**4. (b)** Now find $T_2$ from the block 3 equation:

$$T_2 = m_3 (g - a) = (3.0\ \text{kg})\left(9.81 - 4.9\ \text{m/s}^2\right) = \boxed{15\ \text{N}}$$

**Insight:** Note that the blocks move as if they were a single block of mass 6.0 kg under the influence of a force equal to $m_3 g = 29$ N. Note also that $T_2$ is less than 29 N because the blocks accelerate. It would be zero if $m_3$ fell freely, 29 N if $m_3$ (and the entire system) were at rest.

81. **Picture the Problem**: The force vectors acting on blocks A and B as well as the rope knot are shown in the diagram at right.

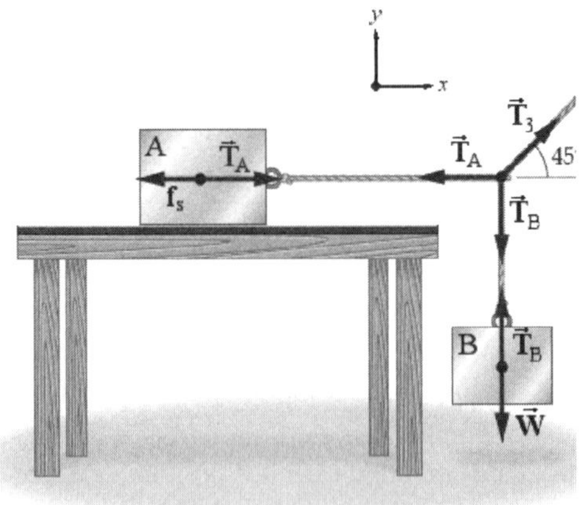


**Strategy:** Write Newton's Second Law for blocks A and B, as well as Newton's Second Law for the rope knot. In all cases the acceleration is zero. Combine the equations to solve for $\vec{\mathbf{f}}_s$, which acts on block A and points toward the left. Let the $x$ direction point to the right for block A and down for block B.

**Solution: 1. (a)** Write Newton's Second Law for block A:

$$\sum_{\text{Block A}} F_x = -f_s + T_A = 0$$

**2.** Write Newton's Second Law for block B:

$$\sum_{\text{Block B}} F_x = -T_B + m_B g = 0$$

**3.** Write Newton's Second Law for the rope knot:

$$\sum_{\text{knot}} F_x = -T_A + T_3 \cos 45° = 0$$
$$\sum_{\text{knot}} F_y = -T_B + T_3 \sin 45° = 0$$

**4.** Divide the $y$ equation for the knot by the $x$ equation:

$$\frac{T_3 \sin 45°}{T_3 \cos 45°} = \tan 45° = 1 = \frac{T_B}{T_A}$$

**5.** Substitute $T_A = T_B$ into

$$T_A = m_B g$$

the equation from step 2:

**6.** Substitute the result into the equation from step 1: $f_s = T_A = m_B g = (2.33 \text{ kg})(9.81 \text{ m/s}^2) = 22.9 \text{ N} = \boxed{23 \text{ N}}$

**7. (b)** As long as mass A is heavy enough that $f_{s,\,max} = \mu_s m_A g = (0.320)(8.82 \text{ kg})(9.81 \text{ m/s}^2) = 27.7 \text{ N} \geq 22.9 \text{ N}$, the friction force is not affected by changes in mass A. It will $\boxed{\text{stay the same}}$ if the mass of block A is doubled.

**Insight:** The minimum mass of block A that will satisfy the criteria of step (b) is 7.29 kg. The answer to (a) is reported with only two significant figures because the angle 45° is only given to two significant figures.

85. **Picture the Problem**: The free-body diagram of the airplane is depicted at right.

**Strategy:** Let the $x$ axis point horizontally from the airplane toward the center of its circular motion, and let the $y$ axis point straight upward. Write Newton's Second Law in both the horizontal and vertical directions and use the resulting equations to find $\theta$ and the tension $T$.

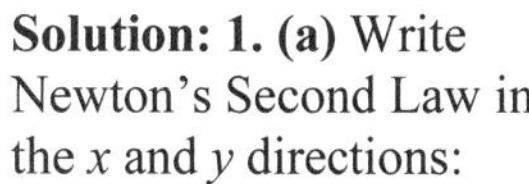

**Solution: 1. (a)** Write Newton's Second Law in the $x$ and $y$ directions:

$$\sum F_x = T\sin\theta = ma_{cp} = mv^2/r$$
$$\sum F_y = T\cos\theta - mg = 0$$

**2.** Solve the $y$ equation for $T$ and substitute the result into the $x$ equation, and solve for $\theta$:

$$T = mg/\cos\theta$$
$$T\sin\theta = \left(\frac{mg}{\cos\theta}\right)\sin\theta = mv^2/r$$
$$\tan\theta = v^2/rg$$
$$\theta = \tan^{-1}\left[\frac{(1.21 \text{ m/s})^2}{(0.44 \text{ m})(9.81 \text{ m/s}^2)}\right] = \boxed{19°}$$

**3. (b)** Calculate the tension from the equation in step 2:

$$T = \frac{mg}{\cos\theta} = \frac{(0.075 \text{ kg})(9.81 \text{ m/s}^2)}{\cos 19°} = \boxed{0.78 \text{ N}}$$

**Insight:** This airplane is pretty small. The toy weighs only 0.74 N = 2.6 ounces and flies in a circle of diameter 2.9 ft.

## Answers to Practice Quiz

**1.** (a) **2.** (b) **3.** (e) **4.** (b) **5.** (d) **6.** (e) **7.** (d) **8.** (a) **9.** (c) **10.** (e) **11.** (c) **12.** (e)

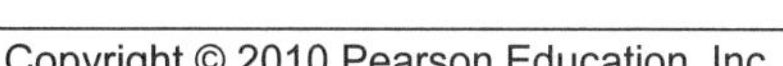

# CHAPTER 7

# WORK AND KINETIC ENERGY

## Chapter Objectives

After studying this chapter, you should

1. understand the value of the concepts of work and kinetic energy.

2. be able to calculate the work done by constant forces and approximate the work of variable forces.

3. be able to determine the kinetic energy of a moving object.

4. know how to calculate the average power delivered when work is done.

## Warm-Ups

1. A weight lifter picks up a barbell and (a) lifts it chest high at a steady speed, (b) holds it for 5 minutes, then (c) puts it down at a steady speed. Rank the amount of work $W$ the weight lifter performs during these three operations. Label the quantities $W_1$, $W_2$, and $W_3$. Justify your ranking order.
2. Estimate the amount of work the engine performed on a 1200-kg car as it accelerated at 1.2 m/s$^2$ over a 150-meter distance.
3. A real-world roller coaster released at point A and coasting without external power would traverse a track somewhat like the figure below. Friction is not negligible in the real world. Does the roller coaster have the same energy at points B and C? Is the total energy conserved during the roller coaster ride? Can you account for all the energy at any point on the track?

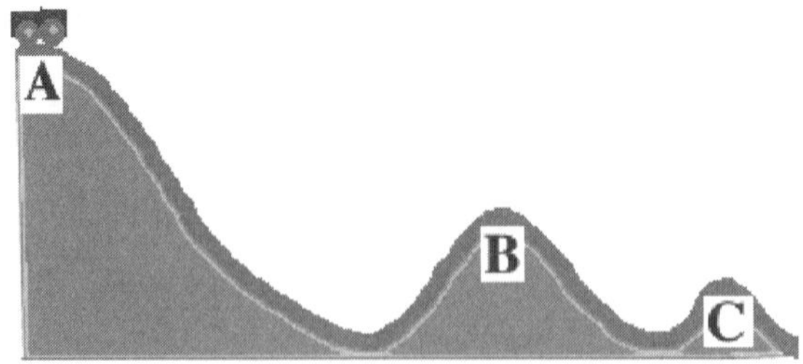


4. A gallon of gasoline contains about $1.3 \times 10^8$ joules of energy. A 2000-kg car traveling at 20 m/s skids to a stop. Estimate how much gasoline it will take to bring the car back to its original speed. To complicate matters further, consider that only about 15% of the energy extracted from gasoline actually propels the car. The rest gets exhausted as heat and unburned fuel.

## Chapter Review

In Chapters 5 and 6, you learned how to analyze mechanical situations by direct use of Newton's laws of motion. In this chapter, you begin to learn about other concepts that are consistent with Newton's laws that can often make such analysis easier to perform. This is especially true when the number of applied forces is large, when they vary with distance or time, or when they are not accurately known.

### 7–1 Work Done by a Constant Force

When a force acts on an object as it undergoes a displacement, we say that the force does a certain amount of **work** on the object. This work can be positive, negative, or zero depending on how the direction of the force relates to the direction of the displacement.

If we consider the case in which a force $\vec{\mathbf{F}}$ and the displacement $\vec{\mathbf{d}}$ are in the same direction, then we can consider the force as acting both through and *with* the displacement to its fullest extent. In this special case, the work done by the force, $W$, will be positive and equal to the product of the magnitude of the force and the magnitude of the displacement: $W = Fd$. This result represents the maximum amount of positive work that the force $\vec{\mathbf{F}}$ can do on the object. The opposite situation occurs when the force and the displacement are in opposite directions (as is often the case with kinetic friction). Here, the force acts through and *against* the displacement to its fullest extent. In this latter case, the work done by the force will be negative $W = -Fd$. A third special case occurs when the direction of $\vec{\mathbf{F}}$ is perpendicular to $\vec{\mathbf{d}}$ (as is the case with the normal force on an object moving across a surface). When $\vec{\mathbf{F}}$ and $\vec{\mathbf{d}}$ are perpendicular, $\vec{\mathbf{F}}$ neither works with nor against the displacement, and the work done by $\vec{\mathbf{F}}$ on the object is zero.

In the most general cases of work done by a constant force, the force and displacement vectors are at arbitrary angles with each other. Consider the following figure:

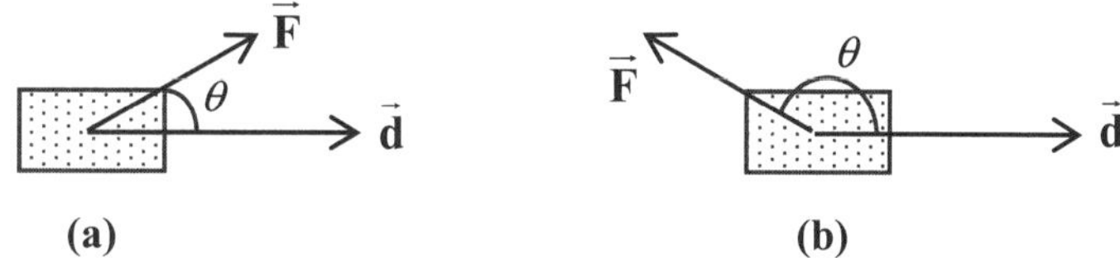


In part (a), we can see that $\vec{\mathbf{F}}$ only partly acts with $\vec{\mathbf{d}}$, and in part (b), it only partly acts against $\vec{\mathbf{d}}$. To determine the work done by $\vec{\mathbf{F}}$ in such cases, we use only that part of $\vec{\mathbf{F}}$ that acts with or against the displacement, that is, we need the component of $\vec{\mathbf{F}}$ along the direction of $\vec{\mathbf{d}}$ — $F_d$. Based on the above diagrams and from what you know about vector components, try to convince yourself that $F_d = F\cos\theta$. In part (a) of the figure, $F_d$ will be positive, and in part (b), it will be negative. The work done by $\vec{\mathbf{F}}$ in this more general case, therefore, is given by

$$W = F_d d = (F\cos\theta)d$$

Notice that this general result for the work done by a constant force includes all three of the special cases discussed in the previous paragraph. As you can see in the preceding equation, the units of work must result from the product of the units of force and displacement. This combination, N·m, when applied to work, is called a **joule** (J).

It is useful to recognize that the work done can also be viewed as the component of the displacement in the direction of the force multiplied by the magnitude of the force. Algebraically, the difference amounts only to a regrouping of the factors involved:

$$W = F(d\cos\theta)$$

Try to convince yourself that the component of $\vec{\mathbf{d}}$ along $\vec{\mathbf{F}}$ really is $d_F = d\cos\theta$. The important concept here is that work depends only on the extent to which the force and the displacement act together (i.e., along the same line).

When several forces act on an object while it is being displaced, each force does work on the object according to the above discussion. The total work done on the object is the sum of all these contributions. An alternative way to approach this calculation is to first determine the total force acting on the object, $\vec{\mathbf{F}}_{total}$, and then calculate the work done by this force. Therefore, we have

$$W_{total} = W_1 + W_2 + W_3 + \cdots = (F_{total}\cos\theta)d$$

where the total force, also called the net force, is given by the vector sum of all the forces

$$\vec{\mathbf{F}}_{total} = \vec{\mathbf{F}}_1 + \vec{\mathbf{F}}_2 + \cdots$$

---

**Example 7–1 Shopping for Groceries** A person pushes a 751-N shopping cart full of groceries down the aisle at the local store. The person applies a force of 112 N at an angle of 40.0 degrees below the horizontal. The cart is pushed the full length of a 15.5-m aisle at constant speed. Determine **(a)** the work done by the shopper, **(b)** the work done by gravity, **(c)** the work done by the normal force of the floor on the cart, and **(d)** the work done by various frictional forces during the cart's motion down the aisle.

**Picture the Problem** The diagram shows the shopping cart, the force exerted by the shopper $\vec{\mathbf{F}}$, the weight $m\vec{\mathbf{g}}$, the normal force $\vec{\mathbf{N}}$, and the displacement $\vec{\mathbf{d}}$.

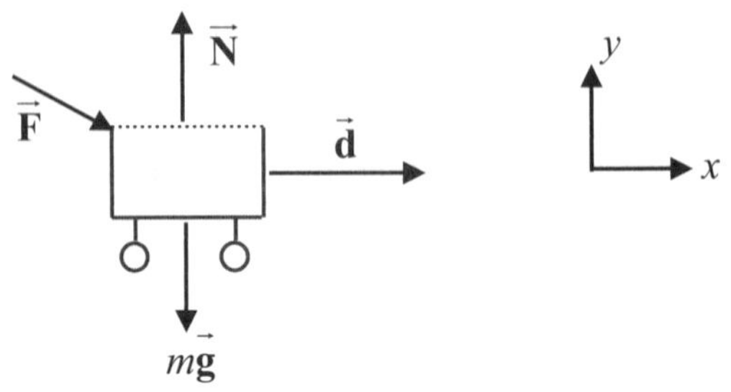

$\vec{\mathbf{N}}$ $\vec{\mathbf{F}}$ $\vec{\mathbf{d}}$ $m\vec{\mathbf{g}}$ $y$ $x$

**Strategy** To calculate the work done by the

individual forces listed, we need to identify the magnitude of each force and its direction relative to the displacement. We can determine the work done by frictional forces from the total work on the cart.

**Solution**

**Part (a)**

1. From the given information, the angle between $\vec{\mathbf{F}}$ and $\vec{\mathbf{d}}$ must be: $\theta_F = 40.0°$

2. The work done by $\vec{\mathbf{F}}$ then is: $W_F = Fd\cos\theta_F = (112\text{ N})(15.5\text{ m})\cos(40.0°)$
$= 1330\text{ J}$

**Part (b)**

3. From the given situation, the angle between the cart's weight and $\vec{\mathbf{d}}$ is: $\theta_g = 90.0°$

4. The work done by gravity is: $W_g = mgd\cos\theta_g = (751\text{ N})(15.5\text{ m})\cos(90.0°) = 0\text{ J}$

**Part (c)**

5. From the given information, the angle between $\vec{\mathbf{N}}$ and $\vec{\mathbf{d}}$ is: $\theta_N = 90.0°$

6. The work done by $\vec{\mathbf{N}}$ then is: $W_N = Nd\cos\theta_N = Nd\cos(90.0°) = 0\text{ J}$

**Part (d)**

7. Because the cart moves with constant speed, the total force on the cart is: $F_{total} = ma = m(0\text{ m/s}^2) = 0$

8. The total work done on the cart then is: $W_{total} = F_{total}d\cos(\theta_{total}) = 0$

9. Because $W_{total} = 0$, the sum of the work done by each force must be zero: $W_{total} = W_F + W_g + W_N + W_{fric} = 0$

10. Because $W_g = W_N = 0$, this implies that: $W_{fric} = -W_F = -1330\text{ J}$

**Insight** Recognizing that both the normal force and the weight of the cart are perpendicular to the displacement, we could have immediately concluded that the work done by these forces is zero. However, the preceding approach shows that even if you don't recognize this, you can arrive at the correct result by performing the general calculation. Notice that the phrase "constant speed" in the statement of the problem was crucial to our ability to solve the problem. Be careful to pay attention to such phrases. Finally, take note that we were able to find the work done by friction forces without knowing anything about those forces. This point has caused many students to get stuck on similar problems. Keep in mind that there are many ways to get at information once you put all the physics together.

**Exercise 7–2 Sliding Down** A 3.5-kg object slides 1.7 m down a ramp that is inclined at 35 degrees to the horizontal. If the coefficient of kinetic friction between the object and the ramp is 0.12, **(a)** how much work is done by gravity, and **(b)** how much work is done by kinetic friction?

**Solution:** The following information is given in the problem:

**Given:** $m = 3.5$ kg, $d = 1.7$ m, $\alpha = 35°$, $\mu_k = 0.12$; **Find: (a)** $W_g$, **(b)** $W_f$

The diagram shows the object on the ramp and the forces acting on it. Notice that the weight has been divided into its components parallel and perpendicular to the incline.

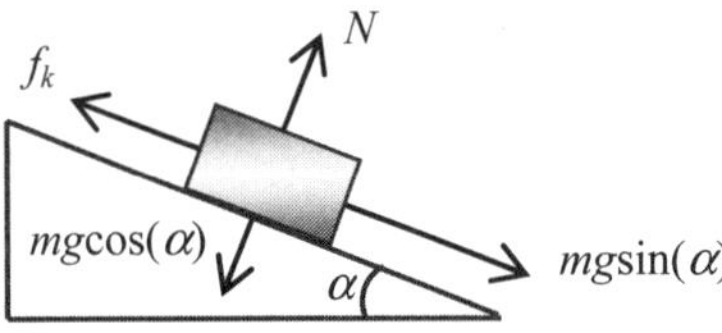


For part (a), we see that $mg\sin(\alpha)$ is the component of the gravitational force in the direction of the displacement. Thus, we can immediately calculate the work done by gravity as

$$W_g = (mg\sin\alpha)d = (3.5\text{ kg})(9.81\text{ m/s}^2)(1.7\text{ m})\sin(35^\circ) = 33\text{ J}$$

For part (b), we need to determine the value of the force of kinetic friction. Since the object does not accelerate in the direction perpendicular to the incline, the two forces along that direction must cancel. This implies that

$$N = mg\cos\alpha$$

The force of friction then is given by

$$f_k = \mu_k N = \mu_k mg\cos\alpha$$

Since $f_k$ always opposes the direction of motion, the angle between $\vec{\mathbf{f}}_k$ and $\vec{\mathbf{d}}$ is 180°. This gives

$$W_f = f_k d\cos(180^\circ) = -\mu_k mgd\cos\alpha = -(0.12)(3.5\text{ kg})(9.81\text{ m/s}^2)(1.7\text{ m})\cos(35^\circ) = -5.7\text{ J}$$

There were two angles in this calculation: the angle of inclination of the ramp and the angle between the force and displacement. A common mistake is to get them confused and use the 35 degrees as the angle between the force and the displacement. Be careful about this situation. Also notice that unlike Example 7–1, this time we determined the force of friction before calculating the work it did. Can you tell why it was more convenient to treat these two cases differently?

---

## Practice Quiz

1. A person pushes an object across a room, on a horizontal floor, by applying a horizontal force. If, instead, he pushed it through the same horizontal distance with the same magnitude of force at an angle $\theta$ below the horizontal, he would do

   **(a)** a greater amount of positive work. **(b)** a lesser amount of positive work. **(c)** zero work.
   **(d)** negative work. **(e)** None of the above.

2. A person pulls a 3.0-kg object across a room, on a horizontal floor, by applying a force of 13 N that makes an angle of 77° above the horizontal. If she does a total of 26 J of work, how far does she pull the object?

   **(a)** 2.0 m **(b)** 230 m **(c)** 3.9 m **(d)** 10 m **(e)** 8.9 m

3. A person pushes an object across a room, on a horizontal floor, by applying a horizontal force. If, instead, he pushed the object through the same horizontal distance with the same magnitude of force at an angle $\theta$ above the horizontal, friction would do

   **(a)** a greater amount of positive work. **(b)** a greater amount of negative work.
   **(c)** a lesser amount of positive work. **(d)** a lesser amount of negative work. **(e)** zero work.

## 7–2 Kinetic Energy and the Work-Energy Theorem

When a nonzero amount of work is done on an object, that work must manifest itself as some sort of observable effect on the object. One such effect is the change in speed that results. However, we cannot say that work is directly converted into speed because work and speed are different types of quantities (work is measured in joules, and speed in m/s); however, work must convert into something closely related to speed. We call that "something" the **kinetic energy,** $K$, of the object. Kinetic energy is given by the equation

$$K = \tfrac{1}{2}mv^2$$

The relationship between the total work done on an object and its kinetic energy is given by the following expression known as the **work-energy theorem:**

$$W_{total} = K_f - K_i = \Delta K$$

The above equation shows that kinetic energy must have the same unit as work, namely, the joule. You can think of kinetic energy as a measure of the amount of work that goes into the motion of an object. Thus, if you see an object with 10 J of kinetic energy, you know that it took at least 10 J of work to get it up to the speed that it has from rest, and it will likewise take at least 10 J of work to bring it to rest.

---

**Exercise 7–3 Skidding to a Stop** While driving at 12.0 m/s in a car of mass 1470 kg, you notice a squirrel running into the street in front of you. You hit the brakes suddenly and come to a stop after skidding for 3.20 m. How much work is done by friction in bringing you to a stop?

**Solution:** The following information is given in the problem:

**Given:** $m = 1470$ kg, $d = 3.20$ m, $v_i = 12.0$ m/s, $v_f = 0$; **Find:** $W_f$

Because we know the mass of the car and its initial and final speeds, we can find the initial and final kinetic energies and therefore the change in kinetic energy:

$$\Delta K = K_f - K_i = \tfrac{1}{2}mv_f^2 - \tfrac{1}{2}mv_i^2 = \tfrac{1}{2}m\left(v_f^2 - v_i^2\right) = \frac{1470\text{ kg}}{2}\left[0 - (12.0\text{ m/s})^2\right] = -1.06\times10^5\text{ J}$$

Since the car stops because of friction only, we know that the total work done is the work done by friction. Therefore,

$$W_f = \Delta K = -1.06\times10^5\text{ J}$$

From the motion alone, we determined the work done by friction without knowing the force at all.

---

## Practice Quiz

4. When a net amount of positive work is done on an object, you expect the object to

   **(a)** speed up. **(b)** slow down. **(c)** stop. **(d)** maintain a constant speed. **(e)** None of these.

5. If a 93-g bullet has a kinetic energy of 4200 J, what is the muzzle velocity of the gun?

   **(a)** 390 m/s **(b)** 45 m/s **(c)** 300 m/s **(d)** 210 m/s **(e)** 77 m/s

## 7–3 Work Done by a Variable Force

When the force that is applied to an object varies with distance, we can use a graphical technique to either approximate or exactly determine the amount of work done by this force. The key to this technique is that any force that varies with distance can be treated as approximately constant over small enough displacements. For these small displacements, we can approximate the amount of work done using the expression for the work done by a constant force. For variable forces, we shall consider only cases for which the force and displacement are along the same direction. As shown in the following graph of force versus displacement on the left, when the force can be treated as constant, the work done by the force equals the area of the rectangle whose width is the displacement $\Delta x$ and whose height is the value of the force $F$.

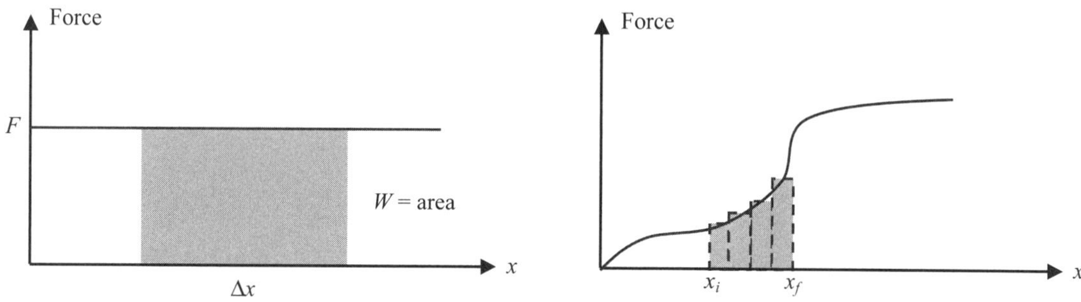


In the graph on the right, we see how small rectangles, representing small displacements, can be used to approximate the area, and therefore the work done, for a given overall displacement from $x_i$ to $x_f$. Notice that the height of the left-hand sides of the rectangles tend to overestimate the force, whereas the right-hand sides tend to underestimate. This compensation helps make this approximation more accurate. The smaller the rectangles, the better the approximation.

A variable force of special interest is the Hooke's law force, $F = -kx$, that exactly describes an ideal spring and approximately describes the behavior of any solid object when it is deformed slightly. Since the force varies linearly with $x$, the area under the curve is that of a simple triangle: area = 1/2 base × height. In this case, we can use the described graphical technique to exactly (not approximately) determine the work done in stretching or compressing the spring. We find that the minimum amount of work *required* to stretch or compress a spring by an amount $x$ from its equilibrium length is

$$W = \tfrac{1}{2}kx^2$$

The work done by the spring while it is being deformed is the opposite of this value $W_{spring} = -\tfrac{1}{2}kx^2$.

**Example 7–4 Spring Cushion** A spring with a force constant of 2000 N/m is used to cushion objects of moderate weight. If a 75-lb object is placed on one of these springs, how much work does the spring do before it balances the weight of the object?

**Picture the Problem** Part (a) of the figure shows the object being balanced by the spring. Part (b) is a free-body diagram of the object. The coordinate system is at the far right.

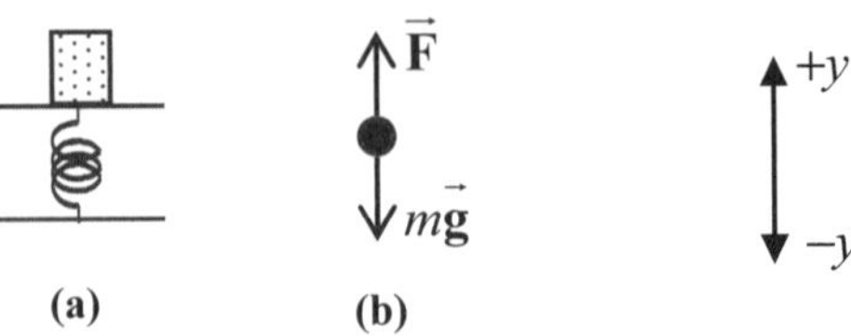


**Strategy** Based on the information given here, we need to know how much the spring compresses in order to calculate the work it does while compressing.

**Solution**

1. Convert the weight to newtons: $\text{weight} = 75\text{ lb}\left(\frac{4.45\text{ N}}{1\text{ lb}}\right) = 333.75\text{ N}$

2. Apply Newton's 2nd law to the vertical direction: $\sum F = ky - mg = 0$

3. Solve this equation for $y$: $ky = mg \Rightarrow y = \frac{mg}{k}$

4. Find the work done by the spring:

$$W = -\frac{1}{2}ky^2 = -\frac{k}{2}\left(\frac{mg}{k}\right)^2$$

$$= -\frac{2000\text{ N/m}}{2}\left(\frac{333.75\text{ N/m}}{2000\text{ N/m}}\right)^2 = -28\text{ J}$$

**Insight** Understand that the work done by the spring is negative because as it compresses downward the spring itself applies an upward force on the object.

## Practice Quiz

6. A spring is held in a compressed position, then released. As the spring uncoils to its equilibrium length it does

   **(a)** negative work. **(b)** positive work. **(c)** zero work. **(d)** None of the above.

7. It requires a minimum of 0.72 J of work to stretch a certain spring 1.5 cm. What is its force constant?

   **(a)** 1.1 N/m **(b)** 0.011 N/m **(c)** 96 N/m **(d)** 6400 N/m **(e)** None of the above.

8. If it requires an amount of work $W$ to stretch a certain spring by an amount $x$, how much work would it take to stretch it by an amount $2x$?

   **(a)** $2W$ **(b)** $\sqrt{2}W$ **(c)** $\frac{1}{2}W$ **(d)** $\frac{1}{4}W$ **(e)** $4W$

## 7–4 Power

From a practical standpoint, the mere fact that a certain amount of work is done is not always good enough. The question of how long it takes to do this work often determines the practical value of certain devices. For example, everyone wants a car that can go from 0 to 60 mi/h, but no one wants it if it takes 5 min. The quantity we use to measure how rapidly work is done is called **power**.

Power is the rate at which work is done. For the applications of interest to us, we will mainly use the average power, which is the amount of work done, $W$, divided by the amount of time, $t$, required to do it:

$$P = \frac{W}{t}$$

The unit of power must be the unit of work divided by the unit of time, or joules per second (J/s). In SI, the unit J/s is given the name **watt** (W). You are probably familiar with this unit, as it is common to give the power rating of light bulbs in watts. For cases when work is being done by a force $\vec{\mathbf{F}}$ on an object moving with velocity $\vec{\mathbf{v}}$ in the same direction, the power delivered by that force is given by

$$P = Fv$$

Verify that the product $Fv$ has the SI unit of watt. Also, note that the above equation is useful even if the force or the speed is not constant. In these cases, the average power can be calculated by substituting the average force or the average speed: $P = F_{av}v$ or $P = Fv_{av}$. If both $F$ and $v$ vary, use $P = F_{av}v_{av}$.

---

**Example 7–5 An Industrial Pulley** The motor of a chain-linked industrial pulley designed to lift heavy weights is rated to deliver 3000 watts of power on average. If the mechanism is only 80% efficient, how long will it take to raise a 55-kg crate a height of 12 m at constant speed?

**Picture the Problem** The diagram shows part of the pulley system lifting the crate with a constant upward velocity $\vec{\mathbf{v}}$.

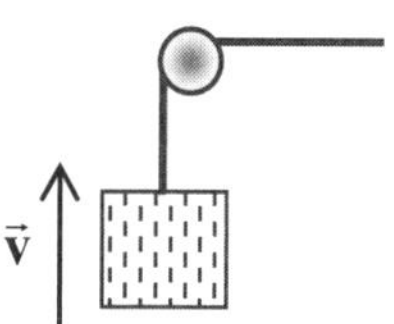

**Strategy** The amount of time it takes is related to how much work is done and the rate at which it is done (the power). Therefore, we will determine the work done and the actual power output.

**Solution**

1. Obtain the expression for the time $t$: $$P_{out} = \frac{W_{pulley}}{t} \Rightarrow t = \frac{W_{pulley}}{P_{out}}$$

2. Since the pulley applies a force in the same direction as the displacement: $$W_{pulley} = F_{pulley}d$$

3. Since its velocity is constant, the force of the pulley must exactly balance the weight: $$F_{pulley} = mg \quad \therefore \quad W_{pulley} = mgd$$

4. Using the fact that the motor is 80% efficient, find the time $t$: $$t = \frac{mgd}{P_{out}} = \frac{mgd}{0.8P} = \frac{(55\text{ kg})(9.81\text{ m/s}^2)(12\text{ m})}{0.8(3000\text{ W})} = 2.7\text{s}$$

**Insight** We could also have solved this problem by finding the velocity and calculating the time for the crate to move the 12-m distance. Try it.

**Exercise 7–6 Test Driving** The manufacturer of a 1500-kg automobile advertises that the engine delivers 175 hp. If all of this power is transferred to the motion of the car with 100% efficiency, what 0-to-60 mi/h time would you expect the car to achieve?

**Solution**: The following information is given in the problem:

**Given:** $m$ = 1500 kg, $P$ = 175 hp, $v_i$ = 0.0 m/s, $v_f$ = 60 mi/h; **Find:** $t$

We have a mix of units here, so let's convert everything to SI.

$$P = 175\text{ hp}\left(\frac{746\text{ W}}{1\text{ hp}}\right) = 1.306\times10^5\text{ W} \quad \text{and} \quad v_f = 60\text{ mi/h}\left(\frac{0.447\text{ m/s}}{1\text{ mi/h}}\right) = 26.82\text{ m/s}$$

The time for the car to reach 60 mi/h depends on the amount of work done: $t = W/P$. The work done by the engine can be determined from the work-energy theorem

$$W = K_f - K_i = K_f = \tfrac{1}{2}mv_f^2$$

So, the expected 0-to-60 time would be

$$t = \frac{mv_f^2}{2P} = \frac{(1500\text{ kg})(26.82\text{ m/s})^2}{2(1.306\times10^5\text{ W})} = 4.1\text{ s}$$

Look up some typical 0-to-60 times. What does this say about the efficiency at which the car converts the engine's power into the motion of the car?

### Practice Quiz

9. If motor A delivers more power than motor B, then

   **(a)** motor A can do more work than motor B.

   **(b)** motor A takes more time to do the same amount of work as motor B.

   **(c)** motor A takes less time to do the same amount of work as motor B.

   **(d)** both motors take the same amount of time to do the same amount of work.

   **(e)** motor A is more efficient than motor B.

10. A lift mechanism delivers a power of 85 W in lifting a 250-N crate at constant speed. What is the speed?

    **(a)** 3.4 m/s **(b)** 0.29 m/s **(c)** 6.0 m/s **(d)** 9.8 m/s **(e)** 0.35 m/s

## Reference Tools and Resources

### I. Key Terms and Phrases

**work** work is done when a force acts through a displacement

**joule** SI unit of work and energy

**kinetic energy** the quantity that measures the amount of work that goes into the motion of an object

**work-energy theorem** the expression $W_{total} = \Delta K$ that relates work and kinetic energy

**power** the rate at which work is done

**watt** SI unit of power equal to one joule per second

### II. Important Equations

| Name/Topic | Equation | Explanation |
|---|---|---|
| work | $W = Fd\cos\theta$ | The work done by a constant force $\vec{\mathbf{F}}$ on an object that undergoes a displacement $\vec{\mathbf{d}}$ |

| | | |
|---|---|---|
| kinetic energy | $K = \frac{1}{2}mv^2$ | The kinetic energy of a mass $m$ moving at speed $v$ |
| work-energy theorem | $W_{total} = \Delta K$ | The relationship between work and kinetic energy change |
| work by a variable force | $W = \frac{1}{2}kx^2$ | The work required to deform a spring an amount $x$ from equilibrium |
| power | $P = \frac{W}{t}$ | The average power delivered when work is done over a time period $t$ |

### III. Know Your Units

| Quantity | Dimension | SI Unit |
|---|---|---|
| work ( $W$ ) | $[M]\cdot[L^2]/[T^2]$ | J |
| kinetic energy ( $K$ ) | $[M]\cdot[L^2]/[T^2]$ | J |
| power ( $P$ ) | $[M]\cdot[L^2]/[T^3]$ | W |

## Puzzle

**WHO IS RIGHT?**

Bill is riding in a railroad car. He throws a ball toward the back wall of the car. The train is moving at a constant 20 m/s to the right. The ball is flying away from Bill at 8 m/s. According to Bill, his 0.145-kg baseball has 4.64 joules of kinetic energy. His brother is standing on the ground disagreeing. According to him the baseball has 10.44 joules of kinetic energy. Who is right? Answer this question in words, not equations, briefly explaining how you obtained your answer.

## Answers to Selected Conceptual Questions

**2.** False. Any force acting on an object can do work. The work done by different forces may add to produce a greater net work, or they may cancel to some extent. It follows that the net work done on an object can be thought of in the following two equivalent ways: (i) The sum of the works done by each individual force; or (ii) the work done by the net force.

**8.** The fact that the boat's velocity is constant means that its kinetic energy is also constant. Therefore, the net work done on the boat is zero. It follows that the net force acting on the boat does no work.

**10.** No. What we can conclude, however, is that the net force acting on the object is zero.

## Solutions to Selected End-of-Chapter Problems and Conceptual Exercises

16. **Picture the Problem**: The force and distance vectors have varying relationships to each other in parts (a), (b), and (c). The appropriate vectors are drawn at right.

(a) 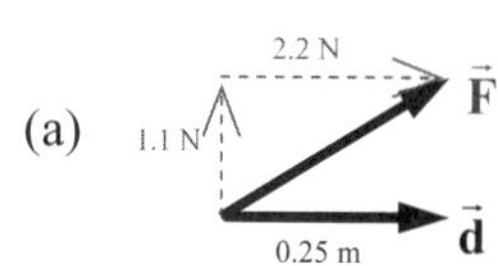


**Strategy:** Multiply the distance by the component of the force that is parallel to the distance. The perpendicular components contribute no work to the total.

**Solution: 1. (a)** Multiply only the $x$-components:

$$W = F_x d_x = (2.2\text{ N})(0.25\text{ m})$$

$$W = \boxed{0.55\text{ J}}$$

(b) 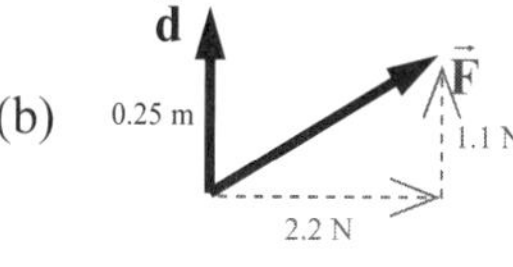


**2. (b)** Multiply only the $y$-components:

$$W = F_y d_y = (1.1\text{ N})(0.25\text{ m})$$

$$W = \boxed{0.28\text{ J}}$$

**3. (c)** Multiply the parallel components and add:

$$W = F_x d_x + F_y d_y$$
$$= (2.2\text{ N})(-0.5\text{ m}) + (1.1\text{ N})(-0.25\text{ m})$$
$$W = \boxed{-1.4\text{ J}}$$

(c) 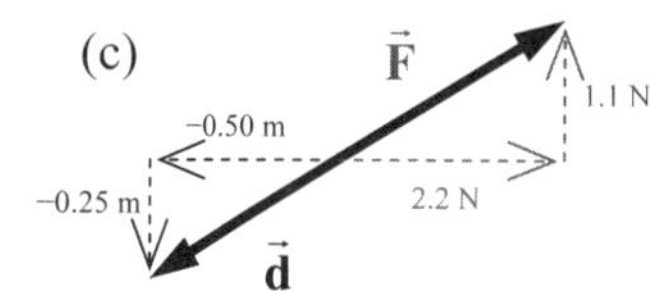


**Insight:** Only the component of the force along the direction of the motion does any work. Multiplying components in this manner is equivalent to calculating the dot product of two vectors $W = \vec{\mathbf{F}} \cdot \vec{\mathbf{d}}$. (See Appendix A.)

29. **Picture the Problem**: The car slows down as it rolls horizontally a distance of 16.0 m through the sand.

**Strategy:** The kinetic energy of the car is reduced by the amount of work done by friction. The change in kinetic energy is related to the initial and final speeds. We can use the speed information from the previous problem to set up a ratio and determine the change in speed for this new case of a shorter sandy section.

**Solution: 1. (a)** Since $\Delta K \propto v^2 \propto Fd$, halving the distance would decrease the speed by $1/\sqrt{2}$ so the speed will decrease by $\boxed{\text{less than 3.5 m/s}}$.

**2. (b)** Create a ratio of the kinetic energies:

$$\frac{\Delta K_{\text{new}}}{\Delta K_{\text{old}}} = \frac{\frac{1}{2}mv_{\text{f, new}}^2 - \frac{1}{2}mv_{\text{i}}^2}{\frac{1}{2}mv_{\text{f, old}}^2 - \frac{1}{2}mv_{\text{i}}^2}$$

**3.** We know $\Delta K_{\text{new}} = \frac{1}{2}\Delta K_{\text{old}}$ because the force is the same but the distance is cut in half:

$$\frac{\Delta K_{\text{new}}}{\Delta K_{\text{old}}} = \frac{v_{\text{f, new}}^2 - v_{\text{i}}^2}{v_{\text{f, old}}^2 - v_{\text{i}}^2} = \frac{1}{2}$$

**4.** Solve for $v_{\text{f, new}}$:

$$v_{\text{f, new}}^2 = \tfrac{1}{2}v_{\text{f, old}}^2 - \tfrac{1}{2}v_{\text{i}}^2 + v_{\text{i}}^2 \Rightarrow v_{\text{f, new}} = \sqrt{\tfrac{1}{2}\left(v_{\text{f, old}}^2 + v_{\text{i}}^2\right)}$$

**5.** Substitute the speeds from the previous question:

$$v_{\text{f, new}} = \sqrt{\tfrac{1}{2}(12\text{ m/s})^2 + \tfrac{1}{2}(19\text{ m/s})^2} = 15.9\text{ m/s thus}$$

$$\Delta v' = 15.9 - 19\text{ m/s} = \boxed{-3.1\text{ m/s}}$$

**Insight:** Kinetic friction always does negative work because the force is always opposite to the direction of motion.

36. **Picture the Problem**: The work done by the force is the area under the force versus position graph.

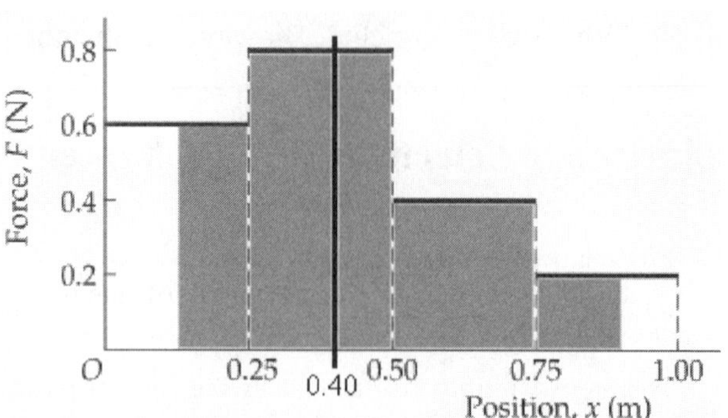


**Strategy:** The initial position is shown by the dark vertical line. Positive work is done on the object if it moves to the right of the line, negative work if it moves to the left. Use the fact that the work is the area under the graph to determine the final position of the object in each case. The shaded area to the right of the dark line represents the positive work done on the object, and the shaded area to the left of the line represents the negative work.

**Solution: 1. (a)** The work done on the object is sufficient to move it past the 0.75 m mark:

$$W = (0.50 - 0.40 \text{ m})(0.8 \text{ N}) + (0.25 \text{ m})(0.4 \text{ m}) + (x - 0.75 \text{ m})(0.2 \text{ N}) = 0.21$$

$$x = \frac{0.21 - 0.08 - 0.10 \text{ J}}{0.2 \text{ N}} + 0.75 \text{ m} = \boxed{0.90 \text{ m}}$$

**2. (b)** The negative work done on the object requires that it be moved left of the 0.25 m mark:

$$W = (0.25 - 0.40 \text{ m})(0.8 \text{ N}) + (x - 0.25 \text{ m})(0.6 \text{ N}) = -0.19 \text{ J}$$

$$x = \frac{-0.19 - (-0.12) \text{ J}}{0.6 \text{ N}} + 0.25 \text{ m} = \boxed{0.13 \text{ m}}$$

**Insight:** The work is positive as long as the object moves from left to right (from small $x$ to large $x$). Therefore the object gains energy as it moves from left to right but loses energy as it is moved from right to left.

39. **Picture the Problem**: The spring is compressed horizontally.

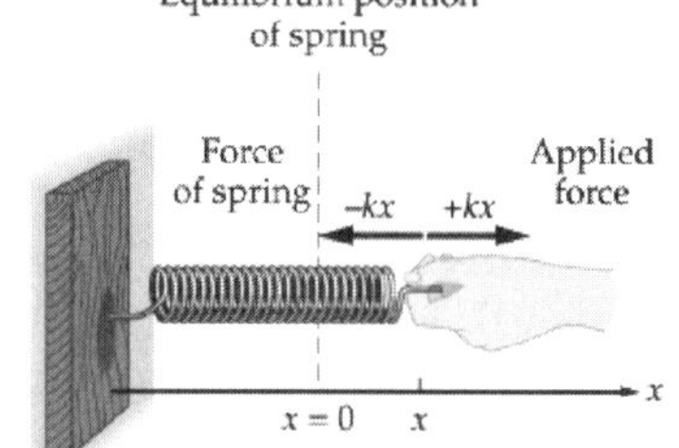


**Strategy:** The work and stretch distance can be used to find the spring constant by applying equation 7-8.

**Solution: 1. (a)** Solve equation 7-8 for $k$:

$$k = \frac{2W}{x^2} = \frac{2(180 \text{ J})}{(0.15 \text{ m})^2}$$

$$= 1.6 \times 10^4 \text{ N/m} = \boxed{16 \text{ kN/m}}$$

**2. (b)** It would take $\boxed{\text{more than 180 J}}$ of work because $W$ is proportional to $x^2$:

$$W = \tfrac{1}{2}kx_2^2 - \tfrac{1}{2}kx_1^2 = \tfrac{1}{2}(1.6\times10^4 \text{ N/m})\left[(0.30 \text{ m})^2 - (0.15 \text{ m})^2\right] = \boxed{540 \text{ J}}$$

**Insight:** The extra work required to stretch the spring an additional 0.15 m can be pictured as the difference in area of two triangles, one with base 0.30 m and one with base 0.15 m, both bounded by the line given by $kx$ (Figure 7-10).

48. **Picture the Problem**: The water is pumped vertically upward.

**Strategy:** The power required is the work required to change the elevation divided by the time. As in Conceptual Checkpoint 7-1 and Example 7-2, the work required to change the elevation of an object is $W = mgh$.

**Solution:** Divide the work required by the time:

$$P = \frac{W}{t} = \frac{mgh}{t} = \frac{(12.0 \text{ lb} \times 4.448 \text{ N/lb})(2.00 \text{ m})}{(1.00 \text{ s})}$$

$$P = 107 \text{ W} \times 1 \text{ hp}/746 \text{ W} = \boxed{0.143 \text{ hp}}$$

**Insight:** A ¼-hp motor would do the trick. Pumping faster than this would require more power.

80. **Picture the Problem**: A car accelerates while traveling on a level road.

**Strategy:** The acceleration can be found from the force and the mass using Newton's Second Law. The force can be determined from the power and the speed.

**Solution: 1. (a)** Combine Newton's Second Law with equation 7-13:

$$a = \frac{F}{m} = \frac{P/v}{m} = \frac{(49\text{ hp} \times 746\text{ W/hp})}{(1300\text{ kg})(14\text{ m/s})} = \boxed{2.0\text{ m/s}^2}$$

**2. (b)** For the same power output the car can exert a smaller force at higher speeds. Therefore the acceleration will $\boxed{\text{decrease}}$ if the speed is doubled.

**3. (c)** Recalculate using the new speed:

$$a = \frac{P}{mv} = \frac{(49\text{ hp} \times 746\text{ W/hp})}{(1300\text{ kg})(28\text{ m/s})} = \boxed{1.0\text{ m/s}^2}$$

**Insight:** The question isn't entirely realistic because car engines generate more power at higher rpm than they do at low speeds, so a real engine would be able to accelerate at a higher rate than that calculated in part (c).

## Answers to Practice Quiz

**1.** (b) **2.** (e) **3.** (d) **4.** (a) **5.** (c) **6.** (b) **7.** (d) **8.** (e) **9.** (c) **10.** (a)

# CHAPTER 8

# POTENTIAL ENERGY AND CONSERVATION OF ENERGY

## Chapter Objectives

After studying this chapter, you should

1. understand the difference between conservative and nonconservative forces.
2. understand the concept of potential energy.
3. be able to apply the principle of the conservation of mechanical energy.
4. know how to handle energy considerations when work is done by nonconservative forces.
5. be able to read information off potential energy curves.

## Warm-Ups

1. Do frictional forces always cause a loss of mechanical energy?
2. A good professional baseball pitcher throws a ball straight up in the air. Estimate how high the ball will go. (A good throw can reach 90 mi/h.)
3. Which potential energy $U(x)$-versus-$x$ graph corresponds to the force $F(x)$-versus-$x$ graph shown?

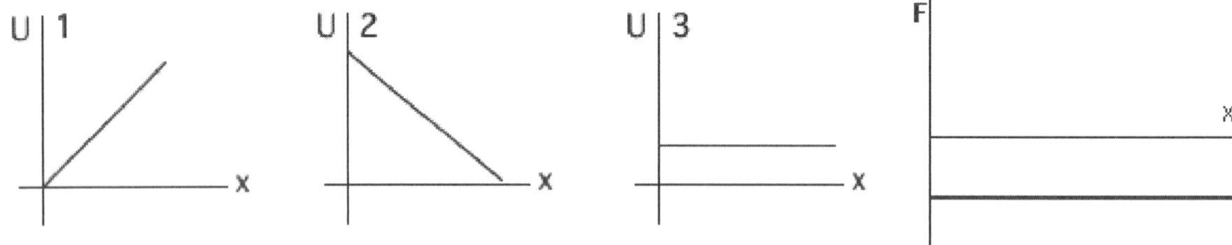


4. Estimate the energy burst you would need to clear a 1-meter hurdle.

## Chapter Review

In this chapter we continue the study of work and energy and encounter another form of energy called potential energy. Most importantly, in this chapter we introduce the principle of the conservation of energy, which is one of the most important principles in science.

## 8–1 Conservative and Nonconservative Forces

Forces are generally divided into two categories depending on properties of the work that a force does. A force can either be a **conservative force** or a **nonconservative force**. A force is called *conservative* if the work it does on any object is independent of the path that object takes during a displacement. Equivalently, a force is conservative if the work it does around any closed path is zero. Any force that does not meet this condition is a *nonconservative* force.

The primary reason for the two classifications is that when work is done by a conservative force, that work is in some sense stored within the system and can be easily recovered, usually as kinetic energy. Gravity is a conservative force. When an object is lifted against gravity, negative work is done by gravity on the object. When that object is released, or lowered back down, gravity does an equal amount of positive work. Kinetic friction is a nonconservative force. If you slide a block across a table, friction does negative work on it. When you slide the block back to its original place, friction does more negative work – the work is not recovered.

## 8–2 Potential Energy and the Work Done by Conservative Forces

In Chapter 7 you were introduced to the idea that when work is done, it goes into changing the kinetic energy of the object, or system, on which the work is done. In that case, it was the *total* work done by all forces acting. A similar concept can be applied to individual forces as well. When work is done by a conservative force, this work goes into changing the **potential energy**, $U$, of the system. You can think of the potential energy as representing the "storage" of this work within the system.

The mathematical definition of potential energy is that the *change in potential energy* equals the negative of the work done by conservative forces

$$\Delta U = -W_c$$

Notice that it is only the change in potential energy that has physical meaning. Particular values of potential energy are defined relative to a chosen reference. This reference can be chosen arbitrarily, although in some cases certain choices are more convenient and usually adopted by convention. Based on the preceding definition, it is clear that the SI unit of potential energy is the same for work and kinetic energy – the joule.

### Gravity

Because of the connection with conservative forces, potential energies are always associated with specific forces. Since gravity is a conservative force, we can define *gravitational potential energy*. Near Earth's surface, where we can treat the force of gravity as constant, the gravitational potential energy is

$$\Delta U = mg(\Delta y)$$

where $\Delta y$ is the change in vertical position. The height at which $U = 0$ can always be chosen to be the initial height. Therefore, we often write the gravitational potential energy as just

$$U = mgy$$

The above equation is written with the understanding that it is only used this way when the reference height has been specified.

**Springs**

Besides gravity, a second conservative force that we have encountered is Hooke's law; therefore, we can define a potential energy to be associated with this force as well. We call this the *spring potential energy*, also often called the *elastic potential energy*. We saw in Chapter 7 that the work done by a spring while being deformed by an amount $x$ from equilibrium is $W = -\frac{1}{2}kx^2$, so the $\Delta U$ associated with this force will be of opposite sign:

$$\Delta U = \tfrac{1}{2}kx^2$$

In the case of Hooke's law, it is almost always convenient to take the equilibrium configuration ($x = 0$) as the reference level for potential energy ($U = 0$). With this understanding, the spring potential energy is written as

$$U = \tfrac{1}{2}kx^2$$

---

**Example 8–1 Rolling Down** A 0.25-kg ball rolls 2.1 m down a ramp that is inclined at 32 degrees to the horizontal. Determine the change in gravitational potential energy.

**Picture the Problem** The diagram shows the ball rolling down the ramp.

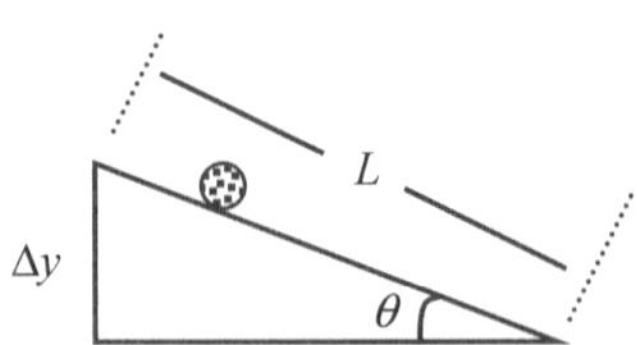


**Strategy** To determine the change in potential energy, we need to find the change in height and use it in the expression for gravitational potential energy.

**Solution**

1. From trigonometry we see that: $|\Delta y| = L\sin\theta$
2. Since the ball rolls downhill: $\Delta y = -L\sin\theta$

**3**. The change in potential energy is:

$$\Delta U = mg\Delta y = -mgL\sin\theta$$
$$= -(0.25 \text{ kg})(9.81 \text{ m/s}^2)(2.1 \text{ m})\sin(32^\circ) = -2.7 \text{ J}$$

**Insight** Notice that we had to be careful to recognize that $\Delta y$ should be negative. The mathematics alone do not give the minus sign.

---

**Exercise 8–2 Hanging from a Spring** A spring of force constant 21 N/cm hangs vertically from a ceiling. A 1.2-kg mass is attached to its end and lowered until it hangs motionless from the end of the spring. If we take the potential energy of the Earth-spring-mass system to be zero initially, what is it in the final configuration?

**Solution** The following information is given in the problem:

**Given:** $m = 1.2$ kg, $k = 21$ N/cm, $U_i = 0$; **Find:** $U_f$

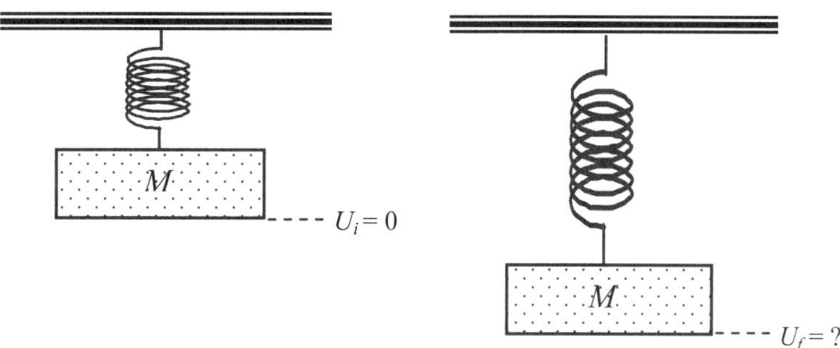


Here we have changes in both the gravitational potential energy and the potential energy of the spring. The change in each depends on how much the spring stretches under the weight of the mass; that is, $x = |\Delta y|$. As we have seen in previous problems, equilibrium between the gravitational force and the Hooke's law force is reached when $kx = mg$. This implies that

$$x = |\Delta y| = \frac{mg}{k}$$

The change in potential energy is then

$$\Delta U = \Delta U_{spring} + \Delta U_{grav} = \tfrac{1}{2}kx^2 + mg\Delta y$$

Inserting the expressions for $x$ and $\Delta y$ (noting that $\Delta y$ is negative), we get

$$\Delta U = \frac{(mg)^2}{2k} - \frac{(mg)^2}{k} = -\frac{(mg)^2}{2k}$$

Because $\Delta U = U_f - U_i = U_f$ we can conclude that

$$U_f = -\frac{(mg)^2}{2k} = -\frac{\left[(1.2\text{ kg})(9.81\text{ m/s}^2)\right]^2}{2(21\text{ N/cm})(100\text{ cm/m})} = -0.033\text{ J}$$

The important thing here was to remember to account for both forms of potential energy.

## Practice Quiz

1. When an object is thrown up into the air, the gravitational potential energy

   **(a)** increases on the way up and decreases on the way down.

   **(b)** decreases on the way up and increases on the way down.

   **(c)** changes the same way going up as it does going down.

   **(d)** remains constant throughout its motion.

   **(e)** None of the above.

2. The potential energy of a spring that is *stretched* by an amount $x$ from equilibrium versus one that is *compressed* an amount $x$ from equilibrium

   **(a)** increases as it is stretched and decreases as it is compressed.

   **(b)** decreases as it is stretched and increases as it is compressed.

   **(c)** changes the same way whether the spring is stretched or compressed.

   **(d)** remains constant throughout its motion.

   **(e)** None of the above.

3. A 2.0-kg object is lifted 2.0 m off the floor, carried horizontally across the room, and finally placed down on a table that is 1.5 m high. What is the overall change in gravitational potential energy for the Earth-object system?

   **(a)** 6.0 J **(b)** 20 J **(c)** 29 J **(d)** 9.8 J **(e)** 59 J

4. If a 0.75-kg block is rested on top of a vertical spring of force constant 45 N/m, how much potential energy is stored in the spring relative to its equilibrium configuration?

   **(a)** 0 J **(b)** 14 J **(c)** 60 J **(d)** 34 J **(e)** 0.60 J

### 8–3 Conservation of Mechanical Energy

Up to this point we have discussed two forms of energy – kinetic and potential. Within any system we refer to the sum of these two forms of energy as the **mechanical energy**, $E = K + U$. The most important thing about this mechanical energy is that for systems in which only conservative forces do work, the total mechanical energy is conserved, that is, remains constant. The energy within the system may change forms between kinetic and potential, but the sum of the two does not change. Three equivalent mathematical statements of this are

$$E = \text{constant}, \qquad E_f = E_i, \qquad K_f + U_f = K_i + U_i$$

---

**Example 8–3 Rolling Down II** In Example 8–1 we had a 0.25-kg ball rolling 2.1 m down a ramp that is inclined at 32° to the horizontal. If the ball starts at rest, calculate its speed at the bottom of the ramp.

**Picture the Problem** The diagram is the same as in Example 8–1.

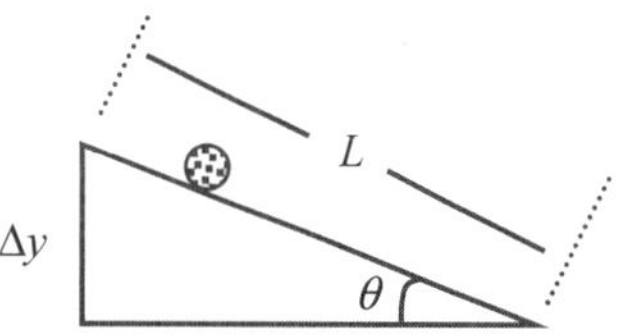


**Strategy** To solve this problem using the conservation of mechanical energy, we shall write expressions for both the initial and final energies and set them equal to each other. We set the reference for gravitational potential energy to be at the bottom of the ramp.

**Solution**

1. Write out the initial mechanical energy. Use the fact that $v_i = 0$ means $K_i = 0$: $E_i = K_i + U_i = 0 + U_i = mg|\Delta y| = mgL\sin(\theta)$

2. Write out the final mechanical energy: $E_f = K_f + U_f = \frac{1}{2}mv_f^2 + 0 = \frac{1}{2}mv_f^2$

3. Use energy conservation, $E_f = E_i$, to solve for $v_f$: $\frac{1}{2}mv_f^2 = mgL\sin(\theta) \Rightarrow v_f = \sqrt{2gL\sin(\theta)}$

4. Calculate the numerical result: $v_f = \sqrt{2(9.81\text{ m/s}^2)(2.1\text{ m})\sin(32^\circ)} = 4.7\text{ m/s}$

**Insight** We already worked out what $\Delta y$ should be in Example 8–1.

---

**Example 8–4 Spring Loaded** At a party, a 0.63-kg ball is going to be shot vertically upward using a spring-loaded mechanism. The spring has a force constant of 188 N/m and is initially compressed, by both the ball and the person loading it, to 45 cm from equilibrium. **(a)** How high will the ball go above the compressed position and **(b)** what will its speed be at half of this height? (Assume the ball leaves contact with the spring when the spring reaches its usual equilibrium position.)

**Picture the Problem** The sketch on the left shows the ball loaded onto the spring, and the sketch on the right shows it after launch.

**Strategy** To determine the height of the ball we can use energy conservation, since only conservative forces are doing work here. Set $U_{\text{grav}} = 0$ in the initial position of the ball.

**Solution**

**Part (a)**

1. Write the total energy for the initial situation where it is all in the spring: $E_i = \frac{1}{2}kx^2$

2. Write the total energy for the final situation where it is all gravitational potential energy: $E_f = mgh$

3. Set these equal to each other and solve for $h$: $mgh = \dfrac{1}{2}kx^2 \Rightarrow h = \dfrac{kx^2}{2mg}$

4. Calculate the numerical result for $h$: $h = \dfrac{(188\text{ N/m})(0.45\text{ m})^2}{2(0.63\text{ kg})(9.81\text{ m/s}^2)} = 3.1\text{ m}$

**Part (b)**

5. Write the total energy for the system when the ball is at half the calculated height in part (a): $E_f = \frac{1}{2}mv^2 + \frac{1}{2}mgh$

6. Set this expression equal to $E_i$ and solve for $v$: $\frac{1}{2}kx^2 = \frac{1}{2}mv^2 + \frac{1}{2}mgh \therefore v = \left(\dfrac{kx^2}{m} - gh\right)^{1/2}$

$$v = \left(\frac{(188\text{ N/m})(0.45\text{ m})^2}{(0.63\text{ kg})} - (9.81\text{ m/s}^2)(3.08\text{ m})\right)^{1/2}$$
$$= 5.5\text{ m/s}$$

**Insight** This problem is a good example of one that would have been very difficult to solve by just using Newton's second law. It demonstrates why the concepts of work and energy are useful.

## Practice Quiz

5. If mechanical energy is conserved within a system,

   **(a)** all objects in the system move at constant speed.

   **(b)** there is no change in the potential energy of the system.

   **(c)** only gravitational forces act within the system.

   **(d)** all the energy is in the form of potential energy.

   **(e)** None of the above.

6. Which of the following statements is most accurate?

   **(a)** Mechanical energy is always conserved.

   **(b)** Mechanical energy is conserved only when potential energy is transformed into kinetic energy.

   **(c)** Mechanical energy is conserved when no nonconservative forces are applied.

   **(d)** Mechanical energy is conserved when no work is done by nonconservative forces.

   **(e)** Mechanical energy is never conserved.

7. A horizontal spring of force constant $k = 100$ N/m is compressed by 5.6 cm, placed against a 0.25-kg ball (that rests on a frictionless, horizontal surface), and then released. How fast will the ball move if it loses contact with the spring when the spring passes its equilibrium point?

   **(a)** 1.3 m/s **(b)** 1.1 m/s **(c)** 1.4 m/s **(d)** 140 m/s **(e)** 2.5 m/s

## 8–4 Work Done by Nonconservative Forces

In the previous section, we discussed how mechanical energy is conserved when only conservative forces do work. It follows that when work is done by nonconservative forces, mechanical energy is not conserved. Instead, the work done by nonconservative forces goes into changing the mechanical energy by an amount equal to the work done:

$$W_{nc} = \Delta E$$

If $W_{nc}$ is positive, then we say that energy is being added to the system. If $W_{nc}$ is negative, then energy is being lost, or dissipated, from the system.

---

**Exercise 8–5 The Luge** In the 2002 Winter Olympic games in Salt Lake City, Utah, Armin Zöggeler of Italy won the gold medal in the luge. The total course length was 1335 m with a vertical drop of 103.5 m. If Armin made a starting leap that gave him an initial speed of 1.00 m/s, and the coefficient of kinetic friction was 0.0222, what was his speed at the finish line?

**Solution:** The luge track is a winding and uneven surface, so a straight inclined plane would not be an accurate picture. However, on average, the results work out to be the same as you would get on a straight inclined plane. Thus, we work the problem as if it were a straight inclined path with the understanding that intermediate results can be considered to be only average values. Try to sketch a diagram for the luge.

**Given:** $L = 1335$ m, $\Delta y = -103.5$ m, $\upsilon_i = 1.00$ m/s, $\mu_k = 0.0222$; **Find:** $\upsilon_f$

In this problem, we cannot apply the conservation of mechanical energy because work is being done by kinetic friction (a nonconservative force), and the amount of work done by friction causes a change in the total mechanical energy. The work done by friction is

$$W_f = -f_k L = -\mu_k NL = -\mu_k mg\cos(\alpha)L$$

where the angle is $\alpha = \sin^{-1}(|\Delta y|/L) = 4.446°$. The minus sign is from cos(180°).

If we take the potential energy to be zero at the bottom of the track, the change in mechanical energy is

$$\Delta E = \Delta K + \Delta U = \tfrac{1}{2}m\upsilon_f^2 - \tfrac{1}{2}m\upsilon_i^2 + mg\Delta y$$

Putting it all together, we have

$$-\mu_k mgL\cos(\alpha) = \tfrac{1}{2}m\upsilon_f^2 - \tfrac{1}{2}m\upsilon_i^2 + mg\Delta y$$

Solving for $v_f$ gives

$$\upsilon_f = \left[\upsilon_i^2 - 2g\ (\Delta y) - 2\mu_k gL\cos\alpha\right]^{1/2}$$

$$= \left[(1.00\ \text{m/s})^2 - 2(9.81\ \text{m/s}^2)(-103.5\ \text{m}) - 2(0.0222)(9.81\ \text{m/s}^2)(1335\ \text{m})\cos(4.446°)\right]^{1/2} = 38.1\ \text{m/s}$$

In this problem, every quantity requires knowledge of Armin's mass, which was not given. Precisely because every term required it, we were able to cancel it in the equation. Don't necessarily think you're stuck if you don't know something that appears to be required; it may cancel out in the long run.

---

**Example 8–6 Assembly Line** A conveyer belt in a factory carries a part down an assembly line to be checked by different workers. At the end of the line the part is boxed and has a total mass of 2.35 kg. The belt then carries it to a 20.0° ramp on which it slides down to the shippers waiting 3.00 meters below. If the coefficient of kinetic friction between the box and the ramp is 0.377, and the box reaches the shippers with a speed of 0.100 m/s, with what speed did the conveyer belt carry the part?

**Picture the Problem** The diagram shows the box moving along the conveyer belt toward the ramp.

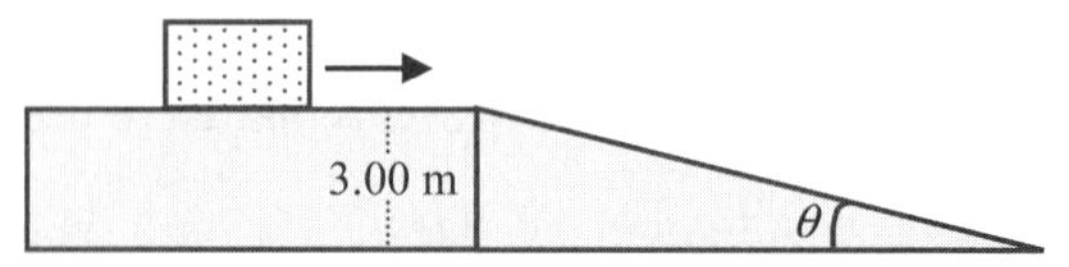

**Strategy** As in the above example, kinetic friction does work here and changes the mechanical energy. To solve this problem, we write out the work done by friction and set it equal to $\Delta E$.

**Solution**

1. The work done by friction is: $W_f = -f_k L = -\mu_k mg\cos(\theta)L$

2. The length of the ramp is ($h$ = height): $L = h/\sin(\theta)$

3. Write the work as: $W_f = -f_k L = -\mu_k mgh\dfrac{\cos(\theta)}{\sin(\theta)} = -\dfrac{\mu_k mgh}{\tan\theta}$

4. Taking $U$ = 0 at the bottom of the ramp, calculate the change in mechanical energy: $\Delta E = \Delta K + \Delta U = \frac{1}{2}mv_f^2 - \frac{1}{2}mv_i^2 - mgh$

5. Set the frictional work equal to $\Delta E$: $-\dfrac{\mu_k mgh}{\tan\theta} = \frac{1}{2}mv_f^2 - \frac{1}{2}mv_i^2 - mgh$

6. Solve for $v_i$:

$$v_i = \left[v_f^2 - 2gh + \frac{2\mu_k gh}{\tan\theta}\right]^{1/2} = \left[v_f^2 - 2gh\left(1 - \frac{\mu_k}{\tan\theta}\right)\right]^{1/2}$$

$$= \left[(0.100\ \text{m/s})^2 - 2(9.81\ \text{m/s}^2)(3.00\ \text{m})\left(1 - \frac{0.377}{\tan(20.0^\circ)}\right)\right]^{1/2}$$

$$= 1.46\ \text{m/s}$$

**Insight** Having worked through this example and the previous one, you should begin to become comfortable with applying the work done by nonconservative forces, especially friction, to work and energy problems.

## Practice Quiz

8. Work done by a nonconservative force
   **(a)** can change only the kinetic energy.
   **(b)** can change only the potential energy.
   **(c)** can change either the kinetic or potential energy, or both.
   **(d)** Work cannot be done by a nonconservative force.
   **(e)** None of the above.

**9**. In which of the following cases does work done by a nonconservative force cause a change in potential energy?

**(a)** A spring launches a ball into the air.

**(b)** A person throws a ball into the air.

**(c)** A ball falls vertically downward with negligible air resistance.

**(d)** A sliding block on a horizontal surface comes to rest.

**(e)** None of the above.

**10**. A block given an initial speed of 5.7 m/s comes to rest due to friction after sliding 2.4 m upward along a surface that is inclined at 30 degrees. What is the coefficient of kinetic friction?

**(a)** 0.22 **(b)** 0.42 **(c)** 0.58 **(d)** 0.50 **(e)** 0.87

## 8–5 Potential Energy Curves and Equipotentials

A plot of the potential energy function of a system, known as a potential energy curve, can provide a lot of useful information about the behavior of the system. Consider the following potential energy curve for a case in which the mechanical energy is conserved.

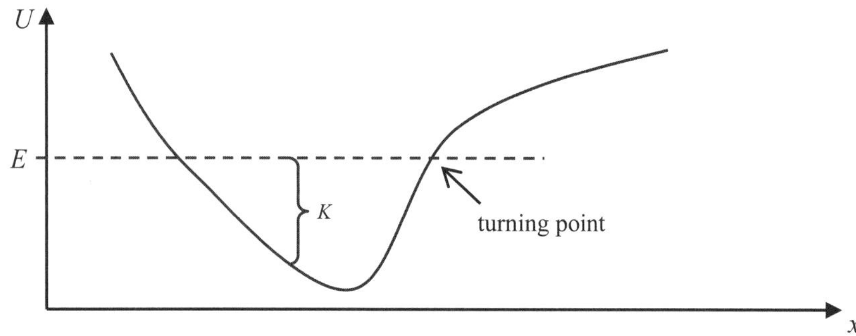


The dashed horizontal line represents the value of the mechanical energy of the system. The difference between the potential energy curve and the mechanical energy must equal the kinetic energy, $K$. Where the horizontal line intersects the curve is where the energy of the system is completely in the form of potential energy, so that $K = 0$; the motion of the system must stop here. These points are called **turning points** because objects within the system stop moving in one direction and begin moving in the opposite direction.

Within the *potential well* between the two turning points, the motion will be oscillatory, going back and forth between the two turning points. Regions of the curve that are higher than $E$ would correspond to negative kinetic energy, which is not possible. These are the *forbidden regions* of the motion. The above plot is of a one-dimensional potential energy function. In many situations, $U$ is a function of three-dimensional space. In these cases it can be useful to plot just the lines of constant potential energy – the **equipotentials**. This type of analysis is used widely in chemistry.

**Example 8–7 Potential Energy Curves** A ball, starting at point $a$, moves under the influence of the potential energy curve shown. Describe the final motion of the ball.

**Picture the Problem** The diagram shows the potential energy curve labeled as $U$, the mechanical energy $E$, and five specific points labeled $a \rightarrow e$.

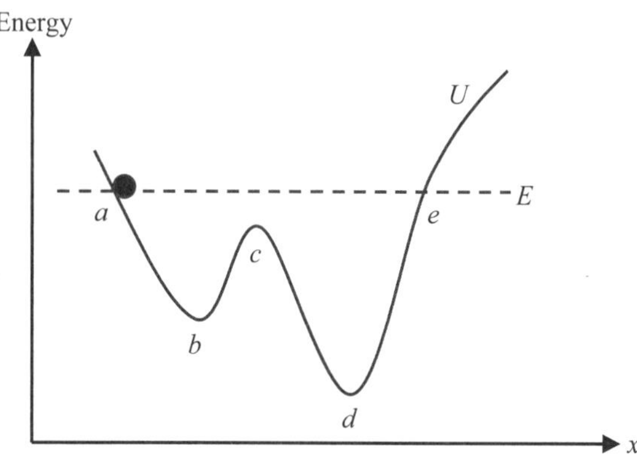


**Strategy/Solution**

At point $a$, $U = E$, so the ball is at rest. It will gain kinetic energy as it moves toward $b$ so its speed increases from $a$ to $b$. As the ball moves from $b$ to $c$, it loses kinetic energy as it gains potential energy, so it slows down; however, it does not lose all its kinetic energy so it does not stop at $c$. In moving from $c$ to $d$ it speeds up again, moving faster at $d$ than it did at $b$. On its way to $e$, the ball gets slower and slower, eventually coming to rest instantaneously at $e$.

After coming to rest at $e$, the ball's motion reverses, and it has the same speeds at the same points along its path. The ball eventually comes back to rest at point $a$. Since there is no work done to change the mechanical energy, this cycle continues with the ball going back and forth between the turning points $a$ and $e$.

**Insight** What would happen if there were a small amount of friction consuming energy from the ball's motion?

**Practice Quiz**

11. When an object reaches a turning point in its motion

    **(a)** it must stop and remain at rest.

    **(b)** it must continue moving past this point.

    **(c)** all of its mechanical energy is in the form of kinetic energy.

    **(d)** none of its mechanical energy is in the form of kinetic energy.

# Reference Tools and Resources

## I. Key Terms and Phrases

**conservative force** any force for which the work done is independent of path

**nonconservative force** any force that is not a conservative force

**potential energy** a representation of the extent to which work is stored in the configuration of a system

**mechanical energy** the sum of the kinetic and potential energies in a system

**turning point** the position on a potential energy curve at which an object will stop and reverse direction

**equipotential** a region in which every point has an equal value of potential energy

## II. Important Equations

| Name/Topic | Equation | Explanation |
|---|---|---|
| potential energy | $\Delta U = -W_c$ | The definition of potential energy |
| gravitational potential energy | $U = mgy$ | The gravitational potential energy, assuming $U = 0$ at $y = 0$ |
| spring potential energy | $U = \frac{1}{2}kx^2$ | The spring potential energy assuming $U = 0$ at $x = 0$ |
| mechanical energy | $E = K + U$ | The definition of mechanical energy |
| work by nonconservative forces | $W_{nc} = \Delta E$ | The work done by nonconservative forces changes the mechanical energy |

## III. Know Your Units

| Quantity | Dimension | SI Unit |
|---|---|---|
| potential energy ( $U$ ) | $[\mathrm{M}]\cdot[\mathrm{L}^2]/[\mathrm{T}^2]$ | J |

## Puzzle

**ENERGY DROP**

Trapeze artist Lea stands at point A in the picture (marked by the square). Under her knees she attaches a taut 10-meter rope, which is tied to the ceiling at the point marked with the dot. She lets herself go. At the bottom of the quarter circle she picks up her twin sister, who is standing at point B. Together they continue on the 10-meter-radius circle until they come to a stop at the platform marked C. Between platforms B and C they rise vertically 2.5 meters. Can you account for all the energy transformations in this stunt?

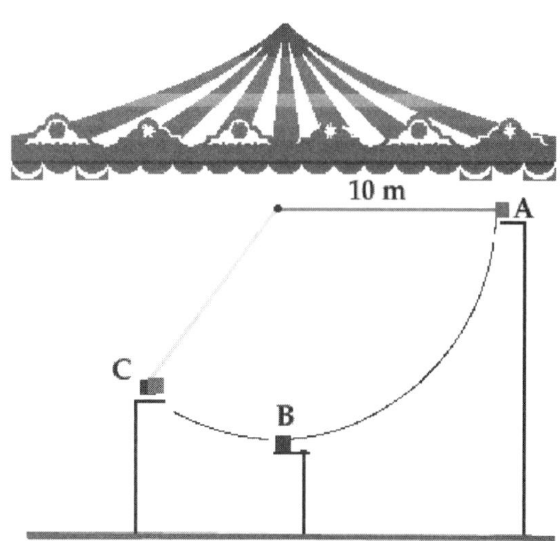


## Answers to Selected Conceptual Questions

**8.** The jumper's initial kinetic energy is largely converted to a compressional, spring-like potential energy as the pole bends. The pole straightens out, converting its potential energy into gravitational potential energy. As the jumper falls, the gravitational potential energy is converted into kinetic energy, and finally, the kinetic energy is converted to compressional potential energy as the cushioning pad on the ground is compressed.

**12.** The total mechanical energy decreases with time if air resistance is present. The energy lost by the ball is transferred to the air molecules.

# Solutions to Selected End-of-Chapter Problems and Conceptual Exercises

2. **Picture the Problem**: The three paths of the object are depicted at right.

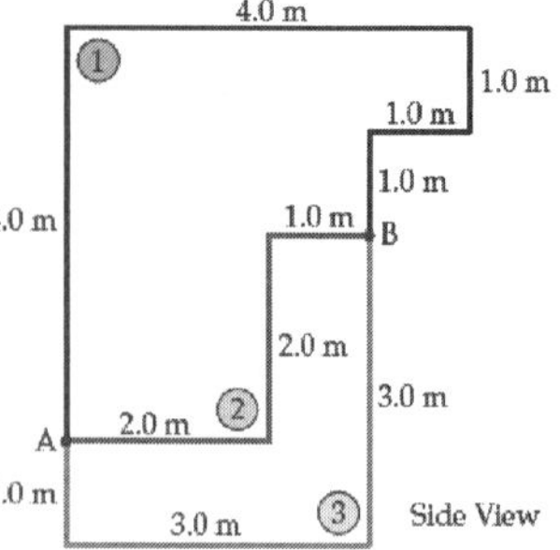


**Strategy:** Find the work done by gravity $W = mgy$ when the object is moved downward, $W = -mgy$ when it is moved upward, and zero when it is moved horizontally. Sum the work done by gravity for each segment of each path.

**Solution: 1.** Calculate the work for path 1:

$$W_1 = mg\left[-y_1 + 0 + y_2 + 0 + y_3\right] = mg\left[-(4.0\text{ m}) + (1.0\text{ m}) + (1.0\text{ m})\right]$$

$$W_1 = (3.2\text{ kg})(9.81\text{ m/s}^2)(-2.0\text{ m}) = \boxed{-63\text{ J}}$$

**2.** Calculate $W$ for path 2:

$$W_2 = mg\left[0 - y_4 + 0\right] = (3.2\text{ kg})(9.81\text{ m/s}^2)[-2.0\text{ m}] = \boxed{-63\text{ J}}$$

**3.** Calculate $W$ for path 3:

$$W_3 = mg\left[y_5 + 0 - y_6\right] = (3.2\text{ kg})(9.81\text{ m/s}^2)\left[(1.0\text{ m}) - (3.0\text{ m})\right] = \boxed{-63\text{ J}}$$

**Insight:** The work is path-independent because gravity is a conservative force.

16. **Picture the Problem**: A graph of the potential energy vs. stretch distance is depicted at right.

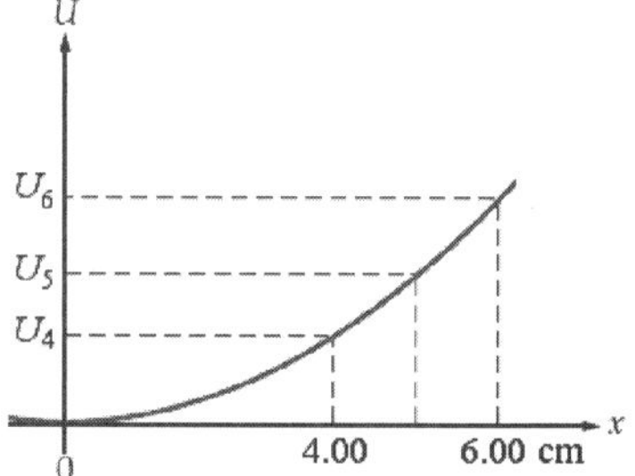


**Strategy:** The work that you must do to stretch a spring is equal to minus the work done by the spring because the force you exert is in the opposite direction from the force the spring exerts. Use equations 8-1 and 8-5 together to find the spring constant and the required work to stretch the spring the specified distance.

**Solution: 1. (a)** Because the stored potential energy in a spring is proportional to the stretch distance squared, the work required to stretch the spring from 5.00 cm to 6.00 cm will be $\boxed{\text{greater than}}$ the work required to stretch it from 4.00 cm to 5.00 cm.

**2. (b)** Use equations 8-1 and 8-5 to find $k$:

$$W_{\text{req}} = -W_{\text{spring}} = -(-\Delta U) = U_5 - U_4 = \tfrac{1}{2}kx_5^2 - \tfrac{1}{2}kx_4^2 = \tfrac{1}{2}k\left(x_5^2 - x_4^2\right)$$

$$k = \frac{2W_{\text{req}}}{x_5^2 - x_4^2} = \frac{2(30.5\text{ J})}{(0.0500\text{ m})^2 - (0.0400\text{ m})^2} = \underline{\underline{6.78\times10^4\text{ N/m}}}$$

**3.** Use $k$ and equations 8-1 and 8-5 to find the new $W_{\text{req}}$:

$$W_{\text{req}} = \tfrac{1}{2}k\left(x_2^{\,2} - x_1^{\,2}\right) = \tfrac{1}{2}\left(6.78\times10^4\text{ N/m}\right)\left[(0.0600\text{ m})^2 - (0.0500\text{ m})^2\right] = \boxed{37.3\text{ J}}$$

**Insight:** Using the same procedure we discover that it would take 44.1 J to stretch the spring from 6.00 cm to 7.00 cm.

35. **Picture the Problem**: The pendulum bob swings from point B to point A and gains altitude and thus gravitational potential energy. See the figure at right.

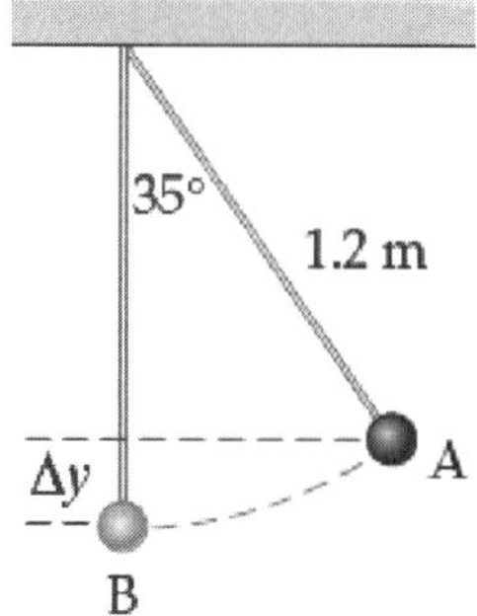


**Strategy:** Use equation 7-6 to find the kinetic energy of the bob at point B. Use the geometry of the problem to find the maximum change in altitude $\Delta y_{max}$ of the pendulum bob, and then use equation 8-3 to find its maximum change in gravitational potential energy. Apply conservation of energy between points B and the endpoint of its travel to find the maximum angle $\theta_{max}$ the string makes with the vertical.

**Solution: 1. (a)** Use equation 7-6 to find $K_B$: $K_B = \frac{1}{2}mv_B^2 = \frac{1}{2}(0.33\text{ kg})(2.4\text{ m/s})^2 = \boxed{0.95\text{ J}}$

**2. (b)** Since there is no friction, mechanical energy is conserved. If we take the potential energy at point B to be zero, we can say that all of the bob's kinetic energy will become potential energy when the bob reaches its maximum height and comes momentarily to rest. Therefore the change in potential energy between point B and the point where the bob comes to rest is $\boxed{0.95\text{ J.}}$

**3. (c)** Find the height change $\Delta y_{max}$ of the pendulum bob: $\Delta y_{max} = L - L\cos\theta_{max} = L(1-\cos\theta_{max})$

**4.** Use equation 8-3 and the result of part (b) to solve for $\theta_{max}$:

$$\Delta U = mg\Delta y_{max} = mgL(1-\cos\theta_{max})$$

$$\theta_{max} = \cos^{-1}\left(1-\frac{\Delta U}{mgL}\right) = \cos^{-1}\left[1-\frac{0.95\text{ J}}{(0.33\text{ kg})(9.81\text{ m/s}^2)(1.2\text{ m})}\right] = \boxed{41°}$$

**Insight:** The pendulum bob cannot swing any farther than 41° because there is not enough energy available to raise the mass to a higher elevation.

50. **Picture the Problem**: The skater travels up a hill (we know this for reasons given below), changing his kinetic and gravitational potential energies, while both his muscles and friction do nonconservative work on him.

**Strategy:** The total nonconservative work done on the skater changes his mechanical energy according to equation 8-9. This nonconservative work includes the positive work $W_{nc1}$ done by his muscles and the negative work $W_{nc2}$ done by the friction. Use this relationship and the known change in potential energy to find $\Delta y$.

**Solution: 1. (a)** The skater has gone $\boxed{\text{uphill}}$ because the work done by the skater is larger than that done by friction, so the skater has gained mechanical energy. However, the final speed of the skater is less than the initial speed, so he has lost kinetic energy. Therefore he must have gained potential energy and has gone uphill.

**2. (b)** Set the nonconservative work equal to the change in mechanical energy and solve for $\Delta y$:

$$W_{nc} = W_{nc1} + W_{nc2} = \Delta E = E_f - E_i$$

$$W_{nc1} + W_{nc2} = (K_f + U_f) - (K_i + U_i) = \frac{1}{2}m(v_f^2 - v_i^2) + mg\Delta y$$

$$\Delta y = \left[W_{nc1} + W_{nc2} - \frac{1}{2}m(v_f^2 - v_i^2)\right]/mg$$

$$= \frac{\left\{(3420\text{ J}) + (-715\text{ J}) - \frac{1}{2}(81.0\text{ kg})\left[(1.22\text{ m/s})^2 - (2.50\text{ m/s})^2\right]\right\}}{(81.0\text{ kg})(9.81\text{ m/s}^2)} = \boxed{3.65\text{ m}}$$

**Insight:** Verify for yourself that if the skates had been frictionless but the skater's muscles did the same amount of work, the skater's final speed would have been 4.37 m/s. He would have sped up if it weren't for friction!

56. **Picture the Problem**: The $U$ vs. $x$ plot for an object is depicted at right.

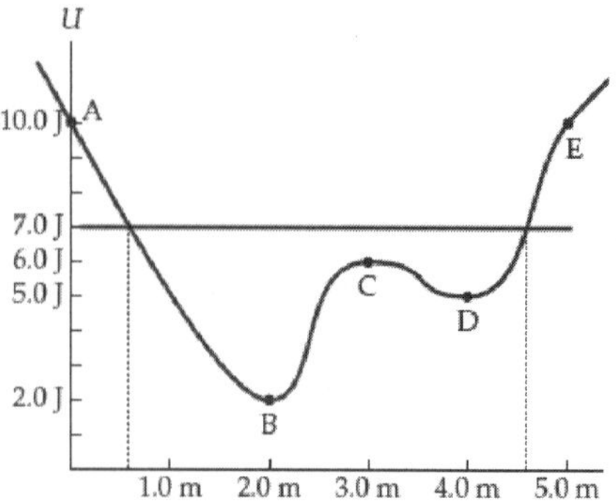


**Strategy:** The object will be at its turning point when its total mechanical energy equals its potential energy. Use the provided information to find the total mechanical energy of the particle, then draw a horizontal line at the corresponding value on the $U$ axis to find the turning points.

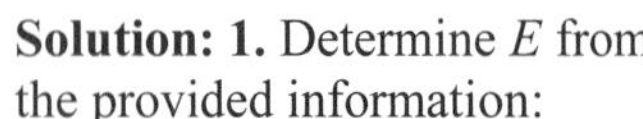

**Solution: 1.** Determine $E$ from the provided information:

$$E = \tfrac{1}{2}mv_C^2 + U_C$$
$$= \tfrac{1}{2}(1.34\text{ kg})(1.25\text{ m/s})^2 + 6.0\text{ J}$$
$$E = \underline{\underline{7.0\text{ J}}}$$

**2.** Draw a horizontal line on the plot at $U = 7.0$ J and determine where the line crosses the potential energy curve. In this case it crosses $\boxed{\text{at approximately } x = 0.6\text{ m and } x = 4.6\text{ m}}$. These are the locations of the turning points.

**Insight:** In order for the turning points to be at A and E, the object needs a speed of 2.36 m/s at point C.

81. **Picture the Problem**: The physical situation is depicted in the figure.

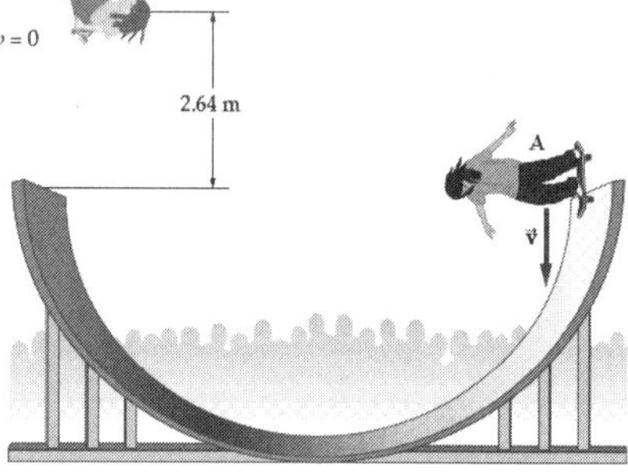


**Strategy:** Use the conservation of mechanical energy and the geometry of the problem to find the speed of the skateboarder at point A. Let $y_A = 0$ and $v_B = 0$.

**Solution:** Set $E_A = E_B$ to find $v_A$:

$$K_A + U_A = K_B + U_B$$
$$\tfrac{1}{2}mv_A^2 + 0 = 0 + mgy_B$$
$$v_A = \sqrt{2gy_B} = \sqrt{2(9.81\text{ m/s}^2)(2.64\text{ m})}$$
$$= \boxed{7.20\text{ m/s}}$$

**Insight:** The solution assumes there is no friction as the skateboarder travels around the half pipe. In real life the skater's speed must exceed 7.20 m/s at point A because friction will convert some of his kinetic energy into heat.

97. **Picture the Problem**: The physical situation is depicted at right.

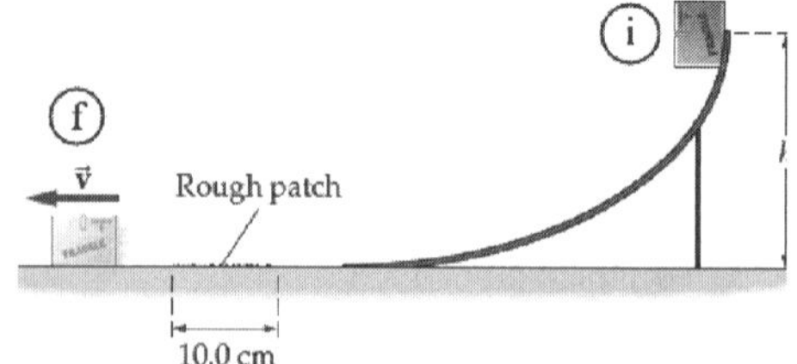


**Strategy:** Use equation 8-9 to relate the nonconservative work of friction on the entire system equal to the energy difference between the initial and final states. Solve the resulting expression for the initial height $h$.

**Solution: 1.** Write equation 8-9 to obtain an expression for $W_{nc}$:

$$W_{nc} = E_f - E_i = (K_f + U_f) - (K_i + U_i)$$
$$= \left(\tfrac{1}{2}mv_f^2 + 0\right) - (0 + mgh)$$

**2.** The nonconservative work is done by friction:

$$W_{nc} = -f_k d = -\mu_k mgd$$

**3.** Substitute the expression from step 2 into step 1 and solve for $v$:

$$-\mu_k mgd = \tfrac{1}{2}mv_f^2 - mgh$$

$$mgh = m\left(\tfrac{1}{2}v_f^2 + \mu_k gd\right)$$

$$h = v_f^2/2g + \mu_k d$$

$$h = \frac{(3.50\text{ m/s})^2}{2(9.81\text{ m/s}^2)} + (0.640)(0.100\text{ m}) = \boxed{0.688\text{ m}}$$

**Insight:** Note that the required $h$ is independent of mass. Friction has an appreciable effect; if there were no friction the initial height would only need to be 0.624 m for the final speed to be 3.50 m/s.

## Answers to Practice Quiz

**1**. (a) **2**. (c) **3**. (c) **4**. (e) **5**. (e) **6**. (d) **7**. (b) **8**. (c) **9**. (b) **10**. (a) **11**. (d)

# CHAPTER 9

# LINEAR MOMENTUM AND COLLISIONS

## Chapter Objectives

After studying this chapter, you should

1. know the definition of *linear momentum* and how it relates to force.

2. know the meaning of *impulse* and how it relates to linear momentum.

3. be able to use linear momentum to analyze elastic and inelastic collisions.

4. be able to determine the location of the center of mass of a system.

5. be able to apply Newton's laws to a system of particles.

6. understand the basic principles behind rocket propulsion.

## Warm-Ups

1. The left ball in the following diagram is chasing after the right ball. The speed of the left ball is 3 m/s, and the speed of the right ball is 2 m/s. After the collision the left ball rebounds at 2 m/s, and the right ball speeds away at 3 m/s. What was the change in velocity experienced by the left ball during the collision? What was the change in velocity experienced by the right ball during the collision? If the mass of the left ball is the same as the mass of the right ball, was momentum conserved during this collision?

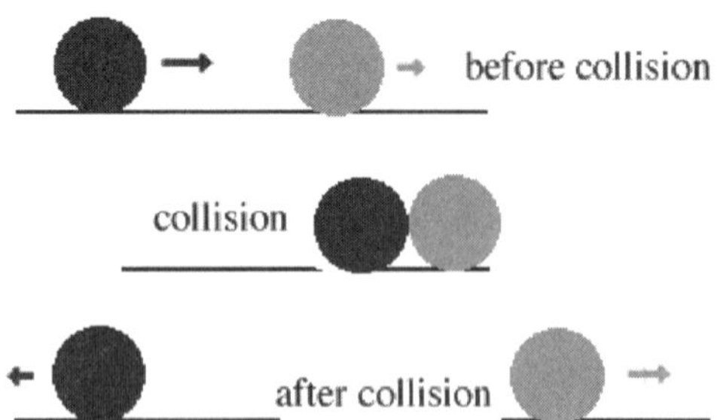

2. Estimate the force on a 145-gram baseball if it hits the bat while moving at 50 miles per hour, stays in contact with the bat for 0.1 second during which time it reverses direction and leaves the bat at 75 miles per hour. Express the result in newtons and in pounds.
3. Suppose the shuttle was launched from a launch pad on the Moon. What changes would we observe? Suppose it was launched from a launch pad in the vacuum of space? Would it work? What changes would we observe?
4. Ball $A$ is propelled forward and collides with ball $B$, which is initially at rest. After the collision, the balls' trajectories, $A'$ and $B'$, are as shown in the diagram. How do the masses of the two balls compare? Briefly explain your answer.

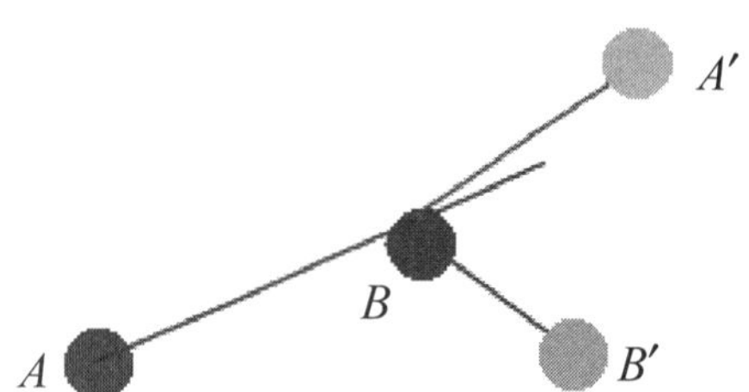


## Chapter Review

In Chapters 7 and 8, we saw that defining concepts beyond just force, velocity, acceleration, and displacement made analyzing some situations much easier, especially the conservation of energy. In this chapter, we do more of the same by introducing the concept of linear momentum. In a way similar to work and energy, the idea of linear momentum often provides a more direct path to understanding physical interactions than working with forces only. Furthermore, as with energy, there is a conservation principle for linear momentum that is of tremendous value in physics.

### 9–1 Linear Momentum

The **linear momentum** of an object is defined as the product of its mass and its velocity

$$\vec{\mathbf{p}} = m\vec{\mathbf{v}}$$

where $\vec{\mathbf{p}}$ is the symbol for linear momentum. Often, we refer to $\vec{\mathbf{p}}$ as just the momentum and leave off the word "linear." Notice that momentum is a vector quantity whose direction is that of the velocity $\vec{\mathbf{v}}$ and whose units are just the mass unit times velocity units, kg·m/s. In one sense, you can think of linear momentum as a measure of the effect the motion of an object has when it interacts with other objects (you'll see this more clearly in the discussion of collisions). In another sense, linear momentum can be viewed as an alternative measure of a moving object's inertia, i.e., its tendency to maintain a constant state of motion (you'll see this more clearly in the discussion of momentum conservation).

For a system of particles, we often speak of the *total momentum* of the system. This total momentum is the vector sum of the momenta of every particle in the system:

$$\vec{\mathbf{p}}_{total} = \vec{\mathbf{p}}_1 + \vec{\mathbf{p}}_2 + \vec{\mathbf{p}}_3 + \cdots$$

We shall see later that the total momentum of a system tells us a lot about how a system behaves as a whole.

---

**Example 9–1 The Total Momentum** Three balls with masses of 2.0 kg, 3.0 kg, and 4.0 kg move with the velocities shown below. Determine the magnitude and direction of the total momentum of the balls.

**Picture the Problem** The diagram shows the three balls and indicates the magnitudes and directions of their velocities.

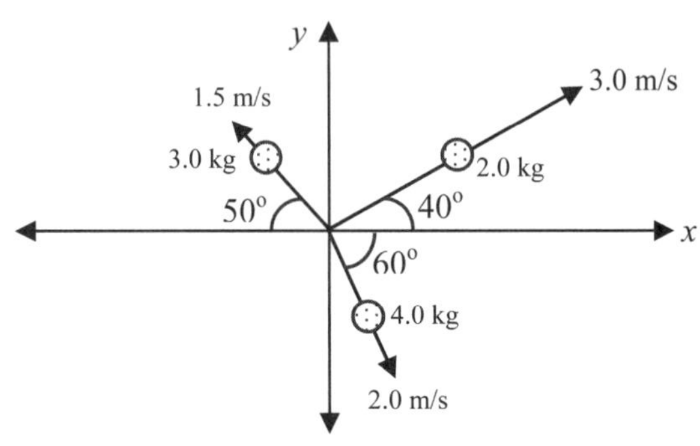


**Strategy** We need to get each particle's momentum and then form a vector sum to get the total. Let's take $m_1 = 2.0$ kg, $v_1 = 3.0$ m/s, $m_2 = 3.0$ kg, $v_2 = 1.5$ m/s, $m_3 = 4.0$ kg, and $v_3 = 2.0$ m/s.

**Solution**

1. Determine the components of $\vec{\mathbf{p}}_1$:

$$p_{1x} = m_1 v_1 \cos(40^\circ) = (2.0\text{ kg})(3.0\text{ m/s})\cos(40^\circ) = 4.60\text{ kg}\cdot\text{m/s}$$

$$p_{1y} = m_1 v_1 \sin(40^\circ) = (2.0\text{ kg})(3.0\text{ m/s})\sin(40^\circ) = 3.86\text{ kg}\cdot\text{m/s}$$

2. Determine the components of $\vec{\mathbf{p}}_2$:

$$p_{2x} = -m_2 v_2 \cos(50^\circ) = -(3.0\text{ kg})(1.5\text{ m/s})\cos(50^\circ) = -2.89\text{ kg}\cdot\text{m/s}$$

$$p_{2y} = m_2 v_2 \sin(50^\circ) = (3.0\text{ kg})(1.5\text{ m/s})\sin(50^\circ) = 3.45\text{ kg}\cdot\text{m/s}$$

3. Determine the components of $\vec{\mathbf{p}}_3$:

$$p_{3x} = m_3 v_3 \cos(-60^\circ) = (4.0\text{ kg})(2.0\text{ m/s})\cos(-60^\circ) = 4.00\text{ kg}\cdot\text{m/s}$$

$$p_{3y} = m_3 v_3 \sin(-60^\circ) = (4.0\text{ kg})(2.0\text{ m/s})\sin(-60^\circ) = -6.93\text{ kg}\cdot\text{m/s}$$

4. Determine the components of $\vec{\mathbf{p}}_{total}$:

$$p_{total,x} = p_{1x} + p_{2x} + p_{3x} = (4.60 - 2.89 + 4.00)\text{ kg}\cdot\text{m/s} = 5.71\text{ kg}\cdot\text{m/s}$$

$$p_{total,y} = p_{1y} + p_{2y} + p_{3y}$$
$$= (3.86 + 3.45 - 6.93)\ \text{kg}\cdot\text{m/s} = 0.380\ \text{kg}\cdot\text{m/}s$$

**5.** Determine the magnitude of $\vec{\mathbf{p}}_{total}$:

$$p_{total} = \left[p_{total,x}^2 + p_{total,y}^2\right]^{1/2}$$
$$\left[(5.71\ \text{kg}\cdot\text{m/s})^2 + (0.380\ \text{kg}\cdot\text{m/s})^2\right]^{1/2} = 5.7\ \text{kg}\cdot\text{m/s}$$

**6.** Determine the direction of $\vec{\mathbf{p}}_{total}$:

$$\theta_{total} = \tan^{-1}\left(\frac{p_{total,y}}{p_{total,x}}\right) = \tan^{-1}\left(\frac{0.380\ \frac{\text{kg}\cdot\text{m}}{\text{s}}}{5.71\ \frac{\text{kg}\cdot\text{m}}{\text{s}}}\right) = 3.8^\circ$$

**Insights** Note that for $\vec{\mathbf{p}}_2$ we put in the signs of the components from our knowledge of their directions instead of using the angle with the $+x$ axis. This way of doing it is common, so be sure you understand it. Also notice that because both components of $\vec{\mathbf{p}}_{total}$ are positive, the 3.8° is the angle from the $+x$ axis, and nothing else needs to be said.

## Practice Quiz

**1.** Two objects have equal velocities but one has twice the mass of the other; the object with larger mass

**(a)** has half the momentum of the other.

**(b)** has twice the momentum of the other.

**(c)** has the same momentum as the other.

**(d)** must move slower than the other.

**(e)** None of the above.

**2.** If two objects have the same mass and speed but move in opposite directions, then

**(a)** they have momenta of equal magnitude and opposite directions.

**(b)** they have equal momenta.

**(c)** each has zero momentum.

**(d)** one must move faster than the other.

**(e)** None of the above.

**3.** A 3.2-kg object has a velocity of 1.8 m/s in the $x$ direction. What is its linear momentum?

**(a)** $5.0\ \text{kg}\cdot\text{m/s}\ \hat{\mathbf{x}}$ **(b)** $1.8\ \text{kg}\cdot\text{m/s}\ \hat{\mathbf{x}}$ **(c)** $0.56\ \text{kg}\cdot\text{m/s}\ \hat{\mathbf{x}}$ **(d)** $5.8\ \text{kg}\cdot\text{m/s}\ \hat{\mathbf{x}}$ **(e)** $1.4\ \text{kg}\cdot\text{m/s}\ \hat{\mathbf{x}}$

## 9–2 Momentum and Newton's Second Law

The form of Newton's second law that we have been using until now, $\vec{\mathbf{F}}_{net} = m\vec{\mathbf{a}}$, applies only to circumstances in which the mass remains constant; however, in the most general cases, as with rockets, the mass may change during the motion. The most general form of Newton's second law is expressed in terms of momentum:

$$\vec{\mathbf{F}}_{net} = \frac{\Delta\vec{\mathbf{p}}}{\Delta t}$$

Thus, force equals the rate at which momentum changes whether the change is in the mass, the velocity, or both. For cases in which mass is constant, this equation reduces to $\vec{\mathbf{F}}_{net} = m\vec{\mathbf{a}}$.

---

**Example 9–2 Newton's Second Law** A 0.25-kg object moves due east at 2.1 m/s. What force is needed to cause it to move due north at 3.6 m/s in 1.52 s?

**Solution** Even though this is a case of constant mass, let's work it in terms of momentum to illustrate that approach.

**Given:** $m = 0.25$ kg, $\vec{\mathbf{v}}_i = 2.1\text{ m/s}\,\hat{\mathbf{x}}$, $\vec{\mathbf{v}}_f = 3.6\text{ m/s}\,\hat{\mathbf{y}}$, $t = 1.52$ s; **Find:** $\vec{\mathbf{F}}$

The initial and final momenta of the object are

$$\vec{\mathbf{p}}_i = m\vec{\mathbf{v}}_i = 0.25\text{ kg}(2.1\text{ m/s})\,\hat{\mathbf{x}} = 0.525\text{ kg}\cdot\text{m/s}\,\hat{\mathbf{x}}$$

$$\vec{\mathbf{p}}_f = m\vec{\mathbf{v}}_f = 0.25\text{ kg}(3.6\text{ m/s})\hat{\mathbf{y}} = 0.900\text{ kg}\cdot\text{m/s}\,\hat{\mathbf{y}}$$

The change in momentum, therefore, is

$$\Delta\vec{\mathbf{p}} = \vec{\mathbf{p}}_f - \vec{\mathbf{p}}_i = \left(-0.525\,\hat{\mathbf{x}} + 0.900\,\hat{\mathbf{y}}\right)\text{kg}\cdot\text{m/s}$$

The force, then, is given by

$$\vec{\mathbf{F}} = \frac{\Delta\vec{\mathbf{p}}}{t} = \frac{\left(-0.525\,\hat{\mathbf{x}} + 0.900\,\hat{\mathbf{y}}\right)\ \text{kg}\cdot\text{m/s}}{1.52\ \text{s}} = \left(-0.35\,\hat{\mathbf{x}} + 0.59\,\hat{\mathbf{y}}\right)\text{N}$$

**Insight** If you are curious, work this out as $\vec{\mathbf{F}} = m\vec{\mathbf{a}}$, and check that you get the same result.

---

## Practice Quiz

4. When does the equation $\vec{\mathbf{F}} = m\vec{\mathbf{a}}$ not accurately describe the dynamics of an object?

   **(a)** when gravity is not present

   **(b)** when more than one force acts on the object

   **(c)** when kinetic friction does work on the object

   **(d)** when the mass of the object is constant

   **(e)** None of the above.

5. If the linear momentum of a 1.6-kg object is decreasing at a rate of 5.0 $\frac{\text{kg}\cdot\text{m/s}}{\text{s}}$, what is the magnitude of the force on the object?

   **(a)** 8.0 N **(b)** 3.4 N **(c)** 5.0 N **(d)** 6.6 N **(e)** 3.1 N

## 9–3 Impulse

As mentioned previously, the concept of momentum is important when two objects interact. In this chapter, the interaction we focus on is called a **collision**. A collision occurs when the forces of interaction between two objects are large for a finite period of time. The average force applied to an object times the amount of time this force is applied is called the **impulse**, $\vec{\mathbf{I}}$

$$\vec{\mathbf{I}} = \vec{\mathbf{F}}_{av}\Delta t$$

The SI unit of impulse is the N·s, which has no special name. The same impulse can be delivered by a weak force acting for a long period of time or a strong force acting for a short period of time. The most common usage of impulse is for the latter case. The concept of impulse is closely related to momentum by what is often called the *impulse-momentum theorem*

$$\vec{\mathbf{I}} = \Delta\vec{\mathbf{p}}$$

The above expression is really just a restatement of Newton's second law in a form that is convenient to describe interactions, like collisions, for which it is difficult to know precise values of the forces involved.

---

**Example 9–3 Ricochet** The velocity of a rock of mass 0.24 kg moving with a speed of 3.33 m/s makes a 60° angle with the normal to a brick wall. The rock is in contact with the wall for only 0.032 s. If the velocity of the rock makes an angle of 40° with the normal to the wall after it strikes and has a magnitude of 2.68 m/s, **(a)** what impulse does the wall apply to the rock, and **(b)** what average force causes this impulse?

**Picture the Problem** The diagram shows the initial and final momenta of the rock as it bounces off the wall.

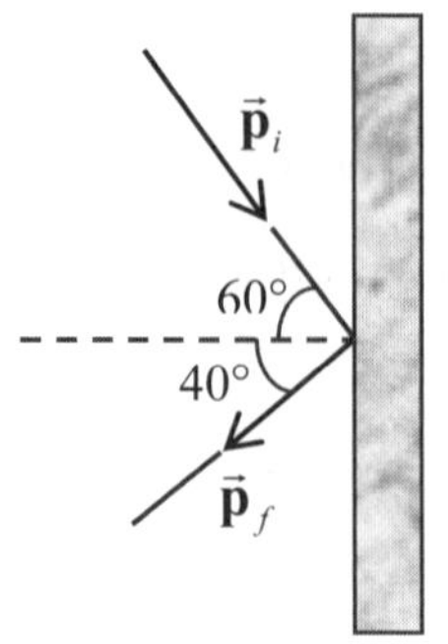


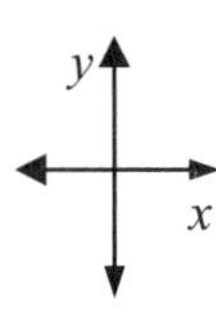


**Strategy** To solve for the impulse, we can find the change in momentum that results from the bounce. Once we know the impulse, we'll use it to get the average force.

**Solution**

**Part (a)**

1. Determine the components of the initial and final momenta of the rock:

$$p_{ix} = mv_i\cos(60^\circ) = 0.24\text{ kg}(3.33\text{ m/s})\cos(60^\circ) = 0.3996\text{ kg}\cdot\text{m/s}$$

$$p_{iy} = -mv_i\sin(60^\circ) = -0.24\text{ kg}(3.33\text{ m/s})\sin(60^\circ) = -0.6921\text{ kg}\cdot\text{m/s}$$

$$p_{fx} = -mv_f\cos(40^\circ) = -0.24\text{ kg}(2.68\text{ m/s})\cos(40^\circ) = -0.4927\text{ kg}\cdot\text{m/s}$$

$$p_{fy} = -mv_f\sin(40^\circ) = -0.24\text{ kg}(2.68\text{ m/s})\sin(40^\circ) = -0.4134\text{ kg}\cdot\text{m/s}$$

2. Determine the impulse as the change in momentum:

$$\vec{\mathbf{I}} = \vec{\mathbf{p}}_f - \vec{\mathbf{p}}_i = (p_{fx} - p_{ix})\hat{\mathbf{x}} + (p_{fy} - p_{iy})\hat{\mathbf{y}} = [(-0.4927 - 0.3996)\hat{\mathbf{x}} + (-0.4134 + 0.6921)\hat{\mathbf{y}}]\text{ kg}\cdot\text{m/s} = (-0.892\,\hat{\mathbf{x}} + 0.279\,\hat{\mathbf{y}})\text{ N}\cdot\text{s}$$

**Part (b)**

3. Use the definition of impulse to obtain an expression for the average force:

$$\vec{\mathbf{I}} = \vec{\mathbf{F}}_{av}t \;\Rightarrow\; \vec{\mathbf{F}}_{av} = \vec{\mathbf{I}}/t$$

4. Calculate the numerical value of the force:

$$\vec{\mathbf{F}}_{av} = \frac{(-0.892\,\hat{\mathbf{x}} + 0.279\,\hat{\mathbf{y}})\text{ N}\cdot\text{s}}{0.032\text{ s}} = (-28\,\hat{\mathbf{x}} + 8.7\,\hat{\mathbf{y}})\text{ N}$$

**Insight** Be sure you understand the signs and relative magnitudes of the components of $\vec{\mathbf{I}}$ and $\vec{\mathbf{F}}_{av}$.

## Practice Quiz

6. During one trial, a force $F$ is applied to a ball for an amount of time $t$. During a second trial, a force of $3F$ is applied for a time of $t/2$. Which of the following is true concerning the magnitude of the change in momentum of the ball?

**(a)** There's a greater change for the second trial.

**(b)** There's a smaller change for the second trial.

**(c)** We get the same nonzero momentum change in each trial.

**(d)** The momentum doesn't change in either trial.

**(e)** None of the above.

7. An object of mass 2.8 kg moves at 1.1 m/s in the $+x$ direction. If an impulse of $1.3\,\mathrm{N\cdot s}\,\hat{\mathbf{x}}$ is applied, what is its final momentum?

   **(a)** 3.1 kg·m/s **(b)** 4.0 kg·m/s **(c)** 5.2 kg·m/s **(d)** 4.4 kg·m/s **(e)** 4.2 kg·m/s

## 9–4 Conservation of Linear Momentum

In the previous section we discussed the fact that an impulse applied to an object causes a change in the object's momentum. Consequently, if there is no impulse, then there is no change in momentum. The net impulse on an object will be zero when the net force on that object is zero. This result leads to the following statement:

*When the net force on an object is zero, its linear momentum is conserved.*

The situation becomes a little more interesting for a system of objects such as the billions of air molecules in an enclosed container such as your bedroom. In this latter case you have billions of particles interacting with one another by collisions. By the law of action and reaction all these particles apply equal and opposite forces to each other that cancel out when the system is considered as a whole, that is, the sum of all forces internal to the system is zero: $\sum \vec{\mathbf{F}}_{int} = 0$.

We are left, therefore, with only the external forces applied to a system to make up the overall net force: $\vec{\mathbf{F}}_{net} = \sum \vec{\mathbf{F}}_{ext}$. By applying Newton's second law, we can see that the net impulse on a system equals the change in its total momentum: $\vec{\mathbf{p}}_{total}$ (or $\vec{\mathbf{p}}_{net}$)

$$\Delta\vec{\mathbf{p}}_{\text{total}} = \left(\sum \vec{\mathbf{F}}_{\text{ext}}\right)\Delta t$$

By reasoning similar to that in the single-particle case, we can see from this expression that

*if the net external force on a system is zero, then the total linear momentum of the system is conserved.*

When the total momentum of a system is conserved, we can often apply this result by setting the final total momentum equal to the initial total momentum:

$$\vec{\mathbf{p}}_{1f} + \vec{\mathbf{p}}_{2f} + \vec{\mathbf{p}}_{3f} + \cdots = \vec{\mathbf{p}}_{1i} + \vec{\mathbf{p}}_{2i} + \vec{\mathbf{p}}_{3i} + \cdots$$

Notice that because momentum is a vector quantity, it, unlike energy (a scalar), must be conserved in both magnitude and direction.

### Practice Quiz

**8.** Which of the following statements is most accurate?

**(a)** The linear momentum of a system is always conserved.

**(b)** The linear momentum of a system is conserved only if no external forces act on the system.

**(c)** The linear momentum of a system is conserved if the net external force on the system is zero.

**(d)** The linear momentum of a system is not conserved.

**(e)** None of the above.

### 9–5 Inelastic Collisions

When studying collisions, we often divide them into two categories according to whether or not the total kinetic energy of the colliding bodies is conserved. If the total kinetic energy is not conserved, the collision is called an **inelastic collision**. In analyzing these types of collisions, we usually just apply the conservation of the total linear momentum to the system: $\vec{\mathbf{p}}_f = \vec{\mathbf{p}}_i$. A special case of an inelastic collision occurs when the colliding objects stick together and emerge from the collision effectively as one object. This latter case is called a *completely* inelastic collision because in this case the system loses the maximum amount of kinetic energy it can lose while still conserving momentum.

---

**Example 9–4 One-dimensional Collision** The driver of a 1485-kg car is driving along a street at 25 mi/h. Another driver, coming from directly behind (and not paying attention), is driving a 1520-kg car at 38 mi/h and hits the slower car. If both drivers slam the brakes at the moment of impact and their fenders catch on each other, with what combined speed do they begin to skid?

**Picture the Problem** The upper sketch shows the heavier and faster car ($m_1$) catching up with the lighter car ($m_2$). The lower sketch shows them stuck together after the collision.

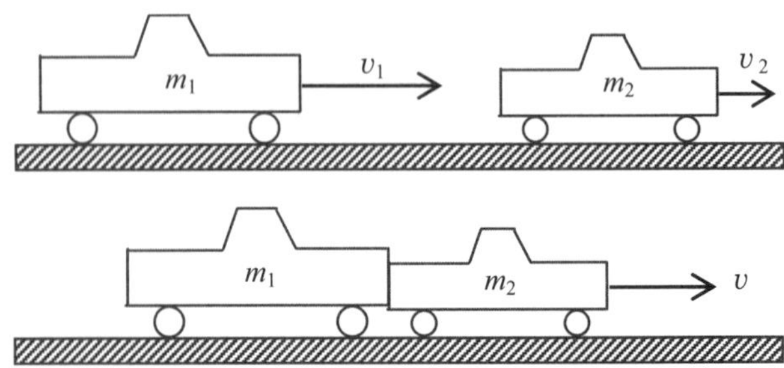

**Strategy** This is a completely inelastic collision. We apply the conservation of momentum by setting the total momenta before and after the collision equal to each other. Take the direction of motion to be the $x$ direction.

**Solution**

**1.** The total momentum before the collision is: $p_{total} = m_1 v_1 + m_2 v_2$

**2.** The total momentum after the collision is: $p_{total} = (m_1 + m_2)v$

**3.** Set them equal: $m_1 v_1 + m_2 v_2 = (m_1 + m_2)v \;\therefore\; v = \dfrac{m_1 v_1 + m_2 v_2}{m_1 + m_2}$

**4.** Calculate the numerical result:

$$v = \frac{1520\text{ kg}\left(38\ \frac{\text{mi}}{\text{h}}\right) + 1485\text{ kg}\left(25\ \frac{\text{mi}}{\text{h}}\right)}{1520\text{ kg} + 1485\text{ kg}}\left(\frac{0.447\text{ m/s}}{1\text{ mi/h}}\right)$$

$$= 14\text{ m/s} \approx 32\text{ mi/h}$$

**Insight** Since this was just a one-dimensional problem, we dispensed with full vector notation.

## Practice Quiz

**9.** A cart is rolling along a horizontal floor when a box is dropped vertically into it. After catching the box, the cart

**(a)** will move more slowly.

**(b)** will move more quickly.

**(c)** will move at the same speed as before.

**(d)** will move in a different direction.

**(e)** will come to a stop as a result of catching the box.

**10.** Two objects of equal mass of 1.7 kg move directly toward each other with equal speed of 2.2 m/s. If they stick together after the collision, their speed will be

**(a)** 0.65 m/s. **(b)** 2.2 m/s. **(c)** 1.1 m/s. **(d)** 3.7 m/s. **(e)** 0 m/s.

## 9–6 Elastic Collisions

A second category of collisions considers those for which the total kinetic energy is conserved. These collisions are called **elastic collisions**. Keep in mind that, as with momentum, it is the *total* kinetic energy of the system, not that of any individual particle, that we are considering

$$K_{1f} + K_{2f} + K_{3f} + \cdots = K_{1i} + K_{2i} + K_{3i} + \cdots$$

In everyday life, few collisions are perfectly elastic. However, many everyday collisions are approximately elastic because only a small fraction of the kinetic energy is lost. Microscopically, perfectly elastic collisions are commonplace. Along with the conservation of kinetic energy, we must apply the conservation of momentum to these collisions as well. Keep in mind that the conservation of kinetic energy discussed here is a very special occurrence; *there is no general principle of the conservation of kinetic energy.*

In one dimension (a head-on collision), we can combine the equations for the conservation of momentum and the conservation of kinetic energy and solve them for the final velocities of the two particles. For particles $m_1$ and $m_2$ with $m_2$ initially at rest and $m_1$ moving with speed $v_0$ directly toward $m_2$, the result is

$$v_{1f} = \left(\frac{m_1 - m_2}{m_1 + m_2}\right) v_0, \quad v_{2f} = \left(\frac{2m_1}{m_1 + m_2}\right) v_0$$

---

**Example 9–5 Head-on Elastic Collision** A 1.3-kg ball initially moving with a speed of 3.1 m/s strikes a stationary 2.2-kg ball head-on. What are the final velocities of the two balls?

**Solution** This is precisely the situation just discussed. Let's identify the masses and velocities.

**Given:** $m_1 = 1.3$ kg, $v_0 = 3.1$ m/s, $m_2 = 2.2$ kg; **Find:** $v_{1f}$, $v_{2f}$

Making direct use of the equations listed above, we have

$$v_{1f} = \frac{1.3\text{ kg} - 2.2\text{ kg}}{1.3\text{ kg} + 2.2\text{ kg}}(3.1\text{ m/s}) = -0.80\text{ m/s}$$

$$v_{2f} = \frac{2(1.3\text{ kg})}{1.3\text{ kg} + 2.2\text{ kg}}(3.1\text{ m/s}) = 2.3\text{ m/s}$$

**Insight** Notice that after the collision, the balls move in opposite directions.

---

**Example 9–6 Elastic Collision in Two Dimensions** Two objects, $m_2$ of mass 1.50 kg moving due south at 15.3 m/s and $m_1$ of mass 1.19 kg moving northeast at 11.7 m/s, collide elastically. After they collide, $m_2$ moves with a velocity of 12.3 m/s at 10.9° north of east. What is the final velocity of $m_1$?

**Picture the Problem** The diagram shows the two masses moving toward each other before colliding as well as the speed and direction of $m_2$ afterward.

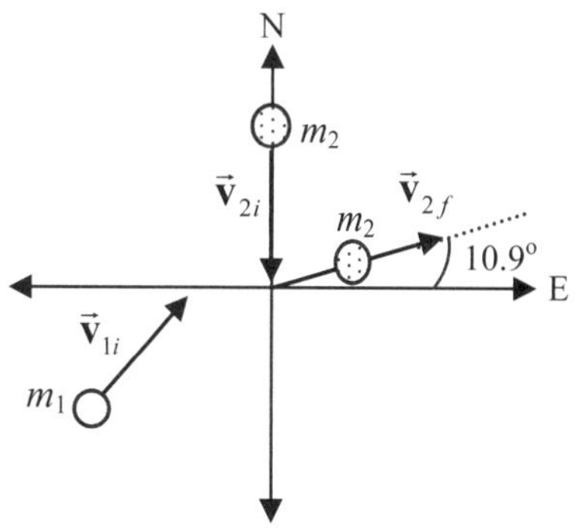

**Strategy**

To perform the analysis we use the conservation of kinetic energy together with the conservation of momentum in both the $x$ and $y$ directions.

**Solution**

1. Apply the conservation of kinetic energy:

$$K_{total}^{before} = K_{total}^{after}$$
$$\therefore \tfrac{1}{2}m_1 v_{1i}^2 + \tfrac{1}{2}m_2 v_{2i}^2 = \tfrac{1}{2}m_1 v_{1f}^2 + \tfrac{1}{2}m_2 v_{2f}^2$$
$$\therefore m_1 v_{1i}^2 + m_2 v_{2i}^2 = m_1 v_{1f}^2 + m_2 v_{2f}^2$$

2. Insert numerical values and solve for $v_{1f}$:

$$514.0\text{ J} = (1.19\text{ kg})v_{1f}^2 + 226.9\text{ J}$$
$$\therefore v_{1f} = \sqrt{\frac{287.1\text{ J}}{1.19\text{ kg}}} = 15.5\text{ m/s}$$

3. Apply the conservation of momentum along the $x$ direction:

$$p_{total,x}^{before} = p_{total,x}^{after}$$
$$\therefore m_1 v_{1i}\cos(45^\circ) = p_{1f,x} + m_2 v_{2f}\cos(10.9^\circ)$$
$$\therefore p_{1f,x} = m_1 v_{1i}\cos(45^\circ) - m_2 v_{2f}\cos(10.9^\circ)$$

4. Evaluate $p_{1f,x}$:

$$p_{1f,x} = 9.845\text{ kg}\cdot\text{m/s} - 18.12\text{ kg}\cdot\text{m/s} = -8.272\text{ kg}\cdot\text{m/s}$$

5. Apply the conservation of momentum along the $y$ direction:

$$p_{total,y}^{before} = p_{total,y}^{after}$$
$$m_1 v_{1i}\sin(45^\circ) - m_2 v_{2i} = p_{1f,y} + m_2 v_{2f}\sin(10.9^\circ)$$
$$p_{1f,y} = (9.845 - 22.95 - 3.489)\text{ kg}\cdot\text{m/s} = -16.59\text{ kg}\cdot\text{m/s}$$

6. Determine the reference angle:

$$\theta_{1f,ref} = \tan^{-1}\left(\frac{p_{1f,y}}{p_{1f,x}}\right) = \tan^{-1}\left(\frac{-16.59\text{ kg}\cdot\frac{\text{m}}{\text{s}}}{-8.272\text{ kg}\cdot\frac{\text{m}}{\text{s}}}\right) = 63.5^\circ$$

7. Since both components are negative, $\theta_{1f,ref}$ is measured from the negative $x$ axis in the 3rd quadrant. Therefore, the final velocity of $m_1$ is: 15.5 m/s at 63.5° south of west

**Insight** This problem had many steps but we started with the conservation of kinetic energy to ensure that we were describing an elastic collision. Also, it was helpful to use the equations for the conservation of momentum in both the $x$ and $y$ directions to make sure that we had the correct orientation of $\vec{\mathbf{v}}_{1f}$.

---

## Practice Quiz

**11.** When a lighter mass undergoes a head-on, elastic collision with a heavier mass that is initially at rest, the lighter mass will

**(a)** stop and come to rest.

**(b)** continue forward at a slower speed.

**(c)** recoil and move in the direction opposite its original motion.

**(d)** cannot answer definitively; it depends on the masses and the initial speed of the lighter mass.

**(e)** None of the above.

## 9–7 Center of Mass

At many times in our study we have treated large objects such as balls, cars, rockets, and so on, as if they were point particles even though they clearly are not. The reason we have been able to do that is because in many situations, systems of particles behave (as a whole) just like point particles. The concept that helps us see this fact most clearly is the **center of mass**.

The center of mass of a system is the average location of mass in that system. This average is a weighted average in that each particle contributes more or less to the average according to its mass. Another way to think of the center of mass of a solid object is as the point on the object at which it can be balanced near Earth's surface where we treat Earth's gravity as constant.

The location of the center of mass can be determined by the following expressions:

$$X_{cm} = \frac{m_1 x_1 + m_2 x_2 + \cdots}{m_1 + m_2 + \cdots} = \frac{\sum m_i x_i}{M}$$

$$Y_{cm} = \frac{m_1 y_1 + m_2 y_2 + \cdots}{m_1 + m_2 + \cdots} = \frac{\sum m_i y_i}{M}$$

In the above equations $X_{cm}$ and $Y_{cm}$ are the $x$ and $y$ coordinates of the center of mass, respectively, $x_i$ and $y_i$ are the $x$ and $y$ coordinates of the individual masses, and $M$ is the total mass of the system. Be aware that the center of mass is just an average location; there does not need to be any mass at that location. For a uniform circular ring, for example, the center of mass is right at the center of the ring, where no mass is located. In fact, we can take it as given that for uniform, symmetric, and continuous distributions of matter, the center of mass is located at the geometric center of the system.

If we examine the motion of the center of mass of a system, we can see the sense in which the system as a whole behaves like a point particle. The velocity of the center-of-mass point is given by $\vec{\mathbf{V}}_{cm} = (\Sigma m\vec{\mathbf{v}}_i)/M$. Slight manipulation of this equation shows that

$$\vec{\mathbf{p}}_{total} = M\vec{\mathbf{V}}_{cm}$$

This is just what we would have for the momentum of a single particle of mass $M$ located at the center of mass. Similarly, the acceleration of the center of mass is given by $\vec{\mathbf{A}}_{cm} = (\Sigma m\vec{\mathbf{a}}_i)/M$. Slight manipulation of this equation and cancellation of the internal forces (by action and reaction) shows that

$$\vec{\mathbf{F}}_{net,ext} = M\vec{\mathbf{A}}_{cm}$$

This is just what we would have for the force on a single particle of mass $M$ located at the center of mass. The two results for the total momentum and force on a system show that when considering the system as a whole, as we have been in many cases such as a moving car, we can treat the system as if it were a single particle with all its mass located at its center of mass.

---

**Example 9–7 Center of Mass** Three objects are located at the following positions in a two-dimensional $(x, y)$ coordinate system: $m_1$ at (1.3 m, 5.4 m), $m_2$ at (–2.2 m, 9.4 m), and $m_3$ at (4.1 m, –0.77 m). If the masses are 10 kg, 15 kg, and 20 kg respectively, what is the position of the center of mass?

**Solution** Sketch a diagram of the three masses on a coordinate system.

**Given:** $m_1 = 10$ kg, $m_2, = 15$ kg, $m_3 = 20$ kg, $x_1 = 1.3$ m, $y_1 = 5.4$ m, $x_2 = -2.2$ m, $y_2 = 9.4$ m, $x_3 = 4.1$ m, $y_3 = -0.77$ m;

**Find:** $X_{cm}$, $Y_{cm}$

We make direct use of the expression for calculating the center of mass:

$$X_{cm} = \frac{\sum m_i x_i}{M} = \frac{m_1x_1 + m_2x_2 + m_3x_3}{m_1 + m_2 + m_3} = \frac{10\text{ kg}(1.3\text{ m}) + 15\text{ kg}(-2.2\text{ m}) + 20\text{ kg}(4.1\text{ m})}{45\text{ kg}} = 1.4\text{ m}$$

$$Y_{cm} = \frac{\sum m_i y_i}{M} = \frac{m_1y_1 + m_2y_2 + m_3y_3}{m_1 + m_2 + m_3} = \frac{10\text{ kg}(5.4\text{ m}) + 15\text{ kg}(9.4\text{ m}) + 20\text{ kg}(-0.77\text{ m})}{45\text{ kg}} = 4.0\text{ m}$$

**Insight** Notice that no mass is actually located at this position.

---

**Practice Quiz**

**12.** For two objects of slightly different mass, the center of mass will be

**(a)** just slightly away from the heavier mass, on the opposite side from the lighter one.

**(b)** between the two, just slightly closer to the lighter mass.

**(c)** very close to the lighter mass, on the opposite side from the heavier one.

**(d)** between the two, just slightly farther away from the lighter object.

**(e)** between the two, very close to the heavier mass.

## *9–8 Systems with Changing Mass: Rocket Propulsion

As mentioned previously, Newton's second law in terms of momentum is a more general form for this law because it is not limited to systems with constant mass. A rocket is a perfect example of a system with changing mass because at a typical launch most of the rocket's mass is fuel, and the amount of this fuel decreases rapidly as it is exhausted. For a rocket whose exhaust is expelled at a speed $v$, relative to the rocket, the magnitude of the forward force exerted on the rocket is

$$F = \left|\frac{\Delta m}{\Delta t}\right| v$$

This force defines the **thrust** of the rocket. Here we can see that attempting to understand rocket propulsion using $\vec{\mathbf{F}} = m\vec{\mathbf{a}}$ would be problematic because this equation assumes that there is just one value of $m$ for the entire acceleration.

---

**Example 9–8 An Accelerating Rocket** At liftoff, an advanced rocket has a total mass (payload + fuel) of $1.8 \times 10^5$ kg. If the rocket burns fuel at a rate of 3000 kg/s with an exhaust velocity of 4500 m/s, what is its acceleration 12.0 seconds after liftoff?

**Picture the Problem** The sketch shows the rocket accelerating upward after liftoff.

**Strategy**
We need to determine the thrust that propels the rocket upward. The net force on the rocket from the thrust and gravity will then allow us to determine the acceleration at that instant.

**Solution**

1. Determine the thrust of the rocket: $F_{thrust} = \left|\frac{\Delta m}{\Delta t}\right| v = (3000\ \text{kg/s})(4500\ \text{m/s}) = 1.35\times10^{7}\ \text{N}$

2. Determine the mass of the rocket after 12 s: $m = m_i - \left|\frac{\Delta m}{\Delta t}\right| t = 1.8\times10^{5}\ \text{kg} - (3000\ \text{kg/s})(12.0\ \text{s})$
$= 1.44\times10^{5}\ \text{kg}$

3. Find the net force at $t = 12.0$ s:
$$F_{net} = F_{thrust} - mg$$
$$= 1.35\times10^{7}\,\text{N} - (1.44\times10^{5}\ \text{kg})(9.81\ \text{m/s}^2)$$
$$= 1.21\times10^{7}\,\text{N}$$

4. Calculate the acceleration of the rocket: $a = \frac{F_{net}}{m} = \frac{1.21\times10^{7}\ \text{N}}{1.44\times10^{5}\ \text{kg}} = 84\ \text{m/s}^2$

**Insight** Notice the roles of the two forms of Newton's second law here. To obtain the thrust on the rocket we need to account for changing mass. To obtain the acceleration we used $F = ma$, but note that this version is applicable only at a particular instant when the mass has a particular value.

**Practice Quiz**

13. The force that thrusts a rocket forward comes from
    **(a)** pushing against the air.
    **(b)** the pull of gravity.
    **(c)** the reaction force from the exhaust.
    **(d)** pushing against the ground at liftoff.
    **(e)** None of the above.

## Reference Tools and Resources

### I. Key Terms and Phrases

**linear momentum** the product of the mass and the velocity of an object

**impulse** the product of force and the amount of time the force acts

**collision** an interaction in which forces are exerted for a finite period of time

**conservation of linear momentum** the principle that the total linear momentum of a system remains constant unless a nonzero external net force is applied

**inelastic collision** a collision in which kinetic energy is not conserved

**elastic collision** a collision in which kinetic energy is conserved

**center of mass** the average location of mass within a system

**thrust** the forward force exerted by the expelled mass in rocket exhaust

## II. Important Equations

| Name/Topic | Equation | Explanation |
|---|---|---|
| linear momentum | $\vec{\mathbf{p}} = m\vec{\mathbf{v}}$ | The definition of linear momentum. |
| Newton's second law | $\vec{\mathbf{F}}_{net} = \dfrac{\Delta\vec{\mathbf{p}}}{\Delta t}$ | The most general form of Newton's second law |
| impulse | $\vec{\mathbf{I}} = \vec{\mathbf{F}}_{av}\Delta t = \Delta\vec{\mathbf{p}}$ | The definition of average impulse and its relationship to linear momentum |
| center of mass | $X_{cm} = \dfrac{m_1x_1 + m_2x_2 + \cdots}{m_1 + m_2 + \cdots} = \dfrac{\sum m_ix_i}{M}$ <br> $Y_{cm} = \dfrac{m_1y_1 + m_2y_2 + \cdots}{m_1 + m_2 + \cdots} = \dfrac{\sum m_iy_i}{M}$ | The $x$ and $y$ coordinates of the center of mass of a system of particles |
| thrust | $F_{thrust} = \left\lvert\dfrac{\Delta m}{\Delta t}\right\rvert v$ | The force exerted by rocket exhaust |

## III. Know Your Units

| Quantity | Dimension | SI Unit |
|---|---|---|
| linear momentum ($\vec{\mathbf{p}}$) | [M]·[L]/[T] | kg·m/s |
| impulse ($\vec{\mathbf{I}}$) | [M]·[L]/[T] | N·s (= kg·m/s) |
| thrust ($F$) | $[\mathrm{M}]\cdot[\mathrm{L}]/[\mathrm{T}^2]$ | N |

## Puzzle

**KNOCK-OFF**

You are standing on a log and a friend is trying to knock you off. He throws a ball at you. You can either catch it or let it bounce off you. Which is more likely to topple you, catching the ball or letting it bounce off? Briefly explain what physics you used to reach your conclusion.

## Answers to Selected Conceptual Questions

**4.** No. Consider, for example, a system of two particles. The total momentum of this system will be zero if the particles move in opposite directions with equal momentum. The kinetic energy of each particle is positive, however, and hence the total kinetic energy is also positive.

**12.** No. Any collision between cars will be at least partially inelastic, due to denting, sound production, heating, and other effects.

**16.** **(a)** Assuming a very thin base, we conclude that the center of mass of the glass is at its geometric center of the glass. **(b)** In the early stages of filling, the center of mass is below the center of the glass. When the glass is practically full, the center of mass is again at the geometric center of the glass. Thus, as water is added, the center of mass first moves downward, then turns around and moves back upward to its initial position.

## Solutions to Selected End-of-Chapter Problems and Conceptual Exercises

20. **Picture the Problem**: The ball rebounds from the player's head in the manner indicated by the figure at right.

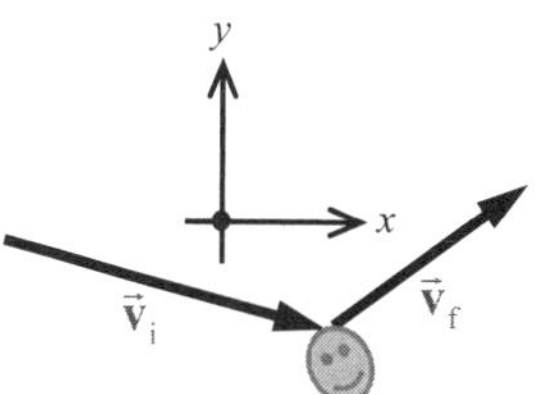


**Strategy:** The impulse is equal to the vector change in the momentum. Analyze the $x$ and $y$ components of $\Delta\vec{\mathbf{p}}$ separately, then use the components to find the direction and magnitude of $\vec{\mathbf{I}}$.

**Solution: 1. (a)** Find $\Delta p_x$: $\Delta p_x = m(v_{fx} - v_{ix}) = (0.43\text{ kg})(5.2 - 8.8\text{ m/s}) = \underline{\underline{-1.5\text{ kg}\cdot\text{m/s}}}$

**2.** Find $\Delta p_y$: $\Delta p_y = m(v_{fy} - v_{iy}) = (0.43\text{ kg})[3.7 - (-2.3)\text{ m/s}] = \underline{\underline{2.6\text{ kg}\cdot\text{m/s}}}$

**3.** Use equation 9-6 to find $\vec{\mathbf{I}}$: $\vec{\mathbf{I}} = \Delta\vec{\mathbf{p}} = (-1.5\text{ kg}\cdot\text{m/s})\hat{\mathbf{x}} + (2.6\text{ kg}\cdot\text{m/s})\hat{\mathbf{y}}$

**4.** Find the direction of $\vec{\mathbf{I}}$: $\theta = \tan^{-1}\left(\frac{I_y}{I_x}\right) = \tan^{-1}\left(\frac{2.58}{-1.55}\right) = -59° + 180° = \boxed{121°}$ from the positive $x$ axis

**5. (b)** Find the magnitude: $I = \sqrt{I_x^2 + I_y^2} = \sqrt{(-1.55\text{ kg}\cdot\text{m/s})^2 + (2.58\text{ kg}\cdot\text{m/s})^2} = \boxed{3.0\text{ kg}\cdot\text{m/s}}$

**Insight:** The ball delivers an equal and opposite impulse to the player's head, which would exert a force of 300 N (67 lb) if the time of collision were 10 milliseconds.

26. **Picture the Problem**: The lumberjack moves to the right while the log moves to the left.

**Strategy:** As long as there is no friction the total momentum of the lumberjack and the log remains zero, as it was before the lumberjack started trotting. Combine vector addition for relative motion (equation 3-8) with the expression from the conservation of momentum to find $v_{\text{L, s}}$ = speed of lumberjack relative to the shore. Let $v_{\text{L, log}}$ = speed of lumberjack relative to the log, and $v_{\text{log, s}}$ = speed of the log relative to the shore.

**Solution: 1. (a)** Write out the equation for relative motion. Let the log travel in the negative direction:

$$\vec{\mathbf{v}}_{\text{L, s}} = \vec{\mathbf{v}}_{\text{L, log}} + \vec{\mathbf{v}}_{\text{log, s}}$$
$$v_{\text{L, s}} = v_{\text{L, log}} - v_{\text{log, s}}$$
$$v_{\text{log, s}} = v_{\text{L, log}} - v_{\text{L, s}}$$

**2.** Write out the conservation of momentum with respect to the shore:

$$\sum \vec{\mathbf{p}} = 0 = m_{\text{L}} v_{\text{L, s}} - m_{\text{log}} v_{\text{log, s}}$$

**3.** Substitute the expression from step 1 into step 2 and solve for $v_{\text{L, s}}$:

$$m_{\text{L}} v_{\text{L, s}} = m_{\text{log}} v_{\text{log, s}} = m_{\text{log}} \left( v_{\text{L, log}} - v_{\text{L, s}} \right)$$
$$v_{\text{L, s}} \left( m_{\text{L}} + m_{\text{log}} \right) = m_{\text{log}} v_{\text{L, log}}$$
$$v_{\text{L, s}} = \frac{m_{\text{log}} v_{\text{L, log}}}{\left( m_{\text{L}} + m_{\text{log}} \right)} = \frac{(380 \text{ kg})(2.7 \text{ m/s})}{(85 + 380 \text{ kg})} = \boxed{2.2 \text{ m/s}}$$

**4. (b)** If the mass of the log had been greater, the lumberjack's speed relative to the shore would have been greater than that found in part (a), because the log would have moved slower in the negative direction.

**5. (c)** Use the expression from step 3 to find the new speed of the lumberjack:

$$v_{\text{L, s}} = \frac{m_{\text{log}} v_{\text{L, log}}}{\left( m_{\text{L}} + m_{\text{log}} \right)} = \frac{(450 \text{ kg})(2.7 \text{ m/s})}{(85 + 450 \text{ kg})} = \boxed{2.3 \text{ m/s}}$$

**Insight:** Taking the argument in (b) to its extreme, if the mass of the log equaled the mass of the Earth, the lumberjack's speed would be exactly 2.7 m/s relative to the Earth (and the log). If the mass of the log were the same as the mass of the lumberjack, the speed of each relative to the Earth would be half the lumberjack's walking speed.

34. **Picture the Problem**: The putty is thrown horizontally, strikes the side of the block, and sticks to it. The putty and the block move together in the horizontal direction immediately after the collision, compressing the spring.

**Strategy:** Use conservation of momentum to find the speed of the putty-block conglomerate immediately after the collision, then use equation 7-6 to find the kinetic energy. Use conservation of energy to find the maximum compression of the spring after the collision.

**Solution: 1. (a)** No, the mechanical energy of the system is not conserved because some of the initial kinetic energy of the putty will be converted to heat, sound, and permanent deformation of material during the inelastic collision.

**2.** Set $\vec{\mathbf{p}}_{\text{i}} = \vec{\mathbf{p}}_{\text{f}}$ and solve for $v_{\text{f}}$:

$$m_{\text{p}} v_{\text{p}} = \left( m_{\text{b}} + m_{\text{p}} \right) v_{\text{f}} \;\Rightarrow\; v_{\text{f}} = \left( \frac{m_{\text{p}}}{m_{\text{b}} + m_{\text{p}}} \right) v_{\text{p}}$$

**3.** Set $E_{\text{after}} = E_{\text{rest}}$ after the collision:

$$K_{\text{after}} + 0 = 0 + U_{\text{rest}}$$
$$\tfrac{1}{2}\left( m_{\text{b}} + m_{\text{p}} \right) \left( \frac{m_{\text{p}}}{m_{\text{b}} + m_{\text{p}}} \right)^2 v_{\text{p}}^2 = \tfrac{1}{2} k x_{\text{max}}^2$$

**4.** Solve the resulting expression for $x_{max}$ : $x_{max} = \sqrt{\left[\dfrac{m_p^2 v_p^2}{k(m_p + m_b)}\right]} = \left[\dfrac{(0.0500 \text{ kg})^2 (2.30 \text{ m/s})^2}{(20.0 \text{ N/m})(0.430 + 0.0500 \text{ kg})}\right]^{\frac{1}{2}}$

$= 0.0371 \text{ m} = \boxed{3.71 \text{ cm}}$

**Insight:** The putty-block conglomerate will compress the spring even farther if $v_p$ is larger or if $m_p$ is larger.

43. **Picture the Problem**: The balls collide elastically along a single direction.

**Strategy:** Set the initial momentum equal to the final momentum and solve for the final speed of the two balls. Then set the initial kinetic energy equal to the final kinetic energy and solve for the final speed of the two balls again.

**Solution: 1. (a)** Set $\vec{\mathbf{p}}_i = \vec{\mathbf{p}}_f$ and solve for $v_f$ : $mv_0 = (2m)v_f \Rightarrow v_f = \boxed{\dfrac{v_0}{2}}$

**2. (b)** Set $K_i = K_f$ and solve for $v_f$ : $\frac{1}{2}mv_0^2 = \frac{1}{2}(2m)v_f^2 \Rightarrow v_f = \boxed{\dfrac{v_0}{\sqrt{2}}}$

**Insight:** If you conserve the momentum and kinetic energy during the collision, assuming the balls that are going into the collision come to rest afterwards, you can show that $m_{in} = m_{out}$ no matter how many balls are sent into the collision.

52. **Picture the Problem**: The geometry of the sulfur dioxide molecule is shown at right.

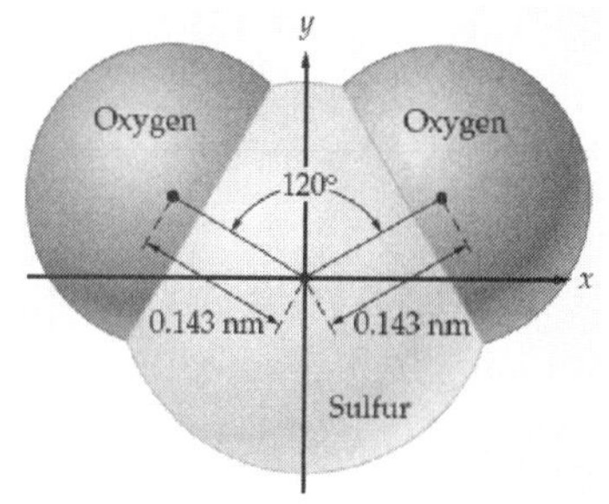


**Strategy:** The center of mass of the molecule will lie somewhere along the $y$ axis because it is symmetric in the $x$ direction. Find $Y_{cm}$ using equation 9-15. Both oxygen atoms will be the same vertical distance $y_O$ from the origin. Let $m$ represent the mass of an oxygen atom, $m_s$ the sulfur atom.

**Solution: 1.** Use equation 9-15 to find $Y_{cm}$ $Y_{cm} = \dfrac{\sum my}{M} = \dfrac{my_O + my_O + m_s y_s}{2m + m_s} = \dfrac{2my_O + 0}{2m + m_s}$

$= \dfrac{2(16 \text{ u})(0.143 \text{ nm})\sin 30°}{2(16 \text{ u}) + 32 \text{ u}} = \underline{\underline{0.036 \text{ nm}}}$

**2.** Recalling that 1 nm = 1 × $10^{-9}$ m, we can write $(X_{cm}, Y_{cm}) = \boxed{(0,\ 3.6 \times 10^{-11} \text{ m})}$

**Insight:** If the angle were to decrease from 120° the center of mass would move upward. For instance, if the bond angle were only 90°, the center of mass would be located at (0, 5.1 × $10^{-11}$ m).

### Answers to Practice Quiz

**1.** (b) **2.** (a) **3.** (d) **4.** (e) **5.** (c) **6.** (a) **7.** (d) **8.** (c) **9.** (a) **10.** (e) **11.** (c) **12.** (d) **13.** (c)

# CHAPTER 10
# ROTATIONAL KINEMATICS AND ENERGY

## Chapter Objectives

After studying this chapter, you should

1. know the meanings of angular position, velocity, and acceleration.
2. be able to describe rotational motion.
3. understand the connection between rotational quantities and corresponding linear quantities.
4. understand and be able to describe rolling motion.
5. understand the concept of the moment of inertia and its role in rotational motion.
6. be able to use rotational kinetic energy and apply the conservation of energy to rotating and rolling objects.

## Warm-Ups

1. When an audio or videotape rewinds, why does the tape wind up faster at the end than at the beginning?
2. Estimate the magnitude of the tangential velocity of an object in your hometown due to the rotation of Earth.
3. A small object is sliding inside a circular frictionless track. At point A, another frictionless track switches it to a smaller circle (see the picture). At all times the tracks can exert only forces that are perpendicular to the motion of the little slug. What happens to the linear velocity of the slug when it switches tracks? What about its angular velocity?

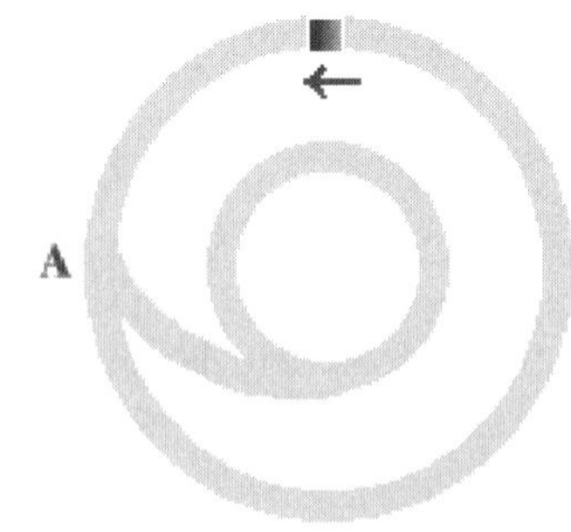


4. A skater is spinning with his arms outstretched. He has a 2-lb weight in each hand. In an attempt to change his angular velocity, he lets go of both weights. Does he succeed in changing his angular velocity? If yes, how does his angular velocity change?

## Chapter Review

In this and the next chapters, we study rotational kinematics and dynamics. Rotational motion is every bit as important as the linear (or translational) motion that we've been studying thus far. As you read these chapters, try to notice how the study of rotational motion parallels that of translational motion. Many of the concepts and mathematical treatments are the same; we only need to alter the physical interpretation of these concepts from that of a translating body to a rotating one.

### 10–1 Angular Position, Velocity, and Acceleration

Just as with linear motion, we describe rotational motion with three basic quantities analogous to displacement, velocity, and acceleration. We shall call these quantities **angular position**, **angular velocity**, and **angular acceleration**.

When we describe the rotation of an object, we think of the object as rotating about an axis. If you consider a line drawn from the axis of rotation to the object, the angular position of the object, denoted by $\theta$, is the angle between that line and an arbitrarily chosen reference line. The reference line plays the same role as the arbitrarily chosen origin of a coordinate system. In SI, the unit of angular measure for rotational motion is the *radian*. Angular measure is actually dimensionless, but radian measure works better in describing motion than do degrees. Angular position plays the same role for rotational motion that position does for translational motion. In keeping with the usual convention, if $\theta$ is measured counterclockwise, it is taken to be a positive angle; if it is measured clockwise, it is taken to be negative.

To describe how rapidly an object rotates about an axis we use the quantity angular velocity, $\omega$. This quantity plays the same role for rotational motion that velocity does for linear motion. Angular velocity is the rate of change of angular position. Often it is useful to use the average rate of change, or average angular velocity:

$$\omega_{av} = \frac{\Delta\theta}{\Delta t}$$

The SI unit of angular velocity is rad/s = $s^{-1}$ (since rad is dimensionless). It is also sometimes useful to use the instantaneous angular velocity:

$$\omega = \lim_{\Delta t \to 0} \frac{\Delta\theta}{\Delta t}$$

To maintain consistency with the sign convention for angular displacement $\Delta\theta$, angular velocity is negative if the object rotates clockwise, and positive if it rotates counterclockwise.

To describe how the angular velocity of an object changes, we use the quantity angular acceleration, $\alpha$. This quantity plays the same role for rotational motion that acceleration, $a$, does for linear motion.

Angular acceleration is the rate of change of angular velocity. Often it is useful to use the average rate of change, or average angular acceleration:

$$\alpha_{\text{av}} = \frac{\Delta\omega}{\Delta t}$$

The SI unit of angular acceleration is rad/s$^2$ = s$^{-2}$ (since rad is dimensionless). It is also sometimes useful to use the instantaneous angular acceleration

$$\alpha = \lim_{\Delta t \to 0} \frac{\Delta\omega}{\Delta t}$$

The sign of $\alpha$ depends on the sign of $\Delta\omega$. If $\Delta\omega$ is negative, then $\alpha$ is negative; if $\Delta\omega$ is positive, then $\alpha$ is positive.

---

**Example 10–1 Turn on the Fan** On a hot day you turn on a small fan to help cool you off. The tip on one of the blades of the fan might go through 1.25 revolutions at an average angular velocity of –7.85 rad/s. **(a)** What is the angular displacement of the tip of the blade, and **(b)** how much time does it take to go through that displacement?

**Picture the Problem** The sketch shows the blades of a fan with one tip indicated by the black dot. In this diagram, the fan is rotating clockwise because $\omega_{\text{av}}$ is negative.

**Strategy** For part (a), we must translate between the number of revolutions and the number of radians to get the SI angular displacement. For part (b), we'll make use of the definition of average angular velocity.

**Solution**

**Part (a)**

1. Convert from revolutions to radians: $\Delta\theta = -1.25 \text{ rev}\left(\frac{2\pi \text{ rad}}{\text{rev}}\right) = -7.85 \text{ rad}$

**Part (b)**

2. Use the definition of $\omega_{\text{av}}$ to solve for $\Delta t$: $\omega_{\text{av}} = \frac{\Delta\theta}{\Delta t} \Rightarrow \Delta t = \frac{\Delta\theta}{\omega_{\text{av}}}$

**3.** Calculate the numerical result: $$\Delta t = \frac{-7.854\ \text{rad}}{-7.85\ \text{rad/s}} = 1.00\ \text{s}$$

**Insight** The minus signs appear by convention only.

## Practice Quiz

1. Which quantity is used to describe how rapidly an object rotates?

   **(a)** angular position **(b)** angular velocity **(c)** angular acceleration **(d)** none of the above

2. If an object is seen to rotate faster and faster, which quantity best describes this aspect of its motion?

   **(a)** angular position **(b)** angular velocity **(c)** angular acceleration **(d)** none of the above

## 10–2 Rotational Kinematics

You may have noticed that the mathematical relationships among the three quantities described in the previous section are exactly the same as the relationships among the corresponding quantities for translational motion. This means that when looking for a way to describe rotational motion, we can follow the prescription already laid out for translational motion. All the equations used in Chapter 2 for describing motion with constant velocity and constant acceleration also apply to rotational motion. The only mathematical difference is that the names of the variables are changed in the following way:

$$x \to \theta,\ \ v \to \omega,\ \ a \to \alpha,\ \ t \to t$$

Of course, along with changing the variables, you should also change the physical picture in your mind of the type of motion you are describing. Nevertheless, the mathematical descriptions are identical.

Thus, to describe rotational motion we make the above replacements and reuse the equations in Chapter 2 for one-dimensional kinematics (assuming $t_0 = 0$, as is customary).

**Motion with Constant Angular Velocity**

$$\Delta\theta = \omega\Delta t,\ \ \omega_f = \omega_i$$

**Motion with Constant Angular Acceleration**

$$\omega = \omega_0 + \alpha t$$

$$\theta = \theta_0 + \tfrac{1}{2}(\omega_0 + \omega)t$$

$$\theta = \theta_0 + \omega_0 t + \tfrac{1}{2}\alpha t^2$$

$$\omega^2 = \omega_0^2 + 2\alpha(\theta - \theta_0)$$

**Example 10–2 Turn on the Fan II** Suppose that the blades of the fan discussed in Example 10–1 rotate counterclockwise with a constant angular acceleration for the first 1.00 seconds. **(a)** What is the instantaneous angular velocity of the blade at $t$ = 1.00 seconds after being turned on, and **(b)** what is the angular acceleration?

**Picture the Problem** The sketch shows the blades of a fan with one tip indicated by the black dot.

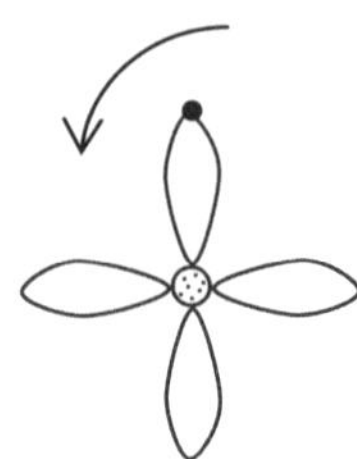

**Strategy** In both parts we make use of the expressions for uniformly accelerated motion. Let's define $\theta_0 = 0$ and $t_0 = 0$. Also, notice that since the fan is just being turned on, $\omega_0 = 0$.

**Solution**

**Part (a)**

1. Choose an appropriate expression for finding $\omega$ based on the given information: $\theta = \theta_0 + \frac{1}{2}(\omega_0 + \omega)t \;\Rightarrow\; \theta = \frac{1}{2}\omega t$

2. Solve for the angular velocity: $\omega = \dfrac{2\theta}{t} = \dfrac{2(7.854\text{ rad})}{1.00\text{ s}} = 15.7\text{ rad/s}$

**Part (b)**

3. Choose an appropriate expression for finding $\alpha$ based on the given information: $\omega = \omega_0 + \alpha t \;\Rightarrow\; \alpha = \dfrac{\omega}{t}$

4. Use the known data to get the numerical result: $\alpha = \dfrac{15.71\text{ rad/s}}{1.00\text{ s}} = 15.7\text{ rad/s}^2$

**Insight** In this example, all the values are positive: for $\theta$ and $\omega$ because the blade is rotating counterclockwise, and for $\alpha$ because the angular speed increases counterclockwise. Try to see how a similar translational problem could have been solved back in Chapter 2. If you think about it, you actually have already done this stuff!

**Exercise 10–3 Turn off the Fan** Now, suppose that the blades of our fan have a top rotation rate of 25 revolutions per second. Once we've cooled down enough we turn off the power, and it takes 6.65 seconds for the blades to come to rest with a constant angular acceleration. **(a)** What is the angular acceleration of the blades, and **(b)** through what angular displacement do the blades turn while coming to rest?

**Solution:** The problem doesn't give us a direction of rotation, so we are free to choose. Let's choose counterclockwise as a default.

**Given:** $\omega_0 = 25$ rev/s, $\omega = 0$, $t = 6.65$ s; **Find:** (a) $\alpha$, (b) $\theta$

First, we'll convert the initial angular velocity to SI units:

$$\omega_0 = 25 \text{ rev/s}\left(\frac{2\pi \text{ rad}}{\text{rev}}\right) = 157 \text{ rad/s}$$

Now, for part (a), we can determine $\alpha$ by using

$$\omega = \omega_0 + \alpha t \quad \Rightarrow \quad \alpha = -\frac{\omega_0}{t} = -\frac{157 \text{ rad/s}}{6.65 \text{ s}} = -24 \text{ rad/s}^2$$

Calculating the answer to part (b) is also a straightforward use of the equations. Most conveniently

$$\theta = \theta_0 + \tfrac{1}{2}(\omega_0 + \omega)t \quad \Rightarrow \quad \theta = \tfrac{1}{2}\omega_0 t = \tfrac{1}{2}(157 \text{ rad/s})(6.65 \text{ s}) = 520 \text{ rad}$$

Notice that because of the original rotation rate, the answers are given to only two significant digits.

## Practice Quiz

3. In comparing linear and rotational motion, angular position is most similar to which linear variable?

   **(a)** $x$ **(b)** $v$ **(c)** $a$ **(d)** $t$ **(e)** none of the above

4. How long would it take an object rotating at a constant speed of 17.0 rad/s to rotate through 235 radians?

   **(a)** 0.0723 s **(b)** 3.4 N **(c)** 5.0 N **(d)** 13.8 s **(e)** 3.1 N

5. If an object rotates at 2.50 rpm, what is the equivalent angular speed in rad/s?

   **(a)** 2.50 **(b)** 15.7 **(c)** 0.262 **(d)** 0.0417 **(e)** 25.0

## 10–3 Connections Between Linear and Rotational Quantities

When an object is moving in a circular path (pure rotation) we can describe its motion using the rotational quantities discussed earlier. However, we can also describe the object's motion using linear quantities. It is very useful to know how to relate these two ways to describe motion.

With pure rotation, the relationship between the angular distance $\theta$ and the linear distance $s$ along the arc is known to be

$$s = r\theta$$

when $\theta$ is measured in radians, and $r$ is the radius of the circular path ($r$ would be the distance of the object from the axis of rotation if the motion were not circular). Notice that the above equation makes it clear that angular measure is dimensionless because both $s$ and $r$ have dimension [L].

The velocity of the object in pure rotation will be tangent to the circular path of its motion. The magnitude of this velocity is called the *tangential speed* of the object, $v_t$. The relationship between $v_t$ and $\omega$ can be determined directly from the preceding expression for the arc length to be

$$v_t = r\omega$$

The relationship between the tangential acceleration and the angular acceleration can similarly be determined from the expression for tangential speed:

$$a_t = r\alpha$$

Because the object is rotating, it will also have an inward-pointing component to its acceleration; the centripetal acceleration discussed in Chapter 6. Recall that $a_{cp}$ depends on the tangential speed $a_{cp} = v_t^2 / r$. Using the above expression for $v_t$ we can get the following relationship between $a_{cp}$ and the rotational quantities:

$$a_{cp} = r\omega^2$$

---

**Example 10–4 Turn off the Fan II** Suppose that the distance from the central axis of the blade unit of the fan to the tip of each blade is 7.00 cm. Given the information for turning off the fan in Example 10–3, calculate **(a)** the linear distance that the tip of the blade traversed while coming to rest, **(b)** the maximum tangential speed of the tip of the blade, **(c)** the tangential component of the acceleration of the tip of the blade, and **(d)** the centripetal acceleration of the tip of the blade.

**Picture the Problem** The sketch shows the blade rotating counterclockwise.

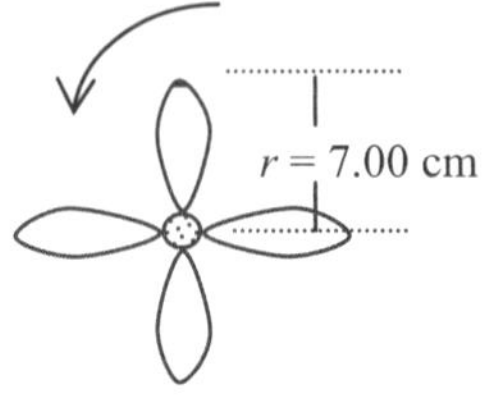

**Strategy** To solve for the desired information, we use the results of Example 10–3 and the known relations between the linear and rotational quantities.

**Solution**

**Part (a)**

1. We know $\theta$ to three significant figures from part (b) in Exercise 10–3, so: $s = r\theta = (0.0700 \text{ m})(522 \text{ rad}) = 37 \text{ m}$

**Part (b)**

2. We know $\omega$ from Exercise 10–3: $v_t = r\omega = (0.0700 \text{ m})(157 \text{ rad/s}) = 11 \text{ m/s}$

**Part (c)**

3. We make direct use of the known result, to three significant figures, from Exercise 10–3: $a_t = r\alpha = (0.0700 \text{ m})(23.6 \text{ rad/s}^2) = 1.7 \text{ m/s}^2$

**Part (d)**

4. Again, we make direct use of the known result: $a_{cp} = r\omega^2 = (0.0700 \text{ m})(157 \text{ rad/s})^2 = 1700 \text{ m/s}^2$

**Insight** Make sure you understand why the units of the final answers don't include the radian.

## Practice Quiz

6. If a wheel of radius 1.3 m rotates at 2.2 rad/s, which is the best way to write the linear speed of a point on the rim of this wheel?

   **(a)** 2.9 rad·m/s **(b)** 1.7 rad·m/s **(c)** 1.7 m/s **(d)** 2.9 m/s **(e)** 0.59 m/s

7. If a wheel of radius 1.3 m rotates at a constant speed of 2.2 rad/s, a point on the rim of the wheel accelerates tangentially at

   **(a)** 6.3 m/s$^2$ **(b)** 2.9 m/s$^2$ **(c)** 1.7 m/s$^2$ **(d)** 3.7 m/s$^2$ **(e)** none of the above

## 10–4 Rolling Motion

Rolling motion represents an important application that combines both rotational and translational motion. This motion also illustrates the importance of understanding the connection between linear and rotational quantities discussed in the previous section.

For a wheel that is free to roll, without slipping, each point on the wheel rotates about the wheel's axle while also being carried forward from the overall translation of the wheel. As a result of this combination of rotational and translational speeds, the point of contact between the wheel and the ground is instantaneously at rest (otherwise there would be slipping), the center of mass of the wheel moves forward at $\upsilon_{cm} = r\omega$ ($r$ – radius, $\omega$ – angular velocity), and the point at the top of the wheel moves at twice this speed: $\upsilon_{top} = 2r\omega$. Armed with just these few facts, much can be understood about rolling motion.

---

**Exercise 10–5 A Bicycle Wheel** The tires of an 18-speed mountain bike have a typical radius of about 34 cm. If you are riding hard and your tires are rotating at an angular velocity of 33 rad/s, **(a)** how fast are you traveling in m/s, and **(b)** how far will you travel after 15 min at this pace?

**Solution:** The following information is given in the problem.

**Given:** $r$ = 34 cm, $\omega$ = 33 rad/s, $t$ = 15 min; **Find:** **(a)** $\upsilon_{trans}$, **(b)** $x$

The key to solving for the speed at which you are moving is to notice that your speed is that of the center of mass of the wheel. Therefore,

$$\upsilon_{trans} = \upsilon_{cm} = r\omega = (0.34\text{ m})(33\text{ rad/s}) = 11\text{ m/s}$$

Since you travel at this constant speed for 15 min, we can determine the distance using

$$x = \upsilon_{cm}t = (11.22\text{ m/s})(15\text{ min})(60\text{ s/min}) = 1.0\times10^4\text{ m}$$

---

### Practice Quiz

8. If a bicycle moves at 6.5 m/s and its wheels have a diameter of 70 cm, how fast are the wheels rotating?

   **(a)** 9.3 rad/s **(b)** 19 rad/s **(c)** 4.5 rad/s **(d)** 2.3 rad/s **(e)** 0.093 rad/s

## 10–5 – 10–6 Rotational Kinetic Energy, Moment of Inertia, and Conservation of Energy

When studying translational motion, we introduced the idea of kinetic energy, which is an energy associated with "mass in motion." Objects that are rotating but not translating also have mass in motion.

We should, therefore, be able to write its kinetic energy in terms of the rotational quantities. In doing this, we find that the rotational kinetic energy of a system is

$$K_{rot} = \tfrac{1}{2}\left(\sum m_i r_i^2\right)\omega^2$$

Comparison between this expression and the translational form, $K = \frac{1}{2}mv^2$, suggests that the quantity $\Sigma mr^2$ plays a role similar to mass. We call this quantity the **moment of inertia,** *I,* of the system (also called *rotational inertia*):

$$I = \sum m_i r_i^2$$

The SI unit of the moment of inertia is kg·m$^2$.

The above expression for the moment of inertia is convenient for calculating *I* when the number of particles in the system is small. However, for systems with so many particles that we should treat the system as a continuous distribution of matter, a more involved calculation is required. Therefore, some of the more common results for the moments of inertia of continuous systems are listed for you in Table 10–1 of your textbook (pg. 298). Notice that *I* depends on the axis about which the object is rotating, so that the same object can have many different values of *I* associated with it. You should become very familiar with Table 10-1.

The conservation of total mechanical energy applies to objects that are rolling without slipping. This is true because, even though friction is present (to prevent slipping), it is *static* friction and therefore does no work. The key to applying energy conservation to rolling objects is to remember that the kinetic energy is taken up by both the rotational and translational motions,

$$K = K_{trans} + K_{rot} = \tfrac{1}{2}mv_{cm}^2 + \tfrac{1}{2}I_{cm}\omega^2$$

Notice here that the translational kinetic energy comes from treating the system as a single particle located at, and moving with, the center of mass of the object. Also, recognize that the moment of inertia in this expression is calculated about an axis passing through the center of mass (perpendicular to the direction of motion).

---

**Example 10–6 The Energy of the Fan** An employee for the manufacturer of our fan needs to estimate the kinetic energy of the fan when the blades are rotating at top speed. She takes, as a working model, **(a)** each blade to be a thin rod of length 7.00 cm and mass 0.113 kg, and as an alternative model, **(b)** each blade as a point mass with all its mass located at its center of mass. Assuming that the central axle is of negligible mass, calculate the rotational kinetic energy in each case. Which do you think is more accurate? **(c)** If she requires this fan to reach top speed 2.50 seconds after being turned on, how powerful a motor should she use?

**Picture the Problem** The sketch shows the blades of the fan rotating counterclockwise.

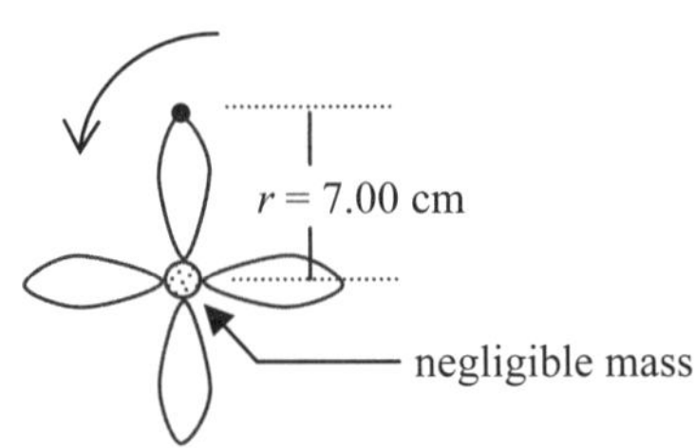


**Strategy** For a continuous thin rod, we make use of Table 10–1 in the text in calculating the moment of inertia. We already know the top angular velocity from previous examples.

**Solution**

**Part (a)**

1. Each blade is a thin rod rotating about an axis through one end; from Table 10–1 the moment of inertia of one blade is:

$$I_1 = \tfrac{1}{3}ML^2 = \tfrac{1}{3}(0.113\text{ kg})(0.0700\text{ m})^2 = 1.846\times10^{-4}\text{ kg}\cdot\text{m}^2$$

2. The total moment of inertia of all 4 blades is:

$$I = 4I_1 = 7.383\times10^{-4}\text{ kg}\cdot\text{m}^2$$

3. The rotational kinetic energy then is:

$$K = \tfrac{1}{2}I\omega^2 = \tfrac{1}{2}\left(7.383\times10^{-4}\text{ kg}\cdot\text{m}^2\right)\left(157\tfrac{\text{rad}}{\text{s}}\right)^2 = 9.1\text{ J}$$

**Part (b)**

4. Each blade is a point mass located at its center 3.50 cm from the axis. The moment of inertia is:

$$I = 4I_1 = 4mr^2 = 4(0.113\text{ kg})(0.035\text{ m})^2 = 5.537\times10^{-4}\text{ kg}\cdot\text{m}^2$$

5. The rotational kinetic energy then is:

$$K = \tfrac{1}{2}I\omega^2 = \tfrac{1}{2}\left(5.537\times10^{-4}\text{ kg}\cdot\text{m}^2\right)\left(157\tfrac{\text{rad}}{\text{s}}\right)^2 = 6.8\text{ J}$$

6. The thin-rod model is a closer approximation to the actual blade and is more accurate.

**Part (c)**

7. The average power of the motor is given by:

$$P = \frac{W}{t}$$

8. By the work-energy theorem:

$$W = \Delta K = K$$

9. The power needed, using part (a), is:

$$P = \frac{K}{t} = \frac{9.099\text{ J}}{2.50\text{ s}} = 3.6\text{ W}$$

**Insight** As can be seen from the differing results of parts (a) and (b), it is not always valid to treat a system as a point particle with its mass located at the center of mass. You have to be careful when doing this. Part (c) is to remind you that the work-energy theorem is equally valid for rotational kinetic energy as for translational kinetic energy that we used to introduce the theorem.

---

**Example 10–7 Gravity Launcher** A solid ball of mass 1.40 kg and radius 0.391 m is released from rest and rolls from the top of a track 10.0 m high down to a height of 0.500 m above the ground, where it is launched vertically upward. How high in the air will the ball go (neglect any friction)?

**Picture the Problem** The sketch shows the odd-shaped track with the ball at rest at the top.

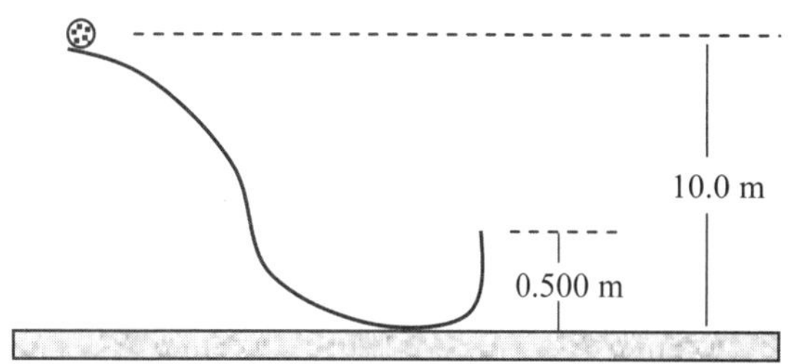


**Strategy** Because the ball rolls without slipping, we can apply mechanical energy conservation. Let's take the bottom of the track (the ground) as the reference level for gravitational potential energy. The greater height will be called $H$ and the lesser $h$.

**Solution**

1. The mechanical energy before the ball is released is in the form of gravitational potential energy only: $E = mgH$ Eq. (1)

2. The final energy, at height $y$, is a combination of potential and rotational kinetic energy: $E = mgy + \frac{1}{2}I\omega^2$ Eq. (2)

3. We need to determine $\omega$ when the ball leaves the track. The energy when it leaves is: $E = mgh + \frac{1}{2}mv_{cm}^2 + \frac{1}{2}I\omega^2$ Eq. (3)

4. From Table 10–1, $I$ is given by: $I = \frac{2}{5}mr^2$

5. The angular speed relates to the linear speed as: $v_{cm} = r\omega$

6. Substituting the results for $I$ and $\omega$ into Eq. (3): $E = mgh + \frac{1}{2}mr^2\omega^2 + \frac{1}{5}mr^2\omega^2 = mgh + \frac{7}{10}mr^2\omega^2$

**7**. Set Eq. (3) equal to Eq. (1) and solve for $\omega$:

$$mgH = mgh + \tfrac{7}{10}mr^2\omega^2 \Rightarrow \omega = \left[\frac{g(H-h)}{\tfrac{7}{10}r^2}\right]^{1/2}$$

$$= \left[\frac{(9.81\text{ m/s}^2)(9.50\text{ m})}{\tfrac{7}{10}(0.391\text{ m})^2}\right]^{1/2} = 29.51\text{ rad/s}$$

**8**. Set Eq. (2) equal to Eq. (1) and solve for the final height, $y$:

$$mgH = mgy + \tfrac{1}{5}mr^2\omega^2 \therefore$$

$$y = H - \frac{r^2\omega^2}{5g} = 10.0\text{ m} - \frac{(0.391\text{ m})^2(29.51\text{ rad/s})^2}{5(9.81\text{ m/s}^2)}$$

$$= 7.29\text{ m}$$

**Insight** Check that you can complete the mathematical steps for items 7 and 8 and understand the choice for $I$.

## Practice Quiz

**9**. Two solid wheels have the same mass, but one has twice the radius of the other. If they both rotate about axes through their centers, the one with the larger radius will have

**(a)** half the rotational inertia of the other one

**(b)** the same rotational inertia as the other one

**(c)** twice the rotational inertia of the other one

**(d)** four times the rotational inertia of the other one

**(e)** eight times the rotational inertia of the other one

**10**. A thin rod of mass 1.0 kg and length 1.0 m rotates about an axis through its center at 1.0 revolutions per second. What is its kinetic energy?

**(a)** 1.6 J **(b)** 1.0 J **(c)** 0.083 J **(d)** 39 J **(e)** 20 J

**11**. A hollow sphere of mass 0.22 kg and radius 0.16 m rolls along the ground with a constant linear speed of 2.0 m/s. What is its kinetic energy?

**(a)** 0.0075 J **(b)** 0.29 J **(c)** 0.070 J **(d)** 0.73 J **(e)** 0.44 J

# Reference Tools and Resources

## I. Key Terms and Phrases

**angular position** the angle measured from a chosen reference line

**angular velocity** the rate of change of angular position

**angular acceleration** the rate of change of angular velocity

**rolling without slipping** a form of rolling motion in which the point of contact between the rolling object and the surface is instantaneously at rest

**moment of inertia** a quantity that represents the inertial property of a rotating object or system

## II. Important Equations

| Name/Topic | Equation | Explanation |
|---|---|---|
| angular velocity | $\omega_{av} = \frac{\Delta\theta}{\Delta t}$ | The average angular velocity of a rotating object |
| angular acceleration | $\alpha_{av} = \frac{\Delta\omega}{\Delta t}$ | The average angular acceleration of a rotating object |
| constant angular acceleration | $\omega = \omega_0 + \alpha t$<br>$\theta = \theta_0 + \frac{1}{2}(\omega_0 + \omega)t$<br>$\theta = \theta_0 + \omega_0 t + \frac{1}{2}\alpha t^2$<br>$\omega^2 = \omega_0^2 + 2\alpha(\theta - \theta_0)$ | The system of equations that describe motion with constant angular acceleration |
| linear quantities | $s = r\theta$<br>$v_t = r\omega$<br>$a_t = r\alpha$<br>$a_{cp} = r\omega^2$ | The 4 equations that relate the linear and angular quantities in rotational kinematics |
| rotational kinetic energy | $K = \frac{1}{2}I\omega^2$ | The rotational kinetic energy of a system |
| moment of inertia | $I = \sum m_i r_i^2$ | The expression for the moment of inertia of a system of particles |

## III. Know Your Units

| Quantity | Dimension | SI Unit |
|---|---|---|
| angular position ($\theta$) | — | rad |
| angular velocity ($\omega$) | $[T^{-1}]$ | rad/s |

| | | |
|---|---|---|
| angular acceleration ($\alpha$) | $[T^{-2}]$ | $rad/s^2$ |
| moment of inertia ($I$) | $[M]\cdot[L^2]$ | $kg\cdot m^2$ |

## Puzzle

**LOOKALIKES**

You are given two identical-looking metal cylinders and a long rope. The cylinders have the same size and shape, and they weigh the same. You are told that one of them is hollow; the other is solid. How would you determine which cylinder is hollow using only the rope and the two cylinders?

## Answers to Selected Conceptual Questions

**2.** Yes. In fact, this is the situation whenever you drive in a circular path with constant speed.

**6.** The moment of inertia of an object changes with the position of the axis of rotation because the distance from the axis to all the elements of mass has been changed. It is not just the shape of an object that matters but the distribution of mass with respect to the axis of rotation.

**10.** **(a)** What determines the winner of the race is the ratio $I/mr^2$, as we see in the discussion just before Conceptual Checkpoint 10-4. This ratio is $MR^2/MR^2 = 1$ for the first hoop and $(2M)R^2/(2MR^2) = 1$ for the second hoop. Therefore, the two hoops finish the race at the same time. **(b)** As in part (a), we can see that the ratio $I/mr^2$ is equal to 1 regardless of the radius. Thus, all hoops, regardless of their mass or radius, finish the race in the same time.

## Solutions to Selected End-of-Chapter Problems and Conceptual Exercises

11. **Picture the Problem**: The propeller rotates about its axis with constant angular acceleration.

**Strategy:** Use the kinematic equations for rotating objects and the given formula to find the average angular speed and angular acceleration during the specified time intervals. By comparison of the formula given in the problem, $\theta = (125\text{ rad/s})t + (42.5\text{ rad/s}^2)t^2$, with equation 10-10, $\theta = \theta_0 + \omega_0 t + \frac{1}{2}\alpha t^2$, we can identify $\omega_0 = 125\text{ rad/s}$ and $\frac{1}{2}\alpha = 42.5\text{ rad/s}^2$.

**Solution: 1. (a)** Use equations 10-3 and 10-10 to find $\omega_{av}$:

$$\omega_{av} = \frac{\Delta\theta}{\Delta t} = \frac{\theta - \theta_0}{t} = \frac{\left[\omega_0 t + \frac{1}{2}\alpha t^2\right] - \theta_0}{t}$$

$$= \frac{\left[(125\text{ rad/s})(0.010\text{ s}) + (42.5\text{ rad/s}^2)(0.010\text{ s})^2\right] - 0}{0.010\text{ s}}$$

$$\omega_{av} = 125\text{ rad/s} = \boxed{1.3\times10^2\text{ rad/s}}$$

**2. (b)** Use equations 10-3 and 10-10 to find $\omega_{av}$:

$$\theta = \left[\omega_0 t + \tfrac{1}{2}\alpha t^2\right] = \left[(125 \text{ rad/s})(1.010 \text{ s}) + (42.5 \text{ rad/s}^2)(1.010 \text{ s})^2\right] = 169.60 \text{ rad}$$

$$\theta_0 = \left[\omega_0 t_0 + \tfrac{1}{2}\alpha t_0^2\right] = \left[(125 \text{ rad/s})(1.000 \text{ s}) + (42.5 \text{ rad/s}^2)(1.000 \text{ s})^2\right] = 167.50 \text{ rad}$$

$$\omega_{av} = \frac{\Delta\theta}{\Delta t} = \frac{\theta - \theta_0}{t - t_0} = \frac{169.60 - 167.50 \text{ rad}}{1.010 - 1.000 \text{ s}} = 210 \text{ rad/s} = \boxed{2.1\times10^2 \text{ rad/s}}$$

**3. (c)** Use equations 10-3 and 10-10 to find $\omega_{av}$:

$$\theta = \left[\omega_0 t + \tfrac{1}{2}\alpha t^2\right] = \left[(125 \text{ rad/s})(2.010 \text{ s}) + (42.5 \text{ rad/s}^2)(2.010 \text{ s})^2\right] = 422.95 \text{ rad}$$

$$\theta_0 = \left[\omega_0 t_0 + \tfrac{1}{2}\alpha t_0^2\right] = \left[(125 \text{ rad/s})(2.000 \text{ s}) + (42.5 \text{ rad/s}^2)(2.000 \text{ s})^2\right] = 420.00 \text{ rad}$$

$$\omega_{av} = \frac{\Delta\theta}{\Delta t} = \frac{\theta - \theta_0}{t - t_0} = \frac{422.95 - 420.0 \text{ rad}}{2.010 - 2.000 \text{ s}} = 295 \text{ rad/s} = \boxed{3.0\times10^2 \text{ rad/s}}$$

**4. (d)** The angular acceleration is $\boxed{\text{positive}}$ because the angular speed is positive and increasing with time.

**5. (e)** Apply equation 10-6 directly:

$$\alpha_{av} = \frac{\omega - \omega_0}{\Delta t} = \frac{210 - 125 \text{ rad/s}}{1.00 - 0.00 \text{ s}} = \boxed{85 \text{ rad/s}^2}$$

**6.** Apply equation 10-6 directly:

$$\alpha_{av} = \frac{\omega - \omega_0}{\Delta t} = \frac{295 - 210 \text{ rad/s}}{2.00 - 1.00 \text{ s}} = \boxed{85 \text{ rad/s}^2}$$

**Insight:** We violated the rules of significant figures in order to report answers with two significant figures. Such problems arise whenever you try to subtract two large but similar numbers to get a small difference. The answers are only known to one significant figure, but we reported two in order to show clearly that the angular acceleration is constant. Of course, we could also have determined from the equation given in the problem that because $\tfrac{1}{2}\alpha = 42.5 \text{ rad/s}^2$, it must be true that $\alpha = 85.0 \text{ rad/s}^2$.

23. **Picture the Problem**: The Earth rotates on its axis, slowing down with constant angular acceleration.

**Strategy:** Equation 10-6 gives an expression for the angular acceleration as a function of rotation rate and time. Determine the difference in rotation rates between 1906 and 2006 by approximating $T + \Delta T \cong T$ because 0.840 s is tiny compared with the time ($3.16 \times 10^7$ s) it takes to complete 365 revolutions. Then use equation 10-6 to find the average angular acceleration over the 100-year time interval.

**Solution: 1.** Find the difference in angular speeds:

$$\omega - \omega_0 = \frac{\theta}{T + \Delta T} - \frac{\theta}{T} = \theta\left[\frac{T - (T + \Delta T)}{T(T + \Delta T)}\right] \cong \theta\left(\frac{-\Delta T}{T^2}\right)$$

$$= (365 \text{ rev} \times 2\pi \text{ rad/rev})\left\{\frac{-(0.840 \text{ s})}{\left[(365 \text{ d})(24 \text{ h/d})(3600 \text{ s/h})\right]^2}\right\}$$

$$\omega - \omega_0 = \underline{\underline{-1.94\times10^{-12} \text{ rad/s}}}$$

**2.** Apply equation 10-6 directly:

$$\alpha_{av} = \frac{\Delta\omega}{\Delta t} = \frac{-1.94\times10^{-12} \text{ rad/s}}{(100 \text{ y} \times 3.16\times10^7 \text{ s/y})} = \boxed{-6.14\times10^{-22} \text{ rad/s}^2}$$

**Insight:** Your first instinct might be to find the angular speed in 1906 assuming a period of 24.000 hrs ($7.272205217\times10^{-5}$ rad/s) and figure out the angular speed in 2006 ($7.272205023\times10^{-5}$ rad/s), but as you can see, attempting to subtract these numbers requires us to ignore the rules for significant figures. Using the approximation outlined above allows us to avoid the subtraction problem and keep three significant figures.

38. **Picture the Problem**: The reel rotates about its axis at a constant rate.

**Strategy:** Calculate the linear speed of the string from the radius of the reel and the angular speed.

**Solution: 1. (a)** Apply equation 10-12 directly: $v_t = r\omega = (3.7\text{ cm})\left(\frac{3.0\text{ rev}}{\text{s}}\right)\left(\frac{2\pi\text{ rad}}{\text{rev}}\right) = \boxed{0.70\text{ m/s}}$

**2. (b)** If the radius of the reel were doubled but the angular speed remained the same, the linear speed would also double as can be seen by an examination of equation 10-12.

**Insight:** If the fish grabbed the lure when there was 10 m of string between it and the reel, it will take 14 seconds to reel in the big one.

51. **Picture the Problem**: The bicycle tire rolls without slipping, increasing its angular speed at a constant rate.

**Strategy:** Because the tire rolls without slipping there is a direct relationship between its linear and angular acceleration (equation 10-14). Use that equation together with the definition of acceleration (equation 3-5) to determine the angular acceleration.

**Solution: 1. (a)** Solve equation 10-14 for $\alpha$: $\alpha = \frac{a}{r} = \frac{1}{r}\frac{\Delta v}{\Delta t} = \frac{(8.90 - 0\text{ m/s})}{(0.360\text{ m})(12.2\text{ s})} = \boxed{2.03\text{ rad/s}^2}$

**2. (b)** If the radius of the tires had been smaller, the angular acceleration would have been greater than 2.03 rad/s$^2$.

**Insight:** The linear acceleration in this case is 0.730 m/s$^2$. If the bike were sliding without friction down an incline, we could say its acceleration equals $g\sin\theta$ and the slope must be $\theta = 4.26°$. However, because the tire rolls we must take torque and moment of inertia into account (see Chapter 11). If we do so we find $\theta$ would still be 4.26° for a hoop-shaped wheel but only 2.13° if the wheel were a solid disk.

75. **Picture the Problem**: The cylinder rolls down the ramp without slipping, gaining both translational and rotational kinetic energy.

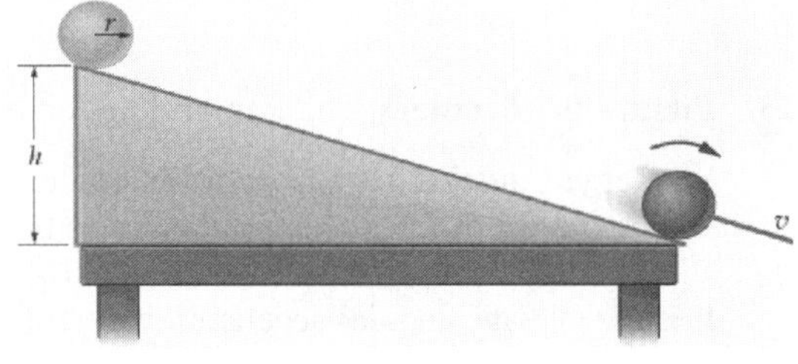


**Strategy:** Use conservation of energy to find total kinetic energy at the bottom of the ramp. Then set that energy equal to the sum of the rotational and translational energies. Because the cylinder rolls without slipping, the equation $\omega = v/r$ can be used to write the expression in terms of linear velocity alone. Use the resulting equation to find expressions for the fraction of the total energy that is rotational and translational kinetic energy.

**Solution: 1. (a)** Set $E_i = E_f$ and solve for $K_f$:

$$U_i + K_i = U_f + K_f$$
$$mgh + 0 = 0 + K_f$$
$$K_f = mgh = (2.0\text{ kg})(9.81\text{ m/s}^2)(0.75\text{ m})$$
$$= 14.7\text{ J} = \boxed{15\text{ J}}$$

**2. (b)** Set $K_f$ equal to $K_t + K_r$:

$$K_f = \tfrac{1}{2}mv^2 + \tfrac{1}{2}I\omega^2 = \tfrac{1}{2}mv^2 + \tfrac{1}{2}\left(\tfrac{1}{2}mr^2\right)(v/r)^2$$
$$= \tfrac{1}{2}mv^2 + \tfrac{1}{4}mv^2 = \tfrac{3}{4}mv^2$$

**3.** Determine $K_r$ from steps 1 and 2:

$$K_r = \tfrac{1}{4}mv^2 = \tfrac{1}{3}\left(\tfrac{3}{4}mv^2\right) = \tfrac{1}{3}K_f = \tfrac{1}{3}(14.7\text{ J}) = \boxed{4.9\text{ J}}$$

**4. (c)** Determine $K_t$ from steps 1 and 2:

$$K_t = \tfrac{1}{2}mv^2 = \tfrac{2}{3}\left(\tfrac{3}{4}mv^2\right) = \tfrac{2}{3}K_f = \tfrac{2}{3}(14.7\text{ J}) = \boxed{9.8\text{ J}}$$

**Insight:** The fraction of the total kinetic energy that is rotational energy depends upon the moment of inertia. If the object were a hoop, for instance, with $I = mr^2$, the final kinetic energy would be half translational, half rotational.

111. **Picture the Problem**: The ball rolls without slipping down the incline, gaining speed, then is launched horizontally off the edge, traveling along a parabolic arc until it hits the floor.

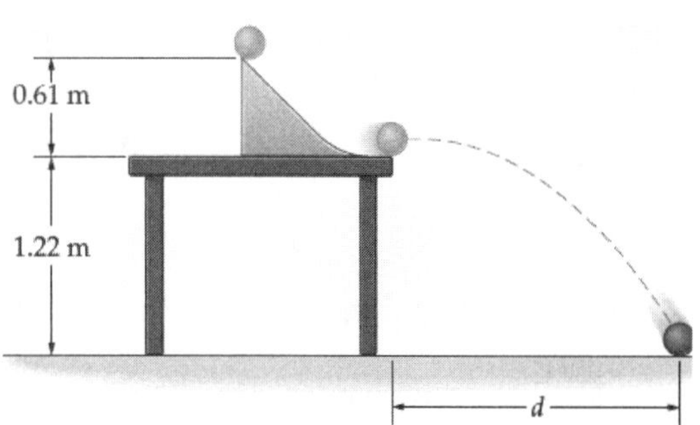


**Strategy:** Use conservation of energy to determine the center of mass speed of the sphere at the bottom of the ramp. Then use equation 4-9 to determine the horizontal range of the ball as it travels through the air. Use the table height together with equation 4-6 to determine the time of fall. Use the angular speed of the ball and the time of fall to find the number of rotations it makes before landing.

**Solution: 1. (a)** Set $E_i = E_f$ and simplify:

$$K_i + U_i = K_f + U_f$$
$$0 + mgh = \tfrac{1}{2}mv^2 + \tfrac{1}{2}I\omega^2 + 0$$
$$= \tfrac{1}{2}mv^2 + \tfrac{1}{2}\left(\tfrac{2}{5}mr^2\right)\left(v/r\right)^2$$
$$mgh = \tfrac{1}{2}mv^2 + \tfrac{1}{5}mv^2 = \tfrac{7}{10}mv^2$$

**2.** Solve the expression from step 1 for $v$:

$$v = \sqrt{\tfrac{10}{7}gh} = \sqrt{\tfrac{10}{7}\left(9.81\text{ m/s}^2\right)\left(0.61\text{ m}\right)} = \underline{\underline{2.92\text{ m/s}}}$$

**3.** Apply equation 4-9 to find $d$:

$$d = v_0\sqrt{\frac{2h_{\text{table}}}{g}} = \left(2.92\text{ m/s}\right)\sqrt{\frac{2\left(1.22\text{ m}\right)}{9.81\text{ m/s}^2}} = \boxed{1.5\text{ m}}$$

**4. (b)** Solve equation 4-6 for $t$:

$$y = y_0 + v_{0y}t - \tfrac{1}{2}gt^2$$
$$0 = h_{\text{table}} + 0 - \tfrac{1}{2}gt^2 \Rightarrow t = \sqrt{\frac{2h_{\text{table}}}{g}} = \sqrt{\frac{2\left(1.22\text{ m}\right)}{9.81\text{ m/s}^2}} = \underline{\underline{0.50\text{ s}}}$$

**5.** Determine the angular speed $\omega$ from equation 10-12:

$$\omega = \frac{v_t}{r} = \frac{2.92\text{ m/s}}{\tfrac{1}{2}\left(0.17\text{ m}\right)} = \underline{\underline{34\text{ rad/s}}}$$

**6.** Find the number of revolutions:

$$\theta = \omega t = \left(34\text{ rad/s}\right)\left(0.50\text{ s}\right) = 17\text{ rad} \times 1\text{ rev}/2\pi\text{ rad} = \boxed{2.7\text{ rev}}$$

**7. (c)** If the ramp were made frictionless, the sphere would slide, not roll. It would therefore store no energy in its rotation, and all of its gravitational potential energy would become translational kinetic energy. It would therefore launch from the table edge with a higher speed and the landing distance $d$ would $\boxed{\text{increase}}$.

**Insight:** If the ramp were frictionless, the launch speed would be $\sqrt{2gh}$ instead of $\sqrt{\tfrac{10}{7}gh}$, an 18% increase in speed and therefore an 18% increase in the landing distance $d$.

## Answers to Practice Quiz

**1.** (b) **2.** (c) **3.** (a) **4.** (d) **5.** (c) **6.** (d) **7.** (e) **8.** (b) **9.** (d) **10.** (a) **11.** (d)

# CHAPTER 11
# ROTATIONAL DYNAMICS AND STATIC EQUILIBRIUM

## Chapter Objectives

After studying this chapter, you should

1. know how to calculate the torque due to a given force about a given axis.
2. understand the role of torque for rotational motion and its relationship to angular acceleration.
3. be able to analyze situations of static equilibrium.
4. know how to calculate angular momentum, understand its relationship to torque, and be able to apply angular momentum conservation.
5. be able to use the rotational forms of work and power.
6. be able to determine the spatial direction of the rotational vector quantities.

## Warm-Ups

1. Two solid disks are linked with a belt. The diameter of the larger disk is twice that of the smaller disk. If the smaller disk is rotating at 500 revolutions per minute (rpm), what is the rotational speed of the larger disk? Express your answer in rpm and in radians per second.

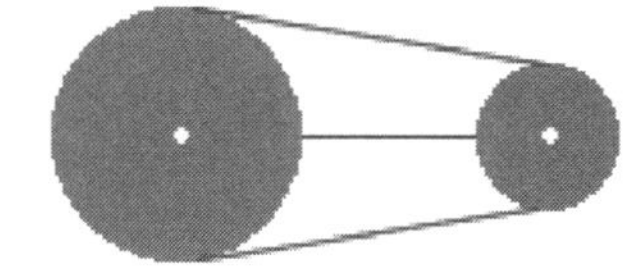

2. Estimate the angular acceleration of a small pebble stuck to a bicycle tire as the bicycle accelerates from rest to 10 mi/h (4.47 m/s) in 2 seconds.
3. A book can be rotated about many different axes. The moment of inertia of the book will depend on the axis chosen. Rank the choices A to C in the sketch in order of increasing moments of inertia.

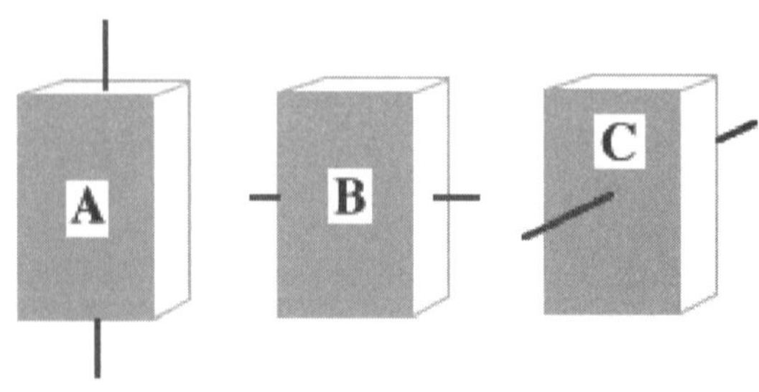

**4**. A hoop and a solid disk are released from rest at the top of an incline and allowed to roll down the incline without slipping. Which of the following is correct?

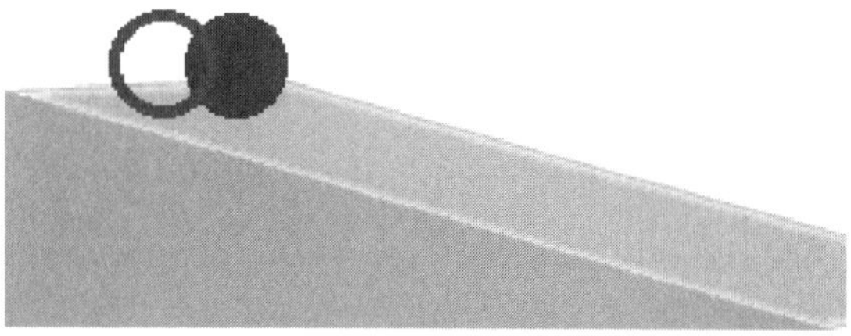

**a**. The hoop has a larger moment of inertia than the disk.

**b**. Gravity is the only force that exerts a torque on the hoop and the disk.

**c**. Both gravity and friction exert a torque on the hoop and the disk.

**d**. Gravity exerts a greater torque on the hoop than on the disk.

**e**. Gravity exerts a greater torque on the disk than on the hoop.

**f**. Gravity torque on the hoop has the same magnitude as gravity torque on the disk.

**g**. If the net torque on the hoop has the same magnitude as the net torque on the disk, the angular acceleration of the hoop is the same as the angular acceleration of the disk.

**h**. If the net torque on the hoop has the same magnitude as the net torque on the disk, the angular acceleration of the hoop is smaller than the angular acceleration of the disk.

**i**. If the angular acceleration is smaller, the tangential acceleration is smaller.

**j**. If the tangential acceleration is smaller, the center-of-mass acceleration is smaller.

**k**. The object with smaller tangential acceleration will take longer to reach the bottom of the incline.

## Chapter Review

In this chapter, we continue the study of rotation by looking at the dynamics of rotational motion. It is here that we introduce rotational versions of concepts such as force, linear momentum, and Newton's second law.

### 11–1 – 11–2 Torque and Angular Acceleration

If you want to cause a nonrotating object to rotate, you must apply a force. However, not just any force will result in a rotation. Crucial to determining the rotation that results from the force are two factors: **(a)** how the applied force is directed relative to the axis of rotation and **(b)** the distance of the point at which the force is being applied from the axis. These two factors combine with the magnitude of the force to form a quantity called **torque**. It is through the application of a torque that an object will begin to rotate. The magnitude of the torque $\tau$ that results from the application of a force $\vec{\mathbf{F}}$ is given by

$$\tau = rF\sin\theta$$

where $r$ is the magnitude of a radial vector $\vec{\mathbf{r}}$ from the axis of rotation to the point of action of the force, and $\theta$ is the smallest angle between $\vec{\mathbf{r}}$ and $\vec{\mathbf{F}}$. The SI unit of torque is the N·m. *Note that in calculations dealing with torque the N·m is **not** called a joule.*

For convenience, the expression for torque is often looked at in two ways. The quantity $F\sin(\theta)$ equals the component of $\vec{\mathbf{F}}$ tangential to a circle of radius $r$ centered on the axis of rotation, $F_t$, so we can write $\tau = rF_t$. Grouped the other way, the quantity $r\sin(\theta)$ equals the component of $\vec{\mathbf{r}}$ perpendicular to the line of force, $r_\perp$, called the **moment arm** (or *lever arm*) of the force, so we can also write $\tau = r_\perp F$.

All the above equations are for the magnitude of the torque only. Torque is a vector quantity, and we'll briefly discuss how to find its spatial direction later. In our applications, torque will be a one-dimensional vector, so we can account for its direction with just an algebraic sign. The convention is as follows:

> *A torque is positive if its tendency is to rotate an object counterclockwise and negative if its tendency is to rotate an object clockwise.*

---

**Example 11–1 Tether Ball** A child is playing tether ball; she has a 2.0-m-long cord with one end connected to a vertical pole. Attached to the other end of the cord is a light ball of mass 0.43 kg. As the ball whirls around in a horizontal circle, the string makes an angle of 35° with the pole. She then hits the ball with a force of 3.6 N that lies in the horizontal plane of the ball's motion and makes an angle of $\phi$ = 30° with the tangent to the ball's path (away from the pole). What magnitude of torque does the force apply to the ball about an axis through the pole?

**Picture the Problem** Figure **(a)** shows the ball connected to the pole by a cord of length $L$. Figure **(b)** is a top view showing the ball's circular path and the force $\vec{\mathbf{F}}$ applied by the child.

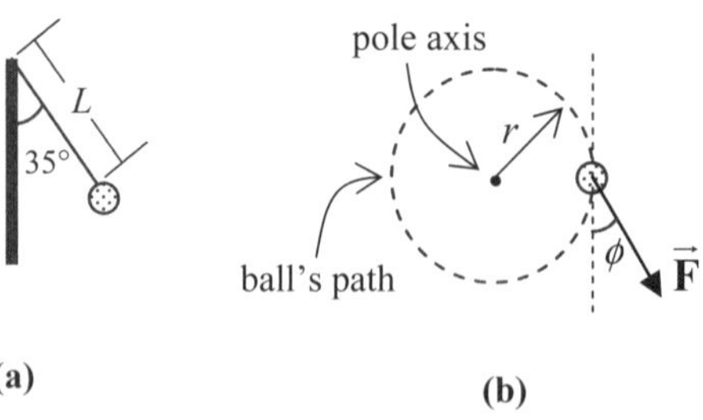


**Strategy** To complete this calculation, we must correctly determine the quantities that are relevant to calculating torque: $r$, $F$, and $\theta$.

**Solution**

1. In figure **(a)**, $r$ equals the horizontal distance between the ball and the pole's axis: $\sin(35^\circ) = \dfrac{r}{L} \quad \therefore \quad r = L\sin(35^\circ)$

2. Since the vector $\vec{\mathbf{r}}$ is perpendicular to the tangent line, the angle $\theta$ can be found from $\phi$: $\theta = 90° - \phi = 90° - 30° = 60°$

3. Directly calculate the magnitude of the torque: $\tau = rF\sin\theta = LF\sin(35^\circ)\sin\theta$
$= (2.0\text{ m})(3.6\text{ N})\sin(35^\circ)\sin(60^\circ) = 3.6\text{ N}\cdot\text{m}$

**Insight** We needed to calculate only the magnitude of the torque in this problem. If we wanted to associate a sign or direction to this torque, according to the sketch it would be negative because the torque contributes to a clockwise rotation of the ball.

Torque is the quantity that plays the role of force for rotational motion. Just as force causes translational acceleration, torque causes angular acceleration. The relationship between $\tau$ and $\alpha$ is very similar to that between force and linear acceleration:

$$\tau = I\alpha$$

Comparing this equation with $F = ma$, we can see how strong the similarity is by recalling that the moment of inertia, $I$, is the rotational quantity that acts like mass does for translational motion. The above equation is referred to as Newton's second law for rotation.

**Exercise 11–2 Engine Specifications** The engine specifications on a Pontiac Grand Am SE state that it supplies 225 ft·lb of torque at 4200 rpm. If this same torque is applied to a solid cylinder of mass 3.3 kg and radius 32 cm that rotates about its central axis, what angular acceleration will result?

**Solution:** The following information is given in the problem.

**Given:** $\tau = 225$ ft·lb, $M = 3.3$ kg, $r = 32$ cm, $\omega = 4200$ rpm; **Find:** $\alpha$

We know that the relationship between torque and angular acceleration is $\tau = I\alpha$. Therefore, if we can determine the moment of inertia for the cylinder, we can use it to determine $\alpha$. From Table 10–1 in the text, we can see that the moment of inertia for a solid cylinder about an axis through its center is

$$I = \tfrac{1}{2}Mr^2$$

Substituting this into Newton's second law for rotation and solving for $\alpha$ yields

$$\alpha = \frac{\tau}{I} = \frac{2\tau}{Mr^2}$$

Inserting numerical values and converting, we obtain the final result:

$$\alpha = \frac{2\tau}{Mr^2} = \frac{2(225\text{ ft}\cdot\text{lb})\left(\dfrac{1\text{ N}\cdot\text{m}}{0.738\text{ ft}\cdot\text{lb}}\right)}{(3.3\text{ kg})(0.32\text{ m})^2} = 1800\text{ rad/s}^2$$

Notice that to answer the question of this problem we did not need the angular velocity. For the engineers working on this car, the angular velocity is important because it is at this angular velocity that the result applies. You should always try to understand the relevance of a calculated result; otherwise, you will greatly diminish its usefulness to you.

## Practice Quiz

**1.** In trial 1, a force $\vec{\mathbf{F}}$ is applied tangentially to the rim of a wheel, causing it to rotate about its center. In trial 2, an equal force is applied tangentially at a point halfway from the center to the rim, also causing a rotation about the center. How do the torques applied in each trial relate to each other?

**(a)** Trial 2 has twice the torque of trial 1.

**(b)** Trial 1 has twice the torque of trial 2.

**(c)** Trial 1 has half the torque of trial 2.

**(d)** The torques are equal.

**2.** If an object has a constant net torque applied to it, this implies that the object must also have

**(a)** constant moment of inertia.

**(b)** constant angular velocity.

**(c)** constant angular acceleration.

**(d)** constant angular displacement.

**(e)** None of the above.

## 11–3 – 11–4 Static Equilibrium and Balance

In many applications, especially in the design of buildings and other structures, objects need to be balanced so that they do not fall down or tip over on people. This means that the objects must be in **static equilibrium**. We have seen examples for which an equilibrium of forces means that an object will not translate; however, to be in complete static equilibrium the object must not rotate either. For the object to be nonrotating, there must be an equilibrium of the torques on the object. The conditions for static equilibrium, then, are that both the net force and the net torque on an object must be zero:

$$\sum F_i = 0, \quad \sum \tau_i = 0$$

When doing an analysis for equilibrium, the center-of-mass concept is often very useful. The weight of an object can exert a torque about any axis of rotation that does not pass through the center of mass. Therefore, with regard to torque, we can treat rigid objects like point masses whose mass is located at the center of mass, as is illustrated by the following example.

---

**Example 11–3 Hanging the Lights** A horizontally suspended 2.7-m track for fluorescent lights is held in place by two vertical beams that are connected to the ceiling. The track and lights jointly weigh 15 N, and each beam is attached 60 cm from opposite ends of the track. What upward force does each beam apply to the track?

**Picture the Problem** The top diagram shows the light (at the bottom) attached to a track that hangs from the ceiling by two vertical beams.

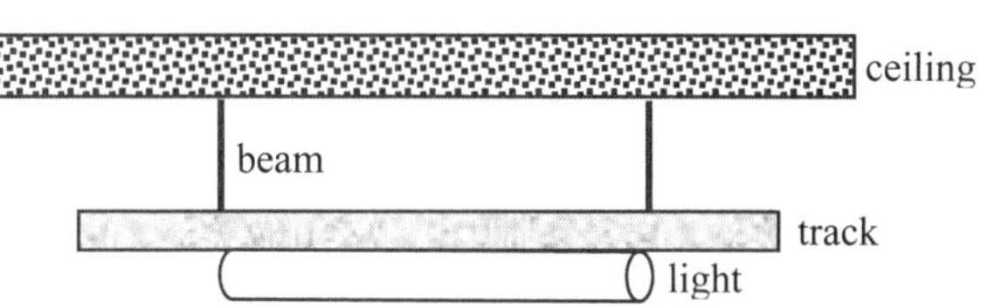


**Strategy** To solve this problem, we will apply the equilibrium conditions. The first step is to draw a free-body diagram of the track (shown in the lower sketch) with the left beam applying a force labeled $F_1$ and the right beam applying a force $F_2$. The axis for calculating torques is located at the left beam. We will calculate torque as force times lever arm.

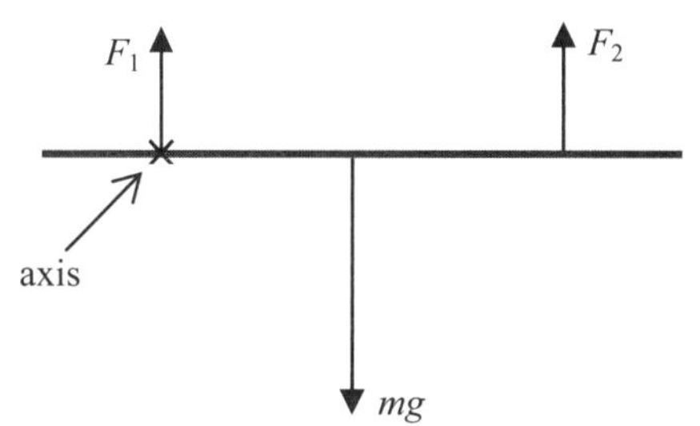


**Solution**

1. Write out the torque due to $F_1$: $\tau_1 = \ell_1 F_1 = (0)F_1 = 0$

2. Write out the torque due to $F_2$: $\tau_2 = \ell_2 F_2 = (2.7\text{ m} - 1.2\text{ m})F_2 = (1.5\text{ m})F_2$

3. Write out the torque due to gravity: $\tau_G = \ell_G mg = (\tfrac{2.7\text{ m}}{2} - 0.60\text{ m})mg = (0.75\text{ m})mg$

4. Apply the equilibrium condition for torque: $\sum \tau_i = \tau_1 + \tau_2 - \tau_G = 0$

$$(1.5\text{ m})F_2 - (0.75\text{ m})mg = 0 \;\therefore\; F_2 = \frac{0.75\text{ m}}{1.5\text{ m}}mg$$

$$F_2 = \tfrac{1}{2}(15\text{ N}) = 7.5\text{ N}$$

5. Apply the equilibrium condition for force: $\sum F_i = 0 \;\therefore\; F_1 + F_2 - mg = 0$

$$F_1 = mg - F_2 = 15\text{ N} - 7.5\text{ N} = 7.5\text{ N}$$

**Insight** Each beam contributes equally to supporting the track. Once you become more used to this type of analysis, you'll be able to solve a problem like this without any calculations. The symmetry of the problem shows that $F_1$ and $F_2$ have equal lever arms about an axis through the center of the track. Because they apply torques of equal magnitude about this axis, they must also apply forces of equal magnitude. Because the lever arms share the 15 N equally, each must be 7.5 N.

---

**Example 11–4 Decorating** A decorative ornament is hung from the end of an 82.3-cm-long horizontal beam attached to a wall as shown below. The beam weighs 12.1 N, and the ornament weighs 7.62 N. **(a)** What is the tension in the cord? **(b)** What are the horizontal and vertical components of the force that the wall directly applies to the beam?

**Picture the Problem** The upper diagram shows the ornament being supported by a horizontal beam attached to a vertical wall. The beam is supported by a cord attached to its end. The lower sketch is a free-body diagram of the beam. The lever arm for the tension $T$ is also shown on the diagram as $\ell_T$.

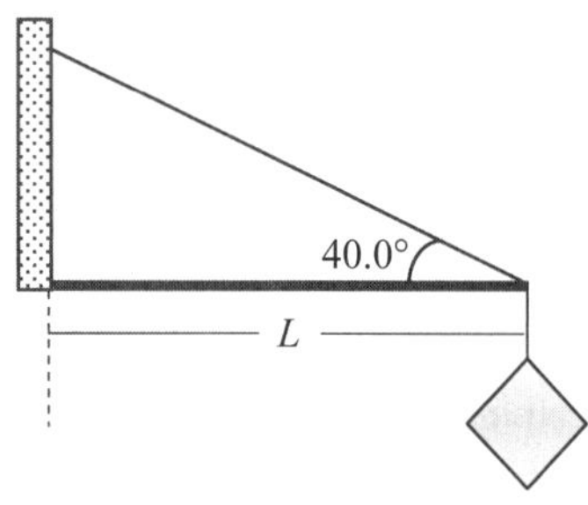


**Strategy** $F_{wall}$ is unknown at this stage. We chose the axis to be at that location to simplify the equation for the net torque. Let's first apply the torque condition and see where it leads us. (When needed we shall choose "up" as the $+y$ direction, and right as the $+x$ direction.)

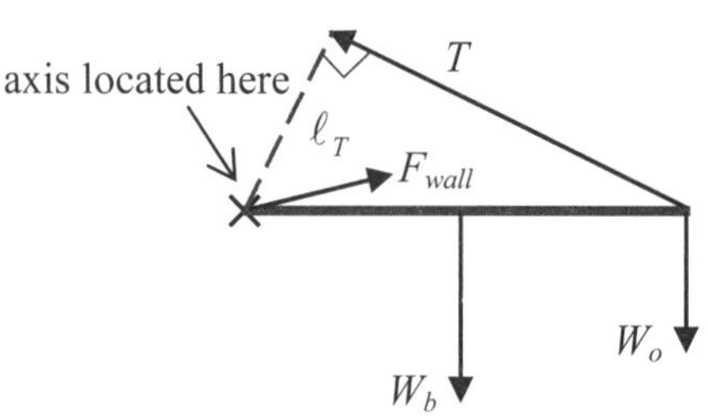


**Solution**

**Part (a)**

1. Write the equilibrium condition for the torques: $\sum \tau_i = T\ell_T - W_b \frac{L}{2} - W_o L = 0$

2. From the geometry of the problem write the lever arm for the tension in terms of $L$: $\sin(40.0^\circ) = \frac{\ell_T}{L} \quad \therefore \quad \ell_T = L\sin(40.0^\circ)$

3. Divide through by $L$: $T\sin(40.0^\circ) - \frac{1}{2}W_b - W_o = 0$

4. Solve for $T$: $T = \frac{W_o + W_b/2}{\sin(40.0^\circ)} = \frac{7.62\text{ N} + (12.1\text{ N})/2}{\sin(40.0^\circ)} = 21.3\text{ N}$

**Part (b)**

**5**. Apply the equilibrium condition for the forces along the $x$ direction:

$$\sum F_x = F_{wall,x} - T_x = 0$$
$$\therefore F_{wall,x} = T_x = T\cos(40.0^\circ)$$
$$F_{wall,x} = (21.27\text{ N})\cos(40.0^\circ) = 16.3\text{ N}$$

**6**. Apply the equilibrium condition for the forces along the $y$ direction:

$$\sum F_y = F_{wall,y} + T_y - W_b - W_o = 0$$
$$F_{wall,y} = W_b + W_o - T_y = W_b + W_o - T\sin(40.0^\circ)$$
$$F_{wall,y} = 12.1\text{ N} + 7.62\text{ N} - (21.27\text{ N})\sin(40.0^\circ)$$
$$= 6.05\text{ N}$$

**Insight** The signs of the results in part (b) tell us that $F_{wall,x}$ is to the right, and $F_{wall,y}$ is upward. We could have discovered these facts without calculations, however, by looking carefully at the equilibrium situation. This will be discussed in more detail in the "Tips" section. Also, the length of the beam does not enter into the calculations.

## Practice Quiz

3. If both the net force and the net torque on an object is zero, the object will
   **(a)** remain motionless.
   **(b)** have constant angular acceleration.
   **(c)** have constant linear acceleration.
   **(d)** have constant angular velocity.
   **(e)** have no forces applied to it.

4. A 1.5-m-long, uniform beam weighing 5.5 N is supported at one end by a fulcrum as shown. What force, $F$, is needed at the other end to maintain static equilibrium?

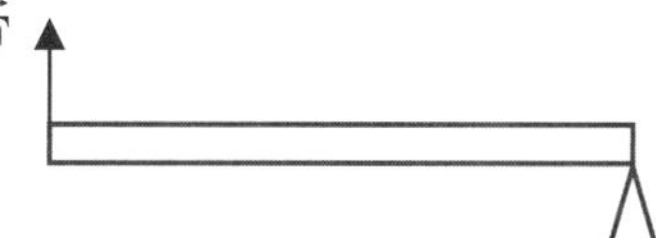


   **(a)** 2.8 N **(b)** 5.5 N **(c)** 3.0 N
   **(d)** 0.75 N **(e)** 8.2 N

## 11–5 Dynamic Applications of Torque

The previous section dealt with static applications of torque. In this section we consider how to handle cases for which the net torque is not zero, resulting in angular acceleration.

**Exercise 11–5 A Design Flaw** Suppose the person who designed the ornament display in Example 11–4 neglected to do the calculation and used a cord that could not withstand the required tension, and it broke. What was the initial angular acceleration of the beam when the cord broke? (Assume the end of the cord that was attached to the beam remained attached.)

**Solution:** The following information is known.

**Given:** $L = 82.3$ cm, $W_b = 12.1$ N, $W_o = 7.62$ N; **Find:** $\alpha$

Since the net torque is not equal to zero in this case, we use Newton's second law, $\tau_{net} = I\alpha$. The moment of inertia is given by both the beam and the ornament: $I = I_b + I_o$. For the beam, we use the results in Table 10–1

$$I_b = \frac{1}{3}M_b L^2$$

and for the ornament we have just $I_o = M_o L^2$. Therefore,

$$I = \left( M_o + \frac{1}{3}M_b \right) L^2$$

The net torque can be calculated from the results of Example 11–4 without the tension,

$$\tau_{net} = -W_b \tfrac{L}{2} - W_o L = -\left(W_o + W_b/2\right)L$$

Using these results in Newton's second law to solve for $\alpha$, we get

$$\alpha = \frac{\tau_{net}}{I} = -\frac{W_o + W_b/2}{\left(M_o + M_b/3\right)L}$$

Making use of the fact that $M = W/g$, we can solve for the angular acceleration,

$$\alpha = -\frac{7.62\text{ N} + (12.1\text{ N})/2}{\left[7.62\text{ N} + (12.1\text{ N})/3\right]\ (0.823\text{ m})/9.81\text{ m/s}^2} = -14.0\text{ rad/s}^2$$

Remember, this result is only the initial angular acceleration of the beam when the cord breaks.

## Practice Quiz

5. A uniform hoop of radius 0.75 m and mass 2.25 kg rotates with an angular acceleration of 44 rad/s$^2$ about an axis through its center. What is the net torque on this hoop?

   **(a)** 0 N·m **(b)** 33 N·m **(c)** 1.3 N·m **(d)** 74 N·m **(e)** 56 N·m

## 11–6 – 11–7 Angular Momentum and Its Conservation

In studying translational motion we found that the concept of linear momentum was important for understanding the behavior of objects, especially systems of particles. For similar reasons, we also have a rotational form of this concept called **angular momentum**, $L$. Just as for linear momentum, there is a conservation principle for angular momentum.

The angular momentum of a rotating object can be calculated as

$$L = I\omega$$

where $I$ is calculated about the axis of rotation. In making the comparison with linear momentum, $p = mv$, we see that angular momentum results by replacing the linear quantities with angular ones: $p \rightarrow L$, $m \rightarrow I$, and $v \rightarrow \omega$. For a particle of mass $m$, the angular momentum can be written in terms of the linear momentum in analogy with how torque is written in terms of force:

$$L = rp\sin\theta$$

where, as with torque, $r$ is the magnitude of a radial vector $\vec{\mathbf{r}}$ from the axis of rotation to the particle, and $\theta$ is the smallest angle between $\vec{\mathbf{r}}$ and $\vec{\mathbf{p}}$. The quantity $p\sin(\theta)$ equals the component of the linear momentum tangential to a circle of radius $r$ centered on the axis of rotation, $p_t$. The quantity $r\sin(\theta)$ equals the component of $\vec{\mathbf{r}}$ perpendicular to the direction of $\vec{\mathbf{p}}$, $r_\perp$. Therefore, we also have

$$L = rp_t = r_\perp p$$

From the above expressions, we can determine that the SI unit of angular momentum is kg·m$^2$/s = J·s.

Recall that with translational motion, the concept of linear momentum led us to a better form for Newton's second law because $F = ma$ works only when mass is constant. The same is true for rotational motion. The expression $\tau = I\alpha$ works only when $I$ remains constant. It is much more common with rotational motion to see situations with varying $I$ than it is to see situations with varying mass in linear motion. The concept of angular momentum solves this problem because the best rotational form of Newton's second law, valid for situations in which $I$ is either constant or changing, is that torque equals the rate of change of angular momentum:

$$\tau = \frac{\Delta L}{\Delta t}$$

---

**Example 11–6 Thrown from a Ledge** A person throws a 0.336-kg object vertically downward from a 7.83-m-high ledge with an initial speed of 2.25 m/s. What average torque is exerted on this object by gravity during its fall to the ground relative to a point $O$ on the ground that is a distance $x = 5.92$ m from where it strikes?

**Picture the Problem** The diagram shows the object at its initial height $h$ with a downward initial velocity $v_i$. The point $O$ relative to which the torque will be calculated is also shown.

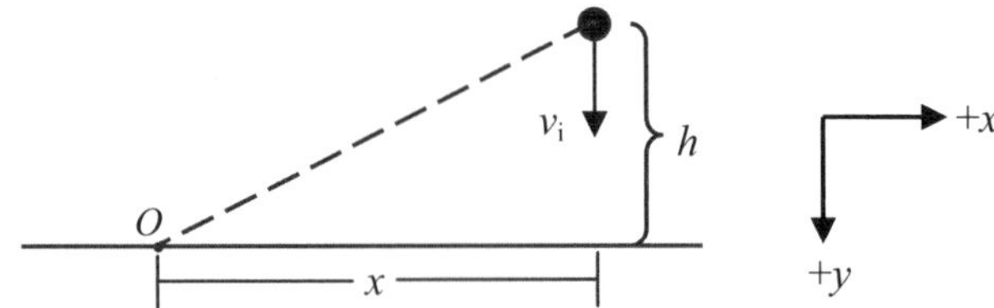


**Strategy** The average torque can be determined from the average rate of change of angular momentum, $\Delta L/\Delta t$, so we can determine $\tau_{av}$ by calculating the angular momentum at the top and the bottom as well as the time it takes the object to reach the ground.

**Solution**

1. Obtain the magnitude of the initial angular momentum:

$$L_i = (r_{i\perp})p_i = xmv_i = (5.92\text{ m})(0.336\text{ kg})(2.25\text{ m/s}) = 4.476\text{ J}\cdot\text{s}$$

2. Obtain the magnitude of the final velocity using the free-fall equations:

$$v_f = \sqrt{v_i^2 + 2gh} = \sqrt{(2.25\tfrac{\text{m}}{\text{s}})^2 + 2(9.81\tfrac{\text{m}}{\text{s}^2})(7.83\text{ m})} = 12.60\text{ m/s}$$

3. The final angular momentum is:

$$L = (r_{f\perp})p_f = xmv_f = (5.92\text{ m})(0.336\text{ kg})(12.60\tfrac{\text{m}}{\text{s}}) = 25.06\text{ J}\cdot\text{s}$$

4. Setting $\Delta t = t$, calculate the time it takes to reach the ground from the equations for free-fall:

$$v_f = v_i + gt \quad \therefore \quad t = (v_f - v_i)/g$$

$$t = \frac{12.60\text{ m/s} - 2.25\text{ m/s}}{9.81\text{ m/s}^2} = 1.055\text{ s}$$

5. Calculate the average torque:

$$\tau_{av} = \frac{L_f - L_i}{t} = \frac{25.06\text{ J}\cdot\text{s} - 4.476\text{ J}\cdot\text{s}}{1.055\text{ s}} = 19.5\text{ N}\cdot\text{m}$$

**Insight** This problem asked only for the average torque, so we used only the total change in angular momentum over a period of time and not the instantaneous rate of change.

---

Newton's second law for rotation, as just discussed, implies that torque causes a change in the angular momentum of a system. Consequently, if there is no net torque on a system, there will be no change in the angular momentum; this is the principle of the conservation of angular momentum. A more complete statement of this principle is the following:

> *If the net external torque on a system is zero, then the total angular momentum of the system is conserved.*

The total angular momentum is the sum of the angular momenta of every object in the system.

---

**Example 11–7 The Merry-go-round** A merry-go-round with a moment of inertia of 750 kg·m$^2$ and a radius $R$ of 2.55 m is rotating with an angular velocity of 9.42 rad/s clockwise (as viewed from above). A 334-N child runs at 2.76 m/s tangent to the rim of the merry-go-round and jumps onto it in the direction opposite its sense of rotation. With what angular velocity does the merry-go-round rotate after the child jumps onto it?

**Picture the Problem** The diagram shows a top view of the rotating merry-go-round and the running child's path before the child jumps on.

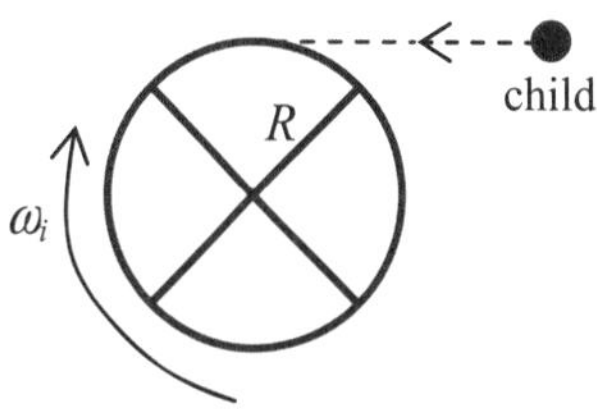


**Strategy** We can solve this problem by using the conservation of angular momentum. We will equate the total angular momentum before the child jumps on with the total angular momentum after the child jumps on.

**Solution**

1. The initial angular momentum at the instant the child is about to jump on is: $L_{total} = L_c - L_{mgr} = Rp_c - I_{mgr}\omega_i$

2. The final angular momentum after the child jumps on is: $L_{total} = I_{total}\omega_f = \left(I_{mgr} + m_c R^2\right)\omega_f$

3. The mass of the child is: { EMBED Equation.3
$m_c = W_c / g = 334\text{ N}/(9.81\text{ m/s}^2) = 34.05\text{ kg}$

4. Set the initial and final angular momenta equal to each other and solve for $\omega_f$:

$$Rp_c - I_{mgr}\omega_i = \left(I_{mgr} + m_c R^2\right)\omega_f$$

$$\omega_f = \frac{Rm_c v_c - I_{mgr}\omega_i}{I_{mgr} + m_c R^2}$$

5. Obtain the numerical result:

$$\omega_f = \frac{(2.55\text{ m})(34.05\text{ kg})(2.76\,\tfrac{\text{m}}{\text{s}}) - (750\text{ kg}\cdot\text{m}^2)(9.42\,\tfrac{\text{rad}}{\text{s}})}{750\text{ kg}\cdot\text{m}^2 + 34.05\text{ kg}(2.55\text{ m})^2}$$
$$= -7.03\text{ rad/s}$$

**Insight** The negative sign on $\omega_f$ results because the rotation is clockwise. The initial angular velocity, $\omega_i$, is also clockwise, and its negative sign was accounted for by subtracting $L_{mgr}$ from $L_c$.

## Practice Quiz

6. A uniform hoop of radius 0.75 m and mass 2.25 kg rotates with an angular velocity of 44 rad/s about an axis through its center. What is the magnitude of its angular momentum?

   **(a)** 0 J·s **(b)** 74 J·s **(c)** 1.3 J·s **(d)** 33 J·s **(e)** 56 J·s

7. If the angular momentum of an object is conserved,

   **(a)** it has zero net torque applied on it.

   **(b)** it has constant angular velocity.

   **(c)** it has constant angular acceleration.

   **(d)** it has constant moment of inertia.

   **(e)** The angular momentum of every object is conserved.

8. If the angular momentum of a system of several objects is conserved,

   **(a)** there is a large net external torque on the system.

   **(b)** no torque acts on any object in the system.

   **(c)** the net external torque on the system is zero.

   **(d)** the system rotates as a rigid body.

   **(e)** None of the above.

## 11–8 Rotational Work and Power

So far we have written rotational forms for several important quantities that we studied in translational motion, namely, displacement, velocity, acceleration, inertia, kinetic energy, force, and momentum. Two quantities that we have not addressed are work and power.

When a force gives rise to a torque that rotates an object, the force acts through a displacement and therefore does work. An analysis of this work shows that it can be written in the following rotational form

$$W = \tau \Delta\theta$$

Once again, you should compare this equation with the one-dimensional linear version, $W = F\Delta x$. Keep in mind that in the above expression, the angular displacement is measured in radians. The average power consumed when a torque does work to rotate an object can also be written in rotational form. Working from the definition of average power, $P = W/\Delta t$, we obtain

$$P = \tau\omega$$

This expression compares directly with $P = Fv$ for the case of translational motion.

---

**Exercise 11–8 Tool Sharpening** To sharpen a tool, the toolmaker uses a grinding wheel of radius 0.695 m with a coarse rim. The tool is pressed against the rim of the wheel with a force of 25.9 N (directed toward the center) as the wheel undergoes exactly 12 rotations. If the coefficient of kinetic friction between the tool and the rim is 0.644, how much work is done by friction in sharpening the tool?

**Solution:** The following information is given in the problem:

**Given:** $N = 25.9$ N, $r = 0.695$ m, # of rev = 12, $\mu_k = 0.644$; **Find:** $W$

In this situation, torque is applied by the force of kinetic friction. This force acts tangentially along the rim of the grinding wheel. The torque about the center of the wheel is then given by

$$\tau = rf_k = r\mu_k N$$

The number of rotations of the wheel corresponds directly to the angular displacement:

$$\Delta\theta = 12\text{ rev}\left(\frac{2\pi\text{ rad}}{\text{rev}}\right) = 75.40\text{ rad}$$

Therefore, the work done by friction is

$$W = \tau\Delta\theta = r\mu_k N\Delta\theta = (0.695\text{ m})(0.644)(25.9\text{ N})(75.40\text{ rad}) = 874\text{ J}$$

---

## Practice Quiz

**9.** If a constant torque of 5.2 N·m rotates an object through 136°, how much work is done by the torque?

**(a)** 5.2 J **(b)** 26 J **(c)** 710 J **(d)** 3.9 J **(e)** 12 J

## *11–9 The Vector Nature of Rotational Motion

Many of the rotational quantities that we have discussed are vector quantities and are therefore associated with particular directions in space. The quantities that must be considered are angular velocity, $\vec{\omega}$; angular acceleration $\vec{\alpha}$; torque $\vec{\tau}$; and angular momentum, $\vec{L}$. The directions of rotating objects are determined by right-hand rules. We need to state only two right-hand rules, and we can use them to get the directions of all the quantities.

**Angular Velocity**

The direction of the angular velocity vector is perpendicular to the plane in which the rotation takes place. To determine the direction in which $\vec{\omega}$ points, we use the following right-hand rule:

> *Curl the fingers of your right hand along the direction of rotation; your extended thumb now indicates the direction of the angular velocity.*

Once we know how to determine the direction of $\vec{\omega}$, we can also determine the direction of angular momentum because the vector equation $\vec{\mathbf{L}} = I\vec{\omega}$ tells us that $\vec{\mathbf{L}}$ must point along $\vec{\omega}$.

**Torque**

The direction of the torque vector is perpendicular to the plane formed by the vectors $\vec{\mathbf{r}}$ and $\vec{\mathbf{F}}$. To determine the side of this plane to which $\vec{\tau}$ points, we use the following right-hand rule:

> *Curl the fingers of your right hand along the direction (in the $\vec{\mathbf{r}}$-$\vec{\mathbf{F}}$ plane) in which the torque tends to rotate the object; your extended thumb now indicates the direction of the torque.*

Once we know how to determine the direction of $\vec{\tau}$, we can also determine the direction of the angular acceleration resulting from this torque because the vector equation $\vec{\tau} = I\vec{\alpha}$ tells us that $\vec{\tau}$ and $\vec{\alpha}$ must point in the same direction.

---

**Example 11–9 Getting the Directions** **(a)** A solid circular disk rotates about a vertical axis as shown. What is the direction of the angular momentum of the disk about its central axis? **(b)** A beam sits on a fulcrum located at its center. If a force, $\vec{\mathbf{F}}$, is then applied to the end of the beam as shown, what is the direction of the torque due to this force about the center of the beam? **(c)** A particle of mass $m$ moves with momentum, $\vec{\mathbf{p}}$, relative to a point $O$ as shown. Determine the direction of the angular momentum of the particle relative to $O$.

**Picture the Problem** The diagram shows the three situations described in the statement of the problem.

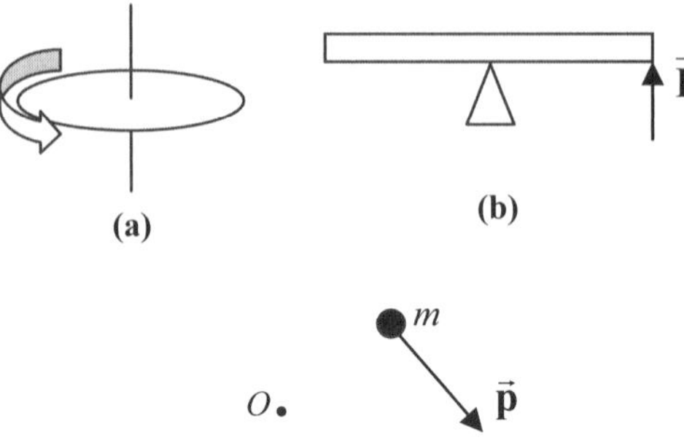


**Strategy** In each case we use the relevant right-hand rule.

**Solution**

**Part (a)**

**1.** Curl the fingers of your right hand along the direction of rotation. Your thumb should indicate that $\vec{\mathbf{L}}$ is along the axis pointing upward.

**Part (b)**

**2.** The vector, $\vec{\mathbf{r}}$, points to the right, so the $\vec{\mathbf{r}}$-$\vec{\mathbf{F}}$ plane is the plane of the page. The force tries to rotate the beam counterclockwise. Curl your fingers this way. Your thumb should indicate that $\vec{\mathbf{L}}$ points directly out of the page (perpendicularly) toward you.

**Part (c)**

**3.** The mass, *m,* can be viewed as instantaneously rotating, in the plane of the page, around point *O* with an angular velocity $\vec{\boldsymbol{\omega}}$. This rotation is clockwise in the figure shown. Curling your fingers clockwise should indicate that $\vec{\mathbf{L}}$ points directly into the page (perpendicularly) away from you.

**Insight** Try reversing the direction of rotation for part (a), $\vec{\mathbf{F}}$ for part (b), and $\vec{\mathbf{p}}$ for part (c), and verify that you will get $\vec{\mathbf{L}}$ pointing in the opposite directions.

## Practice Quiz

**10.** The axis of a rotating sphere is aligned along the north-south direction. If the sphere rotates clockwise as viewed from the southern side, what is the direction of its angular momentum?

**(a)** north **(b)** south **(c)** east **(d)** west **(e)** upward toward the sky

# Reference Tools and Resources

## I. Key Terms and Phrases

**torque** the combination of a force and the distance at which it is applied from an axis that causes angular acceleration

**moment arm** the perpendicular distance from an axis to the line of force for calculating torque

**static equilibrium** the state of motion in which an object neither translates nor rotates

**angular momentum** a vector quantity whose magnitude is equal to $\sum rp\sin\theta$ that represents a rotational analog of linear momentum

**angular momentum conservation** the principle that the total angular momentum of a system remains constant unless a nonzero external net torque is applied

**right-hand rule** the rules for determining the direction of rotational vector quantities

## II. Tips

Sometimes, when you are performing an analysis of a static equilibrium situation, it is not immediately clear in which direction some of the forces are pointing at the time you draw a free-body diagram. In Example 11–4, I drew the direction of the forces exerted by the wall on the beam as being upward and to the right. But, how did I know these directions? You can handle such circumstances by considering different locations for the axis about which torques are being considered. Remember, if there is equilibrium of torques about one axis, then there is equilibrium about every axis.

Let's consider a partial free-body diagram from Example 11–4 (without $F_{wall}$), except this time we place the axis on the right end of the beam instead of on the left end. Now, consider the torques about this new axis. Neither $T$ nor $W_o$ will exert a torque about this axis. The force $W_b$ will try to rotate the beam counterclockwise about this axis. Since we know the beam is in equilibrium, the $y$ component of $F_{wall}$ must be such as to balance the torque of $W_b$. Therefore, we can conclude that $F_{wall,y}$ must point vertically upward. With a little practice you can do this kind of determination in your head fairly quickly. After doing so, you'll have a complete free-body diagram and can solve the numerical problem with the axis placed anywhere you want it.

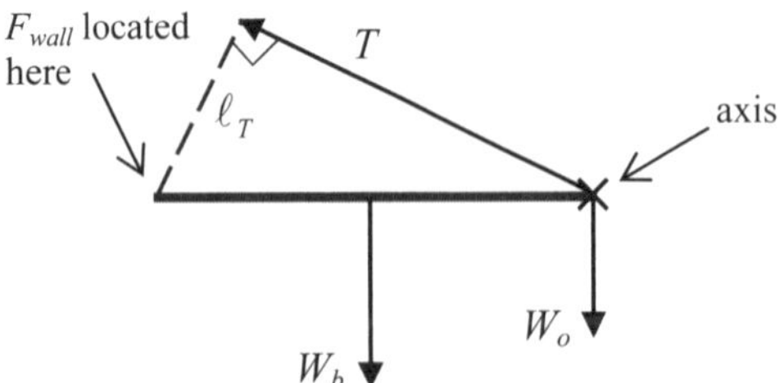

Another tactic is to put the unknown forces in positive directions and solve the equations as you normally would. If your result comes out positive, you guessed correctly; if the result comes out negative, then you should reverse the direction.

## III. Important Equations

| Name/Topic | Equation | Explanation |
|---|---|---|
| torque | $\tau = rF\sin\theta$ | The magnitude of the torque due to a force $\vec{\mathbf{F}}$ |
| rotational dynamics | $\tau = I\alpha$<br>$\tau = \dfrac{\Delta L}{\Delta t}$ | The rotational forms of Newton's second law. The lower equation is more general |
| angular momentum | $L = I\omega$<br>$L = rp\sin\theta$ | The magnitude of the angular momentum. The lower equation applies to point particles |
| work | $W = \tau\Delta\theta$ | Work written in terms of rotational quantities |
| power | $P = \tau\omega$ | Power written in terms of rotational quantities |

## IV. Know Your Units

| Quantity | Dimension | SI Unit |
|---|---|---|
| torque ($\vec{\boldsymbol{\tau}}$) | $[M]\cdot[L^2]/[T^2]$ | N·m |
| angular momentum ($\vec{\mathbf{L}}$) | $[M]\cdot[L^2]/[T]$ | $kg\cdot m^2/s$ (= J·s) |

# Puzzle

### DIZZYBALL

You practice throwing baskets from a spot 15 feet from the pole until you can make a basket every time. The ball leaves your hands the same way every time. The initial velocity of the ball is exactly correct. You visit an amusement park where there is a merry-go-round with a 15-foot-diameter platform. A standard-size basketball post is fastened to the rim of the platform. The platform is rotating clockwise (as seen from above) with a period of 15 seconds.

Step onto a spot at the edge of the platform diametrically opposite the basketball post. Imagine a ground-based coordinate system, with you at the origin, the positive $x$-axis drawn to your right, the $y$-axis drawn away from you toward the post, and the $z$-axis drawn straight up from the spot on which you are standing. When the platform is standing still, you can always make the basket with the $v_y$ and $v_z$ components of the initial velocity you practiced on the ground.

How will you have to adjust the initial velocity of the ball to make a basket on the moving platform? Do you need to add an $x$ component to the velocity? Do you have to adjust the $y$ component and the $z$ components in any way?

## Answers to Selected Conceptual Questions

**2.** As a car brakes, the forces responsible for braking are applied at ground level. The center of mass of the car is well above the ground, however. Therefore, the braking forces exert a torque about the center of mass that tends to rotate the front of the car downward. This, in turn, causes an increased upward force to be exerted by the front springs until the net torque acting on the car returns to zero.

**4.** The force that accelerates a motorcycle is a forward force applied at ground level. The center of mass of the motorcycle, however, is above the ground. Therefore, the accelerating force exerts a torque on the cycle that tends to rotate the front wheel upward.

**12.** The tail rotor on a helicopter has a horizontal axis of rotation, as opposed to the vertical axis of the main rotor. Therefore, the tail rotor produces a horizontal thrust that tends to rotate the helicopter about a vertical axis. As a result, if the angular speed of the main rotor is increased or decreased, the tail rotor can exert an opposing torque that prevents the entire helicopter from rotating in the opposite direction.

## Solutions to Selected End-of-Chapter Problems and Conceptual Exercises

5. **Picture the Problem**: The biceps muscle, the weight of the arm, and the weight of the ball all exert torques on the forearm as depicted at right.

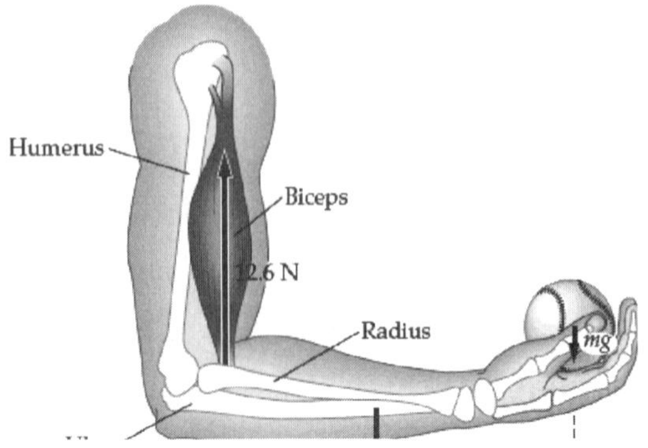


**Strategy:** Use equation 10-3 to determine the torques produced by the biceps muscle, the weight of the forearm, and the weight of the ball. Sum the torques together to find the net torque. According to the sign convention, torques in the counterclockwise direction are positive, and those in the clockwise direction are negative.

**Solution: 1. (a)** Compute the individual torques using equation 10-3 and sum them:

$$\tau_{\text{biceps}} = r_{\perp}F = (0.0275\text{ m})(12.6\text{ N}) = 0.347\text{ N}\cdot\text{m}$$
$$\tau_{\text{forearm}} = r_{\perp}mg = (0.170\text{ m})(1.20\text{ kg})(9.81\text{ m/s}^2) = -2.00\text{ N}\cdot\text{m}$$
$$\tau_{\text{ball}} = r_{\perp}W_{\text{ball}} = (0.340\text{ m})(1.42\text{ N}) = -0.483\text{ N}\cdot\text{m}$$
$$\sum\tau = \tau_{\text{biceps}} + \tau_{\text{forearm}} + \tau_{\text{ball}}$$
$$= +0.347 - 2.00 - 0.483\text{ N}\cdot\text{m} = \boxed{-2.14\text{ N}\cdot\text{m}}$$

**2. (b)** Negative net torque means the clockwise direction; the forearm and hand will rotate downward.

**3. (c)** Attaching the biceps farther from the elbow would increase the moment arm and increase the net torque.

**Insight:** The biceps would need to exert a force of at least 90.3 N in order to prevent the arm from rotating downward (see problem 25).

18. **Picture the Problem**: The object consists of four masses that can be rotated about any of the $x$, $y$, or $z$ axes, as shown in the figure at right.

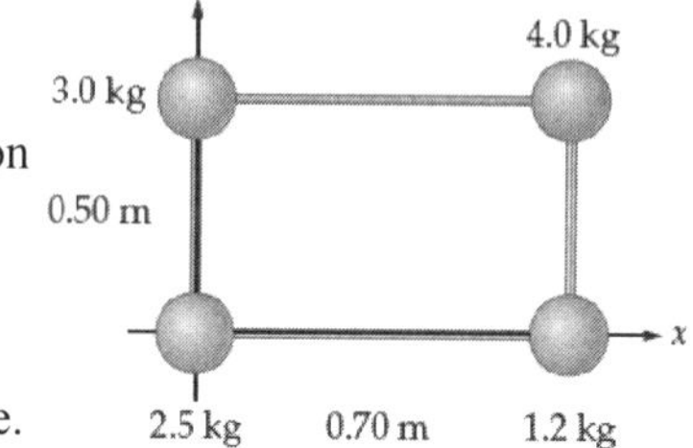


**Strategy:** Calculate the moments of inertia about the $x$, $y$, and $z$ axes using equation 10-18, and then apply equation 11-4 to find the angular acceleration that results from an applied torque of 13 N·m about the various axes. Let $m_1 = 2.5$ kg, $m_2 = 3.0$ kg, $m_3 = 4.0$ kg, and $m_4 = 1.2$ kg.

**Solution: 1. (a)** The angular acceleration will be the greatest when the moment of inertia is the smallest because the torque is the same in each case. The moment of inertia is smallest when the rectangular object is rotated about the $x$ axis and greatest when rotated about the $z$ axis (see below). Therefore we predict the angular acceleration is greatest about the $x$ axis, least about the $z$ axis.

**2. (b)** Calculate $I_x$ using equation 10-18:

$$I_x = m_1r_1^2 + m_2r_2^2 + m_3r_3^2 + m_4r_4^2$$
$$= 0 + (3.0\text{ kg})(0.50\text{ m})^2 + 0 + (4.0\text{ kg})(0.50\text{ m})^2$$
$$I_x = \underline{\underline{1.75\text{ kg}\cdot\text{m}^2}}$$

**3.** Find $\alpha_x$ using equation 11-4:

$$\alpha_x = \frac{\tau}{I_x} = \frac{13\text{ N}\cdot\text{m}}{1.75\text{ kg}\cdot\text{m}^2} = \boxed{7.4\text{ rad/s}^2}$$

**4. (c)** Calculate $I_y$ using equation 10-18:

$$I_y = m_1r_1^2 + m_2r_2^2 + m_3r_3^2 + m_4r_4^2$$
$$= 0 + 0 + (4.0\text{ kg})(0.70\text{ m})^2 + (1.2\text{ kg})(0.70\text{ m})^2$$
$$I_y = \underline{\underline{2.55\text{ kg}\cdot\text{m}^2}}$$

**5.** Find $\alpha_y$ using equation 11-4:

$$\alpha_y = \frac{\tau}{I_y} = \frac{13\text{ N}\cdot\text{m}}{2.55\text{ kg}\cdot\text{m}^2} = \boxed{5.1\text{ rad/s}^2}$$

**6. (d)** Calculate $I_z$ using equation 10-18:

$$I_z = m_1r_1^2 + m_2r_2^2 + m_3r_3^2 + m_4r_4^2$$
$$= 0 + (3.0\text{ kg})(0.50\text{ m})^2 + (4.0\text{ kg})\left((0.50\text{ m})^2 + (0.70\text{ m})^2\right) + (1.2\text{ kg})(0.70\text{ m})^2$$
$$I_z = \underline{\underline{4.3\text{ kg}\cdot\text{m}^2}}$$

**7.** Find $\alpha_z$ using equation 11-4:

$$\alpha_z = \frac{\tau}{I_z} = \frac{13\text{ N}\cdot\text{m}}{4.3\text{ kg}\cdot\text{m}^2} = \boxed{3.0\text{ rad/s}^2}$$

**Insight:** In this case it is a bit difficult to predict the moments of inertia about the $x$ and $y$ axes without calculating anything. That's because, although more mass (7.0 kg) is displaced from the $x$ axis, the masses are at a shorter distance (0.50 m) when compared to the 4.2 kg of mass that are displaced 0.70 m from the $y$ axis. It turns out the difference in distances is what makes $I_x$ smaller than $I_y$. We bent the rules for significant figures a bit in steps 2 and 4 to avoid rounding error.

33. **Picture the Problem**: The rod is supported by the wire and a hinge as depicted in the figure at right.

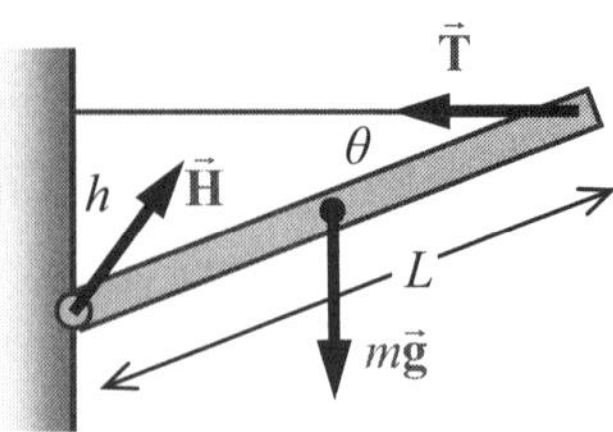


**Strategy:** Write Newton's Second Law for torque (let the hinge be the pivot point) to find the magnitude of the tension in the wire. Then use Newton's Second Law in the $x$ and $y$ directions to determine the components of the hinge force $\vec{H}$.

**Solution: 1. (a)** Set $\sum\tau = 0$ and solve for $T$: $\sum\tau = r_{\perp\text{-wire}}T - r_{\perp\text{-weight}}mg = 0$

$$T = \frac{r_{\perp\text{-weight}}}{r_{\perp\text{-wire}}}mg = \frac{\frac{1}{2}L\cos\theta}{h}mg$$

$$= \frac{\frac{1}{2}(1.2\text{ m})\cos 25°}{0.51\text{ m}}(3.7\text{ kg})(9.81\text{ m/s}^2) = \boxed{39\text{ N}}$$

**2. (b)** Set $\sum F_x = 0$ and solve for $H_x$: $\sum F_x = H_x - T = 0$

$$H_x = T = \boxed{39\text{ N}}$$

**3. (c)** Set $\sum F_y = 0$ and solve for $H_y$: $\sum F_y = H_y - mg = 0$

$$H_y = mg = (3.7\text{ kg})(9.81\text{ m/s}^2) = \boxed{36\text{ N}}$$

**Insight:** In this configuration the horizontal and vertical components of the hinge force $\vec{H}$ are nearly equal so that $\vec{H}$ points a little bit less than 45° above horizontal and to the right. If you put in the numbers you find $\vec{H} = 53$ N at 43° above horizontal.

44. **Picture the Problem**: The books are arranged in a stack as depicted at right, with book 1 on the bottom and book 3 at the top of the stack.

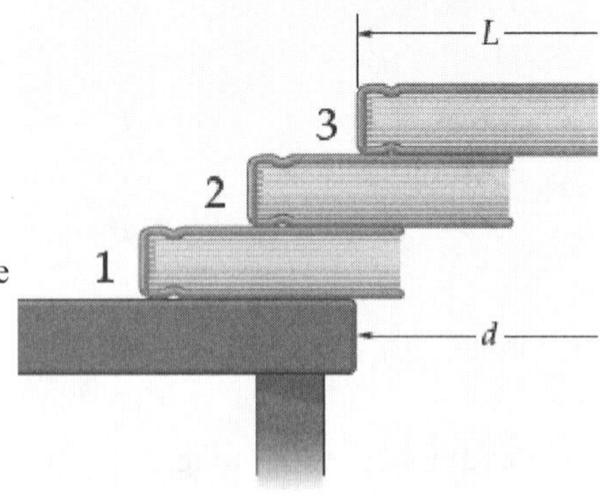


**Strategy:** It is helpful to approach this problem from the top down. The center of mass of each set of books must be above or to the left of the point of support. Find the positions of the centers of mass for successive stacks of books to determine $d$. Measure the positions of the books from the right edge of book 1 (right-hand dashed line in the figure). If the center of mass of the books above an edge is to the right of that edge, there will be an unbalanced torque on the books and they'll topple over. Therefore we can solve the problem by forcing the center of mass to be above the point of support.

**Solution: 1.** The center of mass of book 3 needs to be above the right end of book 2:

$$d_2 = \frac{L}{2}$$

**2.** The result of step 1 means that the center of mass of book 2 is located at $L/2 + L/2 = L$ from the right edge of book 1.

**3.** The center of mass of books 3 and 2 needs to be above the right end of book 1:

$$d_1 = X_{\text{cm},32} = \frac{m(L/2) + m(L)}{2m} = \frac{3}{4}L$$

**4.** The result of step 3 means that the center of mass of book 1 is located at $3L/4 + L/2 = 5L/4$.

**5.** The center of mass of books 3, 2, and 1 needs to be above the right end of the table:

$$d = X_{\text{cm},321} = \frac{m(L/2)+m(L)+m(5L/4)}{3m} = \boxed{\frac{11}{12}L}$$

**Insight:** As we learned in problem 87 of Chapter 9, if you add a fourth book the maximum overhang is $(25/24)L$. If you examine the overhang of each book you find an interesting series: $d = \frac{L}{2}+\frac{L}{4}+\frac{L}{6}+\frac{L}{8} = \frac{25}{24}L$. The series gives you a hint about how to predict the overhang of even larger stacks of books.

52. **Picture the Problem**: The masses and pulley are configured as depicted in the figure at right:

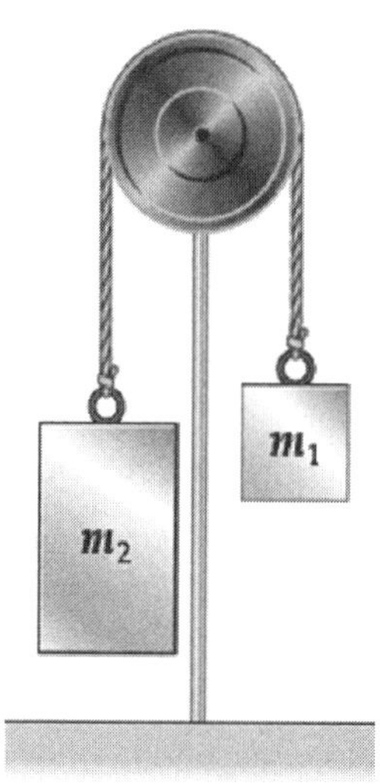

**Strategy:** Write Newton's Second Law in the vertical direction for both $m_1$ and $m_2$, and then write Newton's Second Law for torque about the pulley's rotation axis. Combine the three equations to find the linear acceleration of the masses. Assume the rope does not slip along the rim of the pulley of radius *R*, so that $a = R\alpha$. Let counterclockwise rotation ($m_2$ accelerating downward) be the positive direction for each part of the system. For the disk-shaped pulley the moment of inertia $I = \frac{1}{2}MR^2$.

**Solution: 1.** Write $\sum F_y = m_1 a$. Note that upward is the positive direction for $m_1$:

$$\sum F_y = T_1 - m_1 g = m_1 a$$

**2.** Write $\sum F_y = m_2 a$. Note that downward is the positive direction for $m_2$.

$$\sum F_y = -T_2 + m_2 g = m_2 a$$

**3.** Write $\sum \tau = I\alpha$ for the pulley:

$$\sum \tau = RT_2 - RT_1 = I\alpha = \left(\tfrac{1}{2}MR^2\right)(a/R)$$

$$T_2 - T_1 = \tfrac{1}{2}Ma$$

**4.** Add the two equations from steps 1 and 2, and rearrange:

$$T_1 - m_1 g + (-T_2 + m_2 g) = (m_1 + m_2)a$$

$$(m_2 - m_1)g - (m_1 + m_2)a = T_2 - T_1$$

**5.** Substitute the result from step 3 into step 4 and solve for *a*:

$$(m_2 - m_1)g - (m_1 + m_2)a = \tfrac{1}{2}Ma$$

$$(m_2 - m_1)g = \left(m_1 + m_2 + \tfrac{1}{2}M\right)a$$

$$a = \boxed{\left(\frac{m_2 - m_1}{m_1 + m_2 + \frac{1}{2}M}\right)g}$$

**Insight:** This is similar to the result of Example 6-7 except for the addition of the pulley's mass in the denominator. A massive pulley reduces the acceleration of the system because of the torque required to rotate it.

58. **Picture the Problem**: Jogger 3 runs in a straight line at constant speed in the manner indicated by the figure at right.

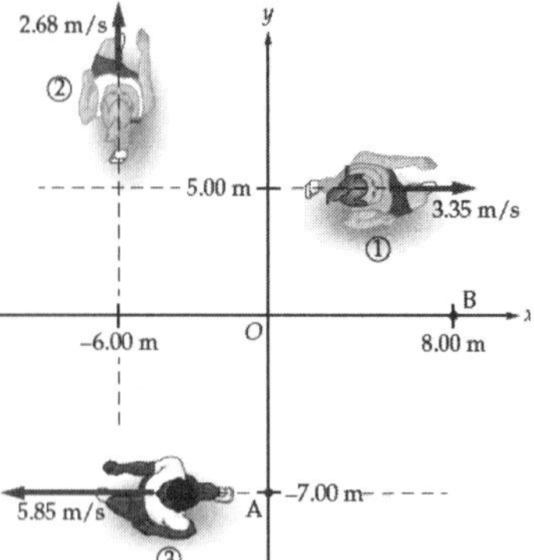


**Strategy:** Use $L = rmv$ (equation 11-12) to find the angular momentum.

**Solution: 1. (a)** The angular momentum increases with the perpendicular distance to the reference point. Since jogger 3's perpendicular distance to point A is zero, his angular momentum is zero with respect to point A. Therefore his angular momentum is greater with respect to $\boxed{\text{point B}}$.

**2. (b)** Jogger 3 has the same perpendicular distance to point B as he has to the origin, O. Therefore, his angular momentum with respect to B is $\boxed{\text{the same}}$ as it is with respect to O.

**3. (c)** Apply equation 11-12 directly: $L_A = r_{\perp,A}mv = (0.00 \text{ m})(62.2 \text{ kg})(5.85 \text{ m/s}) = \boxed{0 \text{ kg}\cdot\text{m}^2/\text{s}}$

**4.** Apply equation 11-12 directly: $L_B = r_{\perp,B}mv = (7.00 \text{ m})(62.2 \text{ kg})(5.85 \text{ m/s}) = \boxed{2.55\times10^3 \text{ kg}\cdot\text{m}^2/\text{s}}$

**5.** Apply equation 11-12 directly: $L_O = r_{\perp,O}mv = (7.00 \text{ m})(62.2 \text{ kg})(5.85 \text{ m/s}) = \boxed{2.55\times10^3 \text{ kg}\cdot\text{m}^2/\text{s}}$

**Insight:** The angular momenta of all 3 joggers have the same sign (they are all clockwise). If you use the right-hand rule introduced in section 11-9, the angular momentum vectors of each point into the page.

73. **Picture the Problem**: A mouse on the freely rotating turntable walks to the rotation axis.

**Strategy:** The moment of inertia of the turntable-mouse system will decrease as the mouse walks toward the axis, but the angular momentum of the system will remain the same because there is no external torque. Use conservation of angular momentum together with equation 11-11 to find the final angular speed of the system.

**Solution: 1. (a)** Angular momentum is conserved as the moment of inertia decreases, so the turntable rotates $\boxed{\text{faster}}$ according to the equation, $I_i\omega_i = I_f\omega_f$.

**2. (b)** Set $L_i = L_f$ and solve for $\omega_f$:

$$I_i\omega_i = I_f\omega_f \Rightarrow \omega_f = \left(\frac{I_i}{I_f}\right)\omega_i = \left(\frac{I + mr^2}{I + 0}\right)\omega_i$$

$$\omega_f = \frac{5.4\times10^{-3}\,\text{kg}\cdot\text{m}^2 + (0.032 \text{ kg})(0.15 \text{ m})^2}{5.4\times10^{-3} \text{ kg}\cdot\text{m}^2}\left(33\tfrac{1}{3} \text{ rev/min}\right) = \boxed{38 \text{ rev/min}}$$

**Insight:** A heavier mouse would have an even larger effect upon the final angular speed because it would create a larger change in the moment of inertia.

84. **Picture the Problem**: The saw blade rotates on its axis and gains rotational kinetic energy due to the torque applied by the electric motor.

**Strategy:** The torque applied through an angular displacement gives the blade its rotational kinetic energy. Use equations 11-17 and 10-17 to relate the kinetic energy to the torque applied by the motor. Then use equation 11-17 again to find the kinetic energy and angular speed after the blade has completed half as many revolutions.

**Solution: 1. (a)** Find $\omega_f$ in units of rad/sec: $\omega_f = 3620 \dfrac{\text{rev}}{\text{min}} \times \dfrac{2\pi \text{ rad}}{\text{rev}} \times \dfrac{1 \text{ min}}{60 \text{ s}} = \underline{\underline{379 \text{ rad/s}}}$

**2.** Set $W = \Delta K$ and solve for $\tau$:

$$W = \tau\Delta\theta = \tfrac{1}{2}I\omega^2 \text{ and } I = \tfrac{1}{2}mr^2$$

$$\tau = \frac{\frac{1}{2}mr^2\omega^2}{2\Delta\theta} = \frac{\frac{1}{2}(0.755 \text{ kg})(0.152 \text{ m})^2(379 \text{ rad/s})^2}{2(6.30 \text{ rev} \times 2\pi \text{ rad/rev})} = \boxed{15.8 \text{ N}\cdot\text{m}}$$

**3. (b)** The time to rotate the first 3.15 revolutions is greater than the time to rotate the last 3.15 revolutions because the blade is speeding up. So more than half the time is spent in the first 3.15 revolutions. Therefore, the angular speed has increased to more than half of its final value. After 3.15 revolutions, the angular speed is $\boxed{\text{greater than}}$ 1810 rpm.

**4. (d)** Set $W = \Delta K$ and solve for $\omega$:

$$\tau\Delta\theta = \tfrac{1}{2}I\omega^2 = \tfrac{1}{4}mr^2\omega^2$$

$$\omega = \sqrt{\frac{4\tau\Delta\theta}{mr^2}} = \sqrt{\frac{4(15.8\text{ N}\cdot\text{m})(3.15\text{ rev}\times 2\pi\text{ rad/rev})}{(0.755\text{ kg})(0.152\text{ m})^2}}$$

$$= (268\text{ rad/s})\left(\frac{60\text{ s}}{\text{min}}\right)\left(\frac{1\text{ rev}}{2\pi\text{ rad}}\right) = \boxed{2560\text{ rev/min}}$$

**Insight:** The angular speed increases linearly upon time ( $\omega = \omega_0 + \alpha t = \alpha t$ ) but depends upon the square root of the angular displacement: $\omega = \sqrt{\omega_0^2 + 2\alpha\Delta\theta} = \sqrt{2\alpha\Delta\theta}$ .

## Answers to Practice Quiz

**1.** (b) **2.** (c) **3.** (d) **4.** (a) **5.** (e) **6.** (e) **7.** (a) **8.** (c) **9.** (e) **10.** (a)

# CHAPTER 12
# GRAVITY

## Chapter Objectives

After studying this chapter, you should

1. understand and be able to use Newton's universal law of gravitation.
2. be able to apply the principle of superposition to gravitational forces and potential energies.
3. know Kepler's laws of orbital motion.
4. know how to apply Kepler's third law to circular orbits.
5. be able to calculate gravitational potential energy and apply it in the conservation of energy.
6. be able to calculate the escape speed from planets and other objects.

## Warm-Ups

1. A planet's orbit around the Sun is an ellipse. Consider points A and B on the ellipse. How does the centripetal force exerted on the planet at point A compare with the centripetal force exerted on the planet at point B? Also compare the potential energies, kinetic energies, and angular momenta at A and B.

2. Estimate the force that the Moon exerts on you when it is directly overhead.
   (You need some data from the text to answer this question.)

3. A space-communication company is planning to take a spy satellite to a spot 35,800 km above Earth's surface and release it into a geosynchronous orbit. (In a geosynchronous orbit, a satellite orbits at the same rate as the points on the surface of Earth below it so as to appear to hover over the same spot.) Is this possible? If yes, how fast must the satellite be moving when it is released?

4. Two different planets are orbiting the same star along two different orbits. The orbit containing points C and D is circular; the orbit containing points A and B is elliptical. Compare the speeds of the planet in the elliptical orbit at points A, B, and E. Compare the speeds of the planet in the circular orbit at points C, D, and E. The planet in the elliptical orbit is to be shifted to the circular orbit as it passes point E. Does it have to speed up or slow down?

## Chapter Review

In this chapter, we study **gravity** – a fundamental force of nature. Of the four fundamental forces, gravity is the most familiar. The other fundamental forces will be discussed later in the text.

### 12–1 – 12–2 Newton's Law of Universal Gravitation and the Attraction of Spherical Bodies

Gravitation is the phenomenon that between every two objects there is a force of attraction. **Newton's law of universal gravitation** describes the behavior of this force. Between any two point masses $m_1$ and $m_2$, the magnitude of the gravitational force on each mass due to the other is given by

$$F = G\frac{m_1 m_2}{r^2}$$

where $r$ is the distance between the two masses, and $G$ is a constant called the *universal gravitational constant*. The value of this constant is

$$G = 6.67\times10^{-11}\ \text{N}\cdot\text{m}^2/\text{kg}^2$$

The force on each mass points directly at the other mass because each mass attracts the other toward it.

For situations involving more than two masses, we apply the **principle of superposition:**

> *The net gravitational force on any given mass due to two or more other masses is the vector sum of the gravitational forces due to each of the other masses individually.*

**Exercise 12–1 Earth and Venus** On average, Earth and Venus are separated by about $4.14 \times 10^{10}$ m. What is the magnitude of the gravitational force between them at this separation?

**Solution**

In this problem, we are given only the distance between Earth and Venus. To calculate the gravitational force between them, we will need to look up their masses. The masses are as follows:

$$M_E = 5.97 \times 10^{24}\text{ kg};\ \ M_V = 4.87 \times 10^{24}\text{ kg}$$

Having all the data we need, the force can now be calculated.

$$F = \frac{GM_E M_V}{r^2} = \frac{(6.67\times10^{-11}\text{ N}\cdot\text{m}^2/\text{kg}^2)(5.97\times10^{24}\text{ kg})(4.87\times10^{24}\text{ kg})}{(4.14\times10^{10}\text{ m})^2} = 1.13\times10^{18}\text{ N}$$

Remember, this is the magnitude of the force exerted on both Earth and Venus.

---

The above law of universal gravitation is stated for point masses. The detailed calculations for extended bodies can become complicated, but Newton figured out that the final result for spherical bodies becomes simple again. Newton showed that two completely separated, uniform spherical bodies attract each other as if they were point masses with all their mass located at their respective centers. This fact explains why the point-mass formula given above works so well for large objects like Earth and the Moon.

Treating the earth as a point mass $M_E$ located at the center of the earth provides further insight into the acceleration due to gravity that we measure near Earth's surface. Objects near the surface are a distance from the center roughly equal to the radius of the earth, $R_E$. Using Newton's law of gravity we can conclude that the acceleration due to gravity near the surface is given by

$$g = \frac{GM_E}{R_E^2} \approx 9.8\text{ m/s}^2$$

With this result, you can also see that the higher you go above the surface, the farther you are from Earth's center, and therefore the weaker the effect of gravity, resulting in a smaller acceleration.

---

**Example 12–2 Martian Gravity** What is the acceleration due to gravity on the surface of Mars?

**Picture the Problem** The sketch is a representation of the planet Mars, with its radius extending from the center to the surface.

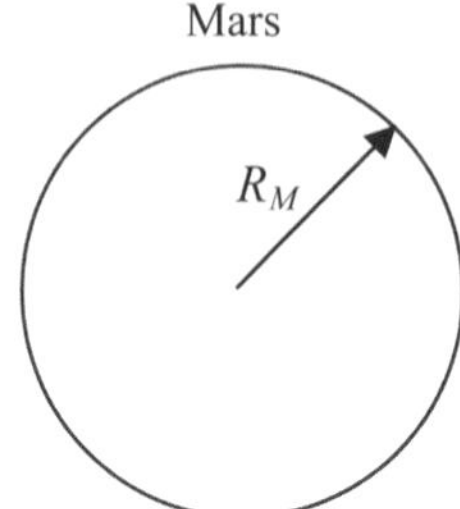


**Strategy** We need to adapt the expression for Earth's surface acceleration of gravity to Mars.

**Solution**

1. Look up the mass of Mars: $M_M = 6.4 \times 10^{23}$ kg

2. Look up the radius of Mars: $R_M = 3.4 \times 10^6$ m

3. Use the mass and radius of Mars in place of those for the Earth in the expression for $g$:

$$g_M = \frac{GM_M}{R_M^2} = \frac{\left(6.67\times10^{-11}\,\text{N}\cdot\text{m}^2/\text{kg}^2\right)\left(6.4\times10^{23}\text{ kg}\right)}{\left(3.4\times10^6\text{ m}\right)^2} = 3.7\text{ m/s}^2$$

**Insight** A similar substitution can be made for any of the spherical astronomical bodies.

## Practice Quiz

1. Two objects are separated by a distance $r$. If the mass of each object is doubled, how does the gravitational force between them change?
   **(a)** It increases to twice as much.
   **(b)** It decreases to half as much.
   **(c)** It increases to four times as much.
   **(d)** It decreases to one-fourth as much.
   **(e)** None of the above.

2. Two objects are separated by a distance $r$. If the distance between them is doubled, how does the gravitational force between them change?
   **(a)** It increases to twice as much.
   **(b)** It decreases to half as much.
   **(c)** It increases to four times as much.
   **(d)** It decreases to one-fourth as much.
   **(e)** None of the above.

3. What is the acceleration due to gravity at a height of 1000 km above Earth's surface?
   **(a)** 9.78 m/s$^2$ **(b)** 4.89 m/s$^2$ **(c)** 3.98 m/s$^2$ **(d)** 7.31 m/s$^2$ **(e)** None of the above

## 12–3 Kepler's Laws of Orbital Motion

One of the great early successes of Newton's work on gravity was his ability to explain accurately the laws of orbital motion that Kepler discovered from his observations of Mars and the other known planets. These laws of orbital motion are summarized as three statements. Kepler's first law is as follows:

*Planets follow elliptical orbits, with the Sun at one focus of the ellipse.*

Kepler's second law is as follows:

*As a planet moves in its orbit, a line from the Sun to the planet sweeps out equal amounts of area in equal amounts of time.*

And finally, the third law relates the amount of time it takes a planet to orbit the Sun, called its **orbital period**, $T$, to the planet's average distance from the Sun, $r$.

*The orbital period of a planet is directly proportional to its average distance from the Sun raised to the 3/2 power. That is,* $T \propto r^{3/2}$.

For the special case of circular orbits, the constant of proportionality is relatively easy to determine. In this case, the result of Kepler's third law can be written in equation form as

$$T = \left( \frac{2\pi}{\sqrt{GM_S}} \right) r^{3/2}$$

Even though the above three laws are stated in terms of planetary orbits about the Sun, they also apply to the orbit of any body around a much more massive body such as the artificial satellites that orbit Earth.

---

**Example 12–3 An Artificial Satellite** If it is required to have a satellite in a circular orbit with an orbital period of two days, at what altitude, $h$, above the ground should it be placed in orbit?

**Picture the Problem** The sketch shows Earth and an artificial satellite a height $h$ above Earth's surface.

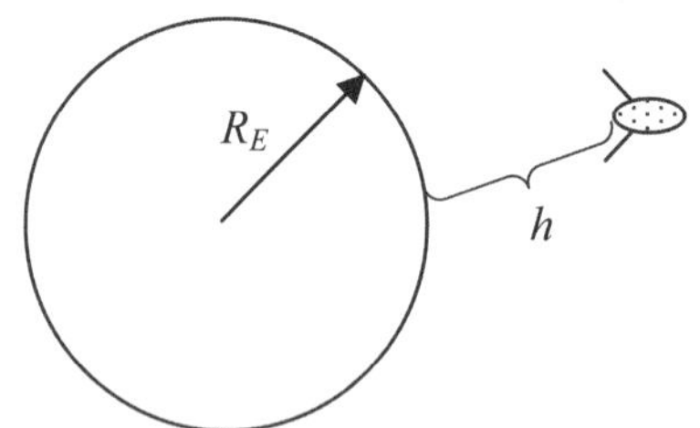


**Strategy** As mentioned above, Kepler's third law applies to satellites orbiting Earth as well, so we can use this law here by replacing $M_S$ with $M_E$.

**Solution**

1. Use the equation for Kepler's third law to solve for $r$:

$$T \left( \frac{2\pi}{\sqrt{GM_E}} \right)^{-1} = r^{3/2} \Rightarrow r = \left( \frac{2\pi / T}{\sqrt{GM_E}} \right)^{-2/3}$$

2. Write the altitude, $h$, in terms of $r$: $h = r - R_E$

3. Determine the period in seconds: $T = 2\,\text{d}\left(\dfrac{24\text{ h}}{\text{d}}\right)\left(\dfrac{3600\text{ s}}{\text{h}}\right) = 172{,}800\text{ s}$

4. Use the preceding results to determine the altitude:

$$h = \left(\frac{2\pi/T}{\sqrt{GM_E}}\right)^{-2/3} - R_E$$

$$= \left(\frac{2\pi/\left(1.728\times10^{5}\text{ s}\right)}{\sqrt{\left(6.67\times10^{-11}\text{ N}\cdot\text{m}^2/\text{kg}^2\right)\left(5.97\times10^{24}\text{ kg}\right)}}\right)^{-2/3} - 6.38\times10^{6}\text{ m}$$

$$= 6.07\times10^{7}\text{ m}$$

**Insight** Since satellites are launched from Earth's surface, calculating the altitude is more practical than calculating the full distance from Earth's center.

## Practice Quiz

4. Which of the following statements is true according to Kepler's laws?
   **(a)** Planets orbit the Sun at constant speed.
   **(b)** Planets move slower when they are closer to the Sun.
   **(c)** Planets move slower when they are farther away from the Sun.
   **(d)** A planet is always the same distance from the Sun.
   **(e)** None of the above.

5. A satellite orbits Earth at a distance $r$ from Earth's center. If you double the distance, by approximately what factor will the new period differ from the old one?
   **(a)** 2.8 **(b)** 1.0 **(c)** 2.0 **(d)** 0.5 **(e)** 4.0

6. A satellite orbits Earth at a distance $r$ from Earth's center. If you double the satellite's mass, by approximately what factor will the new period differ from the old one?
   **(a)** 2.8 **(b)** 1.0 **(c)** 2.0 **(d)** 0.5 **(e)** 4.0

## 12–4 – 12–5 Gravitational Potential Energy and Energy Conservation

In our previous look at gravitational potential energy in Chapter 8, we treated only the approximate case near Earth's surface where the gravitational force on objects is very nearly constant. In this chapter, we see that, in general, the force of gravity behaves differently from the way we considered it before and therefore the gravitational potential energy also behaves differently. For a system of two masses, $m_1$ and $m_2$, separated by a distance $r$, the gravitational potential energy is given by

$$U = -\frac{Gm_1m_2}{r}$$

where it is assumed that $U = 0$ at $r = \infty$ is the chosen reference.

The principle of superposition applies to the gravitational potential energy as a direct consequence of its application to the gravitational force; however, because potential energy is a scalar quantity, we can use simple algebra rather than dealing with vectors:

*The total gravitational potential energy of a system of objects is the algebraic sum of the gravitational potential energies of each pair of objects.*

---

**Example 12–4 A System of Two Masses** In a binary star system, two stars have elliptical orbits about the center of mass of the system. Consider a binary star system consisting of stars of masses $2.3 \times 10^{30}$ kg and $3.1 \times 10^{30}$ kg. At one point in their orbits they are $4.4 \times 10^{16}$ m apart, then they move closer together to become $1.8 \times 10^{15}$ m apart. What is the change in the gravitational potential energy of the system?

**Picture the Problem** The diagram shows two stars separated by a center-to-center distance, $r$.

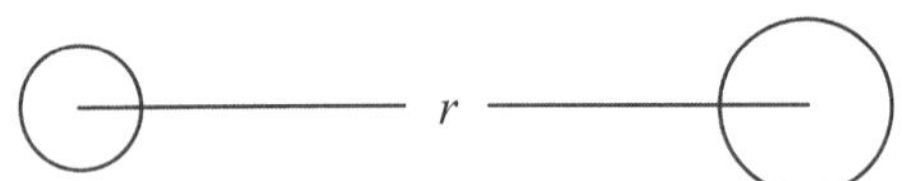


**Strategy** We need to determine the potential energy for each separation and find the difference.

**Solution**

1. The potential energy for the initial separation is:

$$U_i = -\frac{GM_1M_2}{r_i}$$

$$= -\frac{\left(6.67\times10^{-11}\ \frac{\text{N}\cdot\text{m}^2}{\text{kg}^2}\right)\left(2.3\times10^{30}\ \text{kg}\right)\left(3.1\times10^{30}\ \text{kg}\right)}{4.4\times10^{16}\ \text{m}}$$

$$= -1.08\times10^{34}\ \text{J}$$

2. The potential energy for the final separation is:

$$U_f = -\frac{GM_1M_2}{r_f}$$

$$= -\frac{\left(6.67\times10^{-11}\ \frac{\text{N}\cdot\text{m}^2}{\text{kg}^2}\right)\left(2.3\times10^{30}\ \text{kg}\right)\left(3.1\times10^{30}\ \text{kg}\right)}{1.8\times10^{15}\ \text{m}}$$

$$= -2.64\times10^{35}\ \text{J}$$

3. The change in potential energy is:

$$\Delta U = U_f - U_i = -2.64\times10^{35}\,\text{J} + 1.08\times10^{34}\,\text{J}$$
$$= -2.5\times10^{35}\,\text{J}$$

**Insight** As the stars get closer together, the *magnitude* of the potential energy gets larger, but because it is negative there is a decrease in the potential energy ($\Delta U < 0$) rather than an increase.

---

That we can define a gravitational potential energy from Newton's universal law of gravitation means that we can also apply the conservation of mechanical energy to objects that interact only through gravity. If, as is the case with most orbits that we will consider, we can treat one large object $M$ as stationary while a smaller object, $m$, moves, then we can write the mechanical energy, $E$, as

$$E = K + U = \tfrac{1}{2}mv^2 - \frac{GmM}{r}$$

Notice that the kinetic energy is positive, as always, whereas the potential energy is negative. This allows for the case in which the two will cancel out numerically, producing a mechanical energy of zero. This case corresponds to the situation in which the motion of the smaller mass, $m$, has just barely enough kinetic energy to get infinitely far away from the larger mass, $M$, where it will come to rest. In such a case, the speed of $m$ is said to equal the **escape speed**, $v_e$, from $M$. The escape speed can be determined by setting $E = 0$ and solving for $v$; the result of this calculation is

$$v_e = \sqrt{\frac{2GM}{r}}$$

If $m$ is moving slower than $v_e$, it will eventually come to rest momentarily, then fall back toward $M$, speeding up as it gets closer. If $m$ is moving at a speed greater than $v_e$, it will continue moving away from $M$ forever. In this latter case, $m$ will reach infinitely far away from $M$ and still have speed left over.

---

**Example 12–5 A Space Probe Boost** Space probes are often set in orbit around Earth before receiving a *boost* into space. If we have a probe in a circular orbit 3550 km above Earth's surface, how much of a kick in speed (a "$\Delta v$") does it need to achieve escape speed at that position? (Assume that the boost is in the direction of orbital motion.)

**Picture the Problem** The diagram on the right shows a probe in orbit about Earth at height $h$ above the surface.

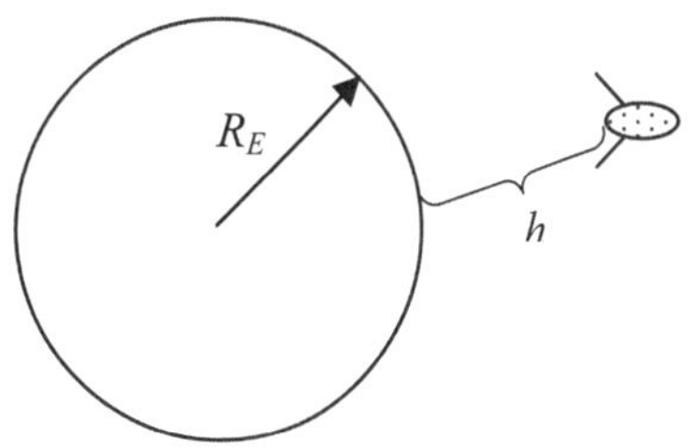


**Strategy** We need to determine the speed that

the probe has in Earth orbit and the escape speed at the position of the probe.

**Solution**

1. In a circular orbit, gravity provides the centripetal force: $\frac{GM_E m}{r^2} = m\frac{v^2}{r}$

2. Solve for the orbital speed of the probe: $v = \sqrt{\frac{GM_E}{r}}$

3. The boost is the difference between the escape speed and the orbital speed: $\Delta v = v_e - v = \sqrt{\frac{2GM_E}{r}} - \sqrt{\frac{GM_E}{r}} = \sqrt{\frac{GM_E}{r}}\left(\sqrt{2}-1\right)$

4. Evaluate the numerical result:

$$\Delta v = \sqrt{\frac{GM_E}{r}}\left(\sqrt{2}-1\right)$$
$$= \sqrt{\frac{\left(6.67\times10^{-11}\ \text{N}\cdot\text{m}^2/\text{kg}^2\right)\left(5.97\times10^{24}\ \text{kg}\right)}{6.38\times10^6\text{m} + 3.55\times10^6\text{m}}}\left(\sqrt{2}-1\right)$$
$$= 2.62\times10^3\ \text{m/s}$$

**Insight** You can tell by examining this problem that, for circular orbits, the $\Delta v$ needed to reach escape speed takes the general form determined here in all cases.

## Practice Quiz

7. Two objects are separated by a distance $r$. If the mass of one object is reduced by half, how does the gravitational potential energy of the system change?
   **(a)** It increases, with $U_f$ having half the magnitude of $U_i$.
   **(b)** It decreases, with $U_f$ having half the magnitude of $U_i$.
   **(c)** It increases, with $U_f$ having twice the magnitude of $U_i$.
   **(d)** It decreases, with $U_f$ having twice the magnitude of $U_i$.
   **(e)** None of the above.

8. Two objects are separated by a distance $r$. If the distance between them is tripled, how does the gravitational potential energy of the system change?
   **(a)** It increases, with $U_f$ having three times the magnitude of $U_i$.
   **(b)** It decreases, with $U_f$ having three times the magnitude of $U_i$.
   **(c)** It increases, with $U_f$ having one-third the magnitude of $U_i$.

**(d)** It decreases, with $U_f$ having one-third the magnitude of $U_i$.

**(e)** It decreases, with $U_f$ having one-ninth the magnitude of $U_i$.

9. Assume that planet X has twice the mass and twice the radius of Earth. The escape speed from the surface of planet X, $v_X$, compared with that from the surface of Earth, $v_E$, is

**(a)** $v_X = 2v_E$ **(b)** $v_X = \sqrt{2}v_E$ **(c)** $v_X = \frac{1}{2}v_E$ **(d)** $v_X = v_E/\sqrt{2}$ **(e)** $v_X = v_E$

## Reference Tools and Resources

### I. Key Terms and Phrases

**gravity** a fundamental force of nature that represents the attraction between objects with mass

**Newton's law of universal gravitation** between any two point masses there is an attractive force directly proportional to the product of the masses and inversely proportional to the square of the distance between them

**principle of superposition** (for gravity) the net result of the gravitational interaction within a system of particles is the sum of the results for interactions between each pair of particles in the system

**orbital period** the amount of time it takes an object to execute one complete orbit

**escape speed** the speed at which a moving object can just barely get infinitely far away from another object; at this speed the mechanical energy of the system is zero

### II. Important Equations

| Name/Topic | Equation | Explanation |
|---|---|---|
| gravitational force | $F = G\dfrac{m_1m_2}{r^2}$ | The magnitude of the gravitational force between two point masses |
| gravitational acceleration | $g = \dfrac{GM_E}{R_E^2}$ | The acceleration due to gravity near Earth's surface |
| circular orbits | $T = \left(\dfrac{2\pi}{\sqrt{GM_S}}\right)r^{3/2}$ | The equation form of Kepler's third law for the case of a circular orbit |

| | | |
|---|---|---|
| potential energy | $U = -\dfrac{Gm_1m_2}{r}$ | The gravitational potential energy of two point masses |
| escape speed | $v_e = \sqrt{\dfrac{2GM}{r}}$ | The escape speed of an object a distance $r$ from an object of mass $M$ |

### III. Know Your Units

| Quantity | Dimension | SI Unit |
|---|---|---|
| universal gravitational constant ( $G$ ) | $\dfrac{[\mathrm{L}^3]}{[\mathrm{M}]\cdot[\mathrm{T}^2]}$ | $\mathrm{N \cdot m^2/kg^2}$ |

## Puzzle

**THE ORBIT PARADOX**

Consider a satellite in a circular orbit about Earth. If NASA scientists want to move the satellite into a higher (circular) orbit, they have to increase the satellite speed, and yet when the satellite is in the new (higher) orbit, its speed is actually slower than it was in the old (lower) orbit. Is this correct? If the answer is yes, can you explain why the satellite slows down? Answer this question in words, not equations, briefly explaining how you obtained your answer.

## Answers to Selected Conceptual Questions

**4.** As the tips of the fingers approach one another, we can think of them as like two small spheres (or we can replace the fingertips with two small marbles if we like). As we know, the net gravitational attraction outside a sphere of mass is the same as that of an equivalent point mass at its center. Therefore, the two fingers simply experience the finite force of two point masses separated by a finite distance.

**8.** Yes. The rotational motion of the Earth is to the east, and therefore if you launch in that direction you are adding the speed of the Earth's rotation to the speed of your rocket.

**12.** Clearly, the Moon orbits both the Sun and the Earth. However, from a larger perspective, it makes more sense to think of the Moon as orbiting the Sun (with everything else in the solar system), with the Earth providing a smaller force that makes the Moon "wobble" back and forth in its solar orbit.

## Solutions to Selected End-of-Chapter Problems and Conceptual Exercises

12. **Picture the Problem**: The 2.0-kg mass is gravitationally attracted to the other three masses.

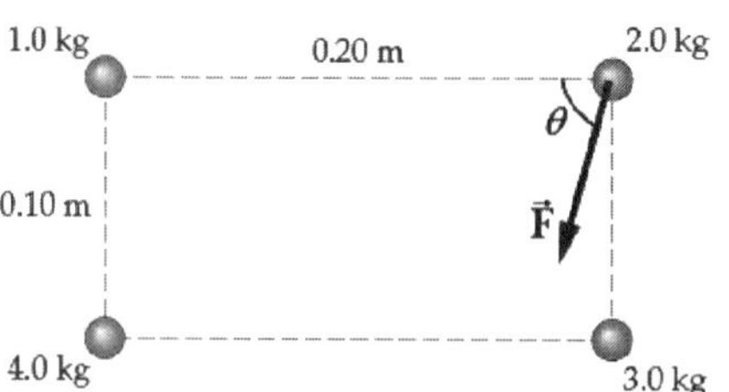


**Strategy:** Add the gravitational forces using the component method of vector addition. Use equation 12-1 and the geometry of the problem to determine the magnitudes of the forces. Let $m_1 = 1.0$ kg, $m_2 = 2.0$ kg, $m_3 = 3.0$ kg, and $m_4 = 4.0$ kg.

**Solution: 1. (a)** Find the $x$ component of the force on $m_2$:

$$F_x = G\frac{m_1 m_2}{r_{12}^2} + G\frac{m_2 m_4}{r_{24}^2}\cos\theta$$

$$= G\frac{m_1 m_2}{r_{12}^2} + G\frac{m_2 m_4}{r_{24}^2}\left(\frac{r_{12}}{r_{24}}\right) = Gm_2\left(\frac{m_1}{r_{12}^2} + \frac{m_4 r_{12}}{r_{24}^3}\right)$$

$$= \left(6.67\times10^{-11}\ \text{N}\cdot\text{m}^2/\text{kg}^2\right)(2.0\ \text{kg})\left\{\frac{1.0\ \text{kg}}{(0.20\ \text{m})^2} + \frac{(4.0\ \text{kg})(0.20\ \text{m})}{\left[(0.20\ \text{m})^2 + (0.10\ \text{m})^2\right]^{3/2}}\right\}$$

$$F_x = \underline{\underline{1.3\times10^{-8}\ \text{N}}}$$

**2.** Find the $y$ component of the force on $m_2$:

$$F_y = G\frac{m_2 m_3}{r_{23}^2} + G\frac{m_2 m_4}{r_{24}^2}\sin\theta$$

$$= G\frac{m_2 m_3}{r_{23}^2} + G\frac{m_2 m_4}{r_{24}^2}\left(\frac{r_{14}}{r_{24}}\right) = Gm_2\left(\frac{m_3}{r_{23}^2} + \frac{m_4 r_{14}}{r_{24}^3}\right)$$

$$= \left(6.67\times10^{-11}\ \text{N}\cdot\text{m}^2/\text{kg}^2\right)(2.0\ \text{kg})\left\{\frac{3.0\ \text{kg}}{(0.10\ \text{m})^2} + \frac{(4.0\ \text{kg})(0.10\ \text{m})}{\left[(0.20\ \text{m})^2 + (0.10\ \text{m})^2\right]^{3/2}}\right\}$$

$$F_y = \underline{\underline{4.5\times10^{-8}\ \text{N}}}$$

**3.** Use the components of $\vec{\mathbf{F}}$ to find its magnitude and direction.

$$F = \sqrt{F_x^2 + F_y^2} = \sqrt{\left(1.3\times10^{-8}\ \text{N}\right)^2 + \left(4.5\times10^{-8}\ \text{N}\right)^2} = \boxed{4.7\times10^{-8}\ \text{N}}$$

$$\theta = \tan^{-1}\left(\frac{F_y}{F_x}\right) = \tan^{-1}\left(\frac{4.5\times10^{-8}\ \text{N}}{1.3\times10^{-8}\ \text{N}}\right) = \boxed{74^\circ\ \text{below horizontal and to the left}}$$

**4. (b)** If the sides of the rectangle are all doubled, all forces are reduced by a factor of $2^2 = 4$; the directions of the forces are unchanged.

**Insight:** Doubling all the masses will quadruple the gravitational force, because the gravitational force depends upon the product of the masses of each object. Therefore, the force would stay exactly the same if we doubled all the masses and doubled the length of the sides of the rectangle.

36. **Picture the Problem**: The satellites travel in circular orbits around the Earth.

**Strategy:** Use equation 12-7 to find the periods of the satellites. Use the mass and radius of the Earth given in the inside back cover of the book. When the altitude $h = R_E$, the orbit radius $r = R_E + h = 2R_E$.

**Solution: 1. (a)** Apply equation 12-7: $T = \left(\dfrac{2\pi}{\sqrt{GM_E}}\right)(2R_E)^{3/2}$

$$= \left[\frac{2\pi}{\sqrt{(6.67\times10^{-11}\ \text{N}\cdot\text{m}^2/\text{kg}^2)(5.97\times10^{24}\ \text{kg})}}\right](2\times6.37\times10^6\ \text{m})^{3/2}$$

$$T = 14{,}300\ \text{s} = \boxed{3.98\ \text{h}}$$

**2. (b)** Repeat for $r = 3R_E$ $T = \left(\dfrac{2\pi}{\sqrt{GM_E}}\right)(3R_E)^{3/2}$

$$= \left[\frac{2\pi}{\sqrt{(6.67\times10^{-11}\ \text{N}\cdot\text{m}^2/\text{kg}^2)(5.97\times10^{24}\ \text{kg})}}\right](3\times6.37\times10^6\ \text{m})^{3/2}$$

$$T = 26{,}300\ \text{s} = \boxed{7.31\ \text{h}}$$

**3. (c)** The periods do not depend on the mass of the satellite because the satellite mass cancels out of the equation as shown on page 390.

**4. (d)** The periods depend inversely on the square root of the mass of the Earth.

**Insight:** The fact that the orbit period of an object is independent of its mass, as discussed in part (c), is ultimately due to the equivalence of gravitational and inertial mass. This equivalence lays the foundation for Einstein's General Theory of Relativity (see section 29-8).

51. **Picture the Problem**: The projectile rises from one Earth radius at launch to two Earth radii at the highest point of its travel.

**Strategy:** Use conservation of energy to determine the initial kinetic energy required to change the distance of the projectile from $R_E$ to $2R_E$ from the center of the Earth. Then use equation 7-6 to find the initial speed of the projectile.

**Solution: 1.** Set $E_i = E_f$ and solve for $K_i = \frac{1}{2}mv_i^2$:

$$K_i + U_i = K_f + U_f$$

$$\tfrac{1}{2}mv_i^2 - G\frac{mM_E}{R_E} = 0 - G\frac{mM_E}{2R_E}$$

$$\tfrac{1}{2}mv_i^2 = \frac{GmM_E}{R_E}\left(1 - \frac{1}{2}\right)$$

**2.** Now multiply by $2/m$ and solve for $v_i$:

$$v_i = \sqrt{\frac{GM_E}{R_E}} = \sqrt{\frac{(6.67\times10^{-11}\ \text{N}\cdot\text{m}^2/\text{kg}^2)(5.97\times10^{24}\ \text{kg})}{6.37\times10^6\ \text{m}}}$$

$$= 7910\ \text{m/s} = \boxed{7.91\ \text{km/s}}$$

**Insight:** It is not practical to achieve a launch speed this large (Mach 23!). Instead, spacecraft are accelerated gradually and over a large distance by the impulse from a rocket motor.

63. **Picture the Problem**: The dumbbell is aligned radially with the center of the Earth, and its center is a distance $r$ from the center of the Earth.

**Strategy:** The gravitational force of attraction is a little bit stronger of the end of the dumbbell that is closest to the Earth and a little weaker on the farthest end. This difference in force tends to stretch the dumbbell along its length. Use Newton's Universal Law of Gravitation (equation 12-1) to find the difference in force between the two ends of the dumbbell.

**Solution: 1.** Use equation 12-1 to find $\Delta F$: $\Delta F = G\dfrac{mM_{\text{E}}}{(r-a)^2} - G\dfrac{mM_{\text{E}}}{(r+a)^2} = GmM_{\text{E}}\left[\dfrac{1}{(r-a)^2} - \dfrac{1}{(r+a)^2}\right]$

**2.** Now if $r >> a$, $\dfrac{1}{(r-a)^2} - \dfrac{1}{(r+a)^2} \approx \dfrac{4a}{r^3}$: $\Delta F = F_{\text{tidal}} = GmM_{\text{E}}\left(\dfrac{4a}{r^3}\right) = \boxed{\dfrac{4GmM_{\text{E}}a}{r^3}}$

**Insight:** The tidal force decreases with $1/r^3$, faster than the magnitude of the gravitational force decreases ($1/r^2$). It is significant only when the masses are large (such as with the Moon and the Earth) or the separation distance is small.

83. **Picture the Problem**: The orbit period of a geosynchronous satellite matches the rotational period of the planet.

**Strategy:** This problem is similar to Active Example 12-1. Set the orbit period equal to the rotational period of the planet and solve for distance $r$ and then the altitude $h$. The rotational period of Mars is $24.6229 \text{ h} \times 3600 \text{ s/h} = 8.86424 \times 10^4 \text{ s}$. Use the mass and radius data given in Appendix C.

**Solution: 1.** Square both sides of equation 12-7 and solve for $r$:

$$T^2 = \frac{4\pi^2 r^3}{GM}$$

$$r = \left(\frac{GMT^2}{4\pi^2}\right)^{1/3}$$

$$= \left[\frac{(6.67\times10^{-11} \text{ N}\cdot\text{m}^2/\text{kg}^2)(0.108\times5.97\times10^{24} \text{ kg})(8.86424\times10^4 \text{ s})^2}{4\pi^2}\right]$$

$$r = 2.05\times10^7 \text{ m}$$

**2.** Solve $r = R + h$ for $h$: $h = r - R = 2.05\times10^7 \text{ m} - 0.3394\times10^7 \text{ m} = \boxed{1.71\times10^7 \text{ m}}$

**Insight:** This distance is less than half of the geosynchronous orbit radius of $4.22\times10^7$ m for the Earth because, although the rotation periods of the two planets are nearly equal, Mars has much less mass than does the Earth.

89. **Picture the Problem**: The space shuttle travels in a circular orbit about the Earth at an altitude of 250 km.

**Strategy:** Use the formula derived in problem 79 to determine the orbital speed of the shuttle. Then solve the relation $v = 2\pi r/T$ for $T$ in order to find the orbit period of the space shuttle.

**Solution: 1. (a)** $\boxed{\text{No}}$, the orbit speed is independent of mass, as seen in problem 79 and elsewhere.

**2. (b)** Use the equation derived in problem 79 to find the orbit speed:

$$v = \sqrt{\frac{GM_{\text{E}}}{R_{\text{E}} + h}} = \sqrt{\frac{(6.67\times10^{-11} \text{ N}\cdot\text{m}^2/\text{kg}^2)(5.97\times10^{24} \text{ kg})}{6.37\times10^6 + 250\times10^3 \text{ m}}}$$

$$= 7760 \text{ m/s} = \boxed{7.76 \text{ km/s}}$$

**3. (c)** Solve $v = 2\pi r/T$ for $T$:

$$T = \frac{2\pi r}{v} = \frac{2\pi\left(6.37\times10^{6} + 250\times10^{3}\ \text{m}\right)}{7.76\times10^{3}\ \text{m/s}} = 5360\ \text{s} = \boxed{1.49\ \text{h}}$$

**Insight:** If the shuttle were to be boosted to a higher orbit, its speed would be smaller and its orbit period greater.

## Answers to Practice Quiz

**1**. (c) **2**. (d) **3**. (d) **4**. (c) **5**. (a) **6**. (b) **7**. (a) **8**. (c) **9**. (e)

# CHAPTER 13
# OSCILLATIONS ABOUT EQUILIBRIUM

## Chapter Objectives

After studying this chapter, you should

1. know why periodic motion usually occurs.
2. know the characteristic quantities that describe periodic motion.
3. be able to analyze simple harmonic motion and understand its connection to uniform circular motion.
4. be able to apply the conservation of energy to systems undergoing simple harmonic motion.
5. understand the basic principles of both simple and physical pendulums.
6. gain good familiarity with the properties of damped and driven oscillations.

## Warm-Ups

1. Consider a mass on a spring, oscillating under the influence of a nonconservative retarding force, such as air drag. How will the retarding force affect the period of the oscillations? In a sentence or two, justify your answer.
2. When you push a child on a swing, your action is most effective when your pushes are timed to coincide with the natural frequency of the motion. You are swinging a 30-kg child on a swing suspended from 5-meter cables. Estimate the optimum time interval between your pushes. Repeat the estimate for a 15-kg child.
3. Suppose you are told that the period of a simple pendulum depends only on the length of the pendulum $L$ and the acceleration due to gravity $g$. Use dimensional analysis to show that the square of the period is proportional to the ratio $L/g$.
4. You want to construct a poor man's amusement ride by mounting a seat on a large spring. Estimate the spring constant that will give you a ride with a period of 10 seconds.

## Chapter Review

In this chapter, we study what happens when objects in equilibrium (Chapter 11) are disturbed slightly away from this equilibrium. The most common result of such disturbances is that objects oscillate about (around) the equilibrium position (or configuration) from which it was disturbed. Here we study this oscillatory motion.

## 13–1 – 13–2 Periodic and Simple Harmonic Motion

An object's motion is referred to as **periodic motion** when it repeats itself over and over again. When the motion comes exactly back to its original state (of position, velocity, and acceleration), we say that the object has gone through a **complete cycle** or **complete oscillation.** The amount of time it takes for one complete cycle to occur is called the **period** of the motion, $T$.

The period is one way to characterize the speed of the periodic motion; another way is through the **frequency,** $f$. Frequency is a measure of how frequently the motion repeats; it is most commonly quoted as the number of oscillations (or *cycles*) per second. The frequency relates directly to the period by

$$f = \frac{1}{T}$$

Since period measures an amount of time, its SI unit is the second; therefore, the SI unit of frequency must be the inverse second, $s^{-1}$. When dealing with frequencies only, the special name hertz (Hz) is given to $s^{-1}$. One hertz refers to one oscillation cycle per second.

The above discussion applies to any type of periodic motion. One type that is very important in physics is called **simple harmonic motion** (SHM). This type of periodic motion occurs as a result of a Hooke's law force. Hence, SHM is the motion experienced by an object on the end of a spring and, to a close approximation, by media that are disturbed slightly away from equilibrium.

The position of an object undergoing SHM changes sinusoidally with time. This fact means that it can be described using a sine or cosine function (we'll use cosine)

$$x = A\cos\left(\frac{2\pi}{T}t\right)$$

In the above equation, $T$ is the period and $A$ is the **amplitude** of the motion. The amplitude represents the maximum distance the object gets from equilibrium as it moves back and forth. The above expression assumes that the object starts its motion at $x = A$. The argument of the cosine function is an angle in radians; therefore, your calculator should be in radian mode when using the above expression.

---

**Exercise 13–1 Simple Harmonic Motion** An object undergoes simple harmonic motion with a frequency of 3.50 Hz and an amplitude of 0.850 m. If its equilibrium position is $x = 0$, where is it **(a)** when the motion starts and **(b)** 1.055 seconds later? **(c)** Write the complete equation of the motion.

**Solution** We are given the following information.

**Given**: $f = 3.50$ Hz, $A = 0.850$ m; **Find**: (a) $x_i$ (b) $x_f$ (c) Write the equation.

**(a)** The start of the motion is when we begin timing, so right at the beginning, $t = 0$. Because the problem does not state the initial conditions of the motion, we use the expression for position versus time in SHM:

$$x = A\cos(0) = A = 0.850 \text{ m}.$$

**(b)** To get the position at $t = 1.055$ s, again we use the general expression for SHM together with the relationship between frequency and period:

$$x = A\cos\left(\frac{2\pi}{T}t\right) = A\cos(2\pi ft) = (0.850 \text{ m})\cos\left(2\pi[3.50 \text{ s}^{-1}][1.055 \text{ s}]\right) = -0.300 \text{ m}$$

**(c)** Here, we only need to insert all the numerical values into the expression for $x$. First, let's find the period:

$$T = \frac{1}{f} = \frac{1}{3.50 \text{ s}^{-1}} = 0.286 \text{ s}$$

Now that we have the period, we can write out the full formula for the motion:

$$x = A\cos\left(\frac{2\pi}{T}t\right) = (0.850 \text{ m})\cos\left(\frac{2\pi}{0.286 \text{ s}}t\right)$$

In part (c), we also could have used the given value for the frequency instead of first finding the period.

## Practice Quiz

1. *Most generally*, simple harmonic motion occurs
   **(a)** when a mass is attached to the end of a spring
   **(b)** when an object moves back and forth
   **(c)** when the motion of an object repeats itself
   **(d)** when an object is acted upon by a Hooke's law force
   **(e)** None of the above

2. If the period of an oscillating body is 2.3 seconds, what is its frequency?
   **(a)** 2.3 Hz **(b)** 0.43 Hz **(c)** 2.3 s **(d)** 0.43 s **(e)** 2.7 Hz

3. An object oscillates with SHM according to the equation $x = (0.15 \text{ m})\cos[(3.7 \text{ rad/s})t]$. What is the amplitude of this motion?
   **(a)** 0.15 m **(b)** 3.7 m **(c)** 1.7 m **(d)** 0.94 m **(e)** 0.59 m

**4.** An object oscillates with SHM according to the equation $x = (0.15 \text{ m})\cos[(3.7 \text{ rad/s})t]$. What is the period of this motion?

**(a)** 0.15 s **(b)** 3.7 s **(c)** 1.7 s **(d)** 0.94 s **(e)** 0.59 s

## 13–3 – 13–4 Connections Between Uniform Circular Motion and Simple Harmonic Motion, and The Period of a Mass on a Spring

There is a basic connection between simple harmonic motion and uniform circular motion that proves to be very useful in understanding SHM. The connection between the two types of motion is the following:

*For an object undergoing uniform circular motion, the projection of its motion onto any diameter of its path executes simple harmonic motion.*

To see this connection more directly, consider the figure below, which shows the circular path of an object of mass, *m,* undergoing uniform circular motion. The radius of the path is *r*, and its center is at the origin of an *x*-*y* coordinate system.

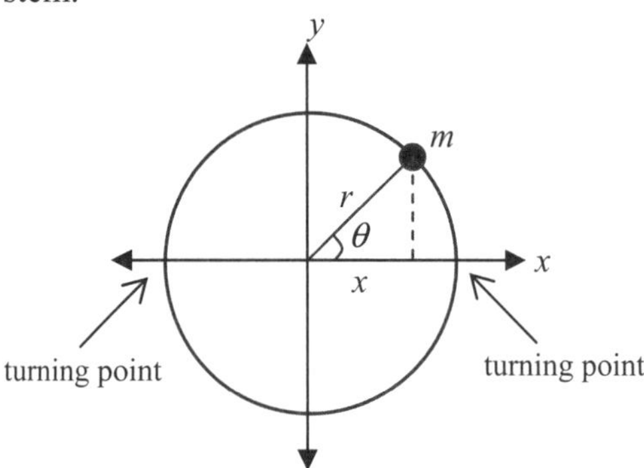


As the object moves around in a circular path, its projection onto the *x*-axis oscillates back and forth along the central diameter of the path between the two turning points. From the geometry shown in the diagram, we can see that the position of the projection is given by $x = r\cos\theta$. In uniform circular motion, the angular position is given by $\theta = \omega t$. Also, the maximum displacement of the projection from the central position (the amplitude, *A*) equals the radius. Therefore, we see that the position can be written as $x = A\cos(\omega t)$. For constant angular velocity (uniform circular motion) the angular velocity is $\omega = 2\pi/T$, where *T* is the period of rotation; therefore, we can finally write

$$x = A\cos\left(\frac{2\pi}{T}t\right),$$

which is exactly the same expression that describes simple harmonic motion.

Because of the connection between uniform circular motion and SHM, *any* simple harmonic motion can be viewed as a projection of uniform circular motion. In such cases, the path is called the *reference circle* of the motion, and $\omega$ is called the **angular frequency**. Angular frequency relates to frequency, $f$, by

$$\omega = 2\pi f$$

Following a similar approach as above, we can also use the connection between uniform circular motion and SHM to determine other relationships valid for SHM. The velocity oscillation in SHM can be determined to be

$$v = -A\omega \sin(\omega t)$$

This expression also establishes that the maximum speed in SHM, which occurs at equilibrium, is

$$v_{\text{max}} = A\omega$$

Similarly, we find that the acceleration oscillation in SHM is given by

$$a = -A\omega^2 \cos(\omega t)$$

and the maximum acceleration, which occurs at the turning points, is given by

$$a_{\text{max}} = A\omega^2$$

---

**Example 13–2 Kinematics of SHM** An object, starting from its maximum displacement at rest, undergoes simple harmonic oscillations of amplitude 5.3 cm and frequency 1.7 Hz. After 0.82 sec have passed, determine its **(a)** position, **(b)** velocity, and **(c)** acceleration.

**Picture the Problem** The diagram shows the line along which the object oscillates with the object out at its initial position.

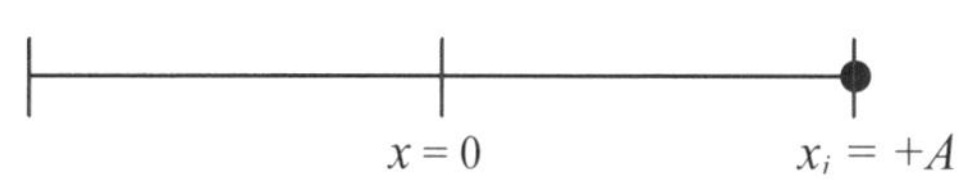


**Strategy** To solve this problem, we make use of the preceding expressions for describing SHM.

**Solution**

**Part (a)**

1. Determine the angular frequency: $\omega = 2\pi f = 2\pi(1.7\text{ Hz}) = 10.68\text{ rad/s}$

2. Find the position of the object: $x = A\cos(\omega t) = (5.3\text{ cm})\cos\left[\left(10.6814\,\tfrac{\text{rad}}{\text{s}}\right)(0.82\text{ s})\right] = -4.2\text{ cm}$

**Part (b)**

3. Use the known quantities to determine the velocity of the object: $v = -A\omega\sin(\omega t) = -(5.3\text{ cm})\left(10.6814\,\tfrac{\text{rad}}{\text{s}}\right)\sin\left[\left(10.6814\,\tfrac{\text{rad}}{\text{s}}\right)(0.82\text{s})\right] = -35\text{ cm/s}$

**Part (c)**

4. Use the known quantities to determine the acceleration of the object:

$$a = -A\omega^2 \cos(\omega t)$$
$$= -(5.3\text{ cm})(10.6814\,\tfrac{\text{rad}}{\text{s}})^2 \cos[(10.6814\,\tfrac{\text{rad}}{\text{s}})(0.82\text{s})]$$
$$= 480\text{ cm/s}^2$$

**Insight** Take note of what the signs of these results mean. From part (a), we know that the object is 4.2 cm to the left of the equilibrium position. Part (b) tells us that it is moving in the negative direction (toward the turning point at $x = -A$). Part (c) tells us that it is slowing down, because the acceleration is positive (toward the central position).

---

Recall that Hooke's law is the force law that leads to simple harmonic motion

$$F = ma = -kx$$

Using the results for $a$ and $x$ in the preceding expression, we obtain the relation

$$\omega = \sqrt{\frac{k}{m}}$$

This gives us a direct relationship between the physical system ($k$ and $m$) and the motion ($\omega$). Since $\omega = 2\pi f = 2\pi / T$, we can also write the above expression in terms of the period:

$$T = 2\pi\sqrt{\frac{m}{k}}$$

Notice that $T$ does not depend on the amplitude of the oscillation.

---

**Example 13–3 Oscillations on a Spring** Consider an industrial spring of force constant 225 N/cm that supports an object of mass 5.03 kg. If the system is disturbed and begins to oscillate, find the period, frequency, and angular frequency of the oscillation.

**Picture the Problem** The sketch shows the spring supporting the object.

**Strategy** We need only to use the relationships between the given quantities.

**Solution**

1. Convert the force constant:

$$k = 225\text{ N/cm}\left(\frac{100\text{ cm}}{\text{m}}\right) = 22500\text{ N/m}$$

**2.** Calculate the period of the motion: $T = 2\pi\sqrt{\frac{m}{k}} = 2\pi\sqrt{\frac{5.03 \text{ kg}}{22500 \text{ N/m}}} = 0.0939 \text{ s}$

**3.** Calculate the frequency of the motion: $f = \frac{1}{T} = \frac{1}{0.09394 \text{ s}} = 10.6 \text{ Hz}$

**4.** Calculate the angular frequency of the motion: $\omega = 2\pi f = 2\pi(10.64 \text{ Hz}) = 66.9 \text{ Hz}$

**Insight** In the calculations for $f$ and $\omega$, we used an extra figure in the previously obtained results.

---

For the case of a mass on a vertical spring, as in the above example, the force on the mass comes from both the spring and from gravity. The motion, however, is still simple harmonic. The effect of gravity is essentially to shift the equilibrium position of the oscillation from the unstretched length of the spring alone to the new equilibrium length of the mass-spring system given by $y_0 = mg/k$. The mass then oscillates up and down about this new equilibrium position.

### Practice Quiz

5. If the frequency, $f$, increases by a factor of 2, by what factor does the angular frequency increase?

   **(a)** 1/2 **(b)** 2π **(c)** 4 **(d)** 1/2π **(e)** 2

6. When the position of an object undergoing SHM is at its maximum displacement from equilibrium, the speed of the object is

   **(a)** at its maximum.
   **(b)** zero.
   **(c)** half its maximum value.
   **(d)** twice its maximum value.
   **(e)** None of the above.

7. If a mass of 2.6 kg oscillates with a period of 1.73 seconds on the end of a vertical spring, what is the force constant of the spring?

   **(a)** 4.5 N/m **(b)** 34 N/m **(c)** 7.7 N/m **(d)** 0.029 N/m **(e)** 1.5 N/m

## 13–5 Energy Conservation in Oscillatory Motion

Because a force obeying Hooke's law is conservative, the mechanical energy in SHM is conserved

$$E = K + U = \tfrac{1}{2}mv^2 + \tfrac{1}{2}kx^2 = \text{constant}$$

The energy oscillates between kinetic and potential. The energy is all in the form of potential energy when the mass is at its amplitude ($x = A$), where $v = 0$. Therefore, the mechanical energy is given by

$$E = \tfrac{1}{2}kA^2$$

Similarly, the energy is all kinetic energy when the mass is at the central position ($x = 0$), where $v = v_{max}$; therefore, we can also write the mechanical energy as $E = \tfrac{1}{2}mv_{max}^2$. Since $v_{max} = A\omega$, we have

$$E = \tfrac{1}{2}mA^2\omega^2$$

---

**Example 13–4 Using the Energy** A spring of force constant 142 N/m is hung vertically. An object of mass 1.25 kg is attached to the end of this spring. The mass is then pulled down an additional 8.37 cm from the equilibrium position and released. Use energy conservation to determine the object's speed when it is halfway between its central position and its amplitude.

**Picture the Problem** The sketch shows the mass on the end of a vertical spring.

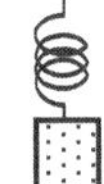

**Strategy** We must identify the most appropriate expressions for the mechanical energy of the system based on the given information.

**Solution**

1. Since the distance the mass is pulled down will be the amplitude $A$, and we know $k$, one useful expression of the energy is: $E = \tfrac{1}{2}kA^2$

2. The energy expression involving velocity at an arbitrary position is: $E = \tfrac{1}{2}mv^2 + \tfrac{1}{2}kx^2$

3. Use $x = A/2$, and equate the two expressions: $\tfrac{1}{2}kA^2 = \tfrac{1}{2}mv^2 + \tfrac{1}{2}k\left(\dfrac{A}{2}\right)^2 = \tfrac{1}{2}mv^2 + \tfrac{1}{8}kA^2$

4. Solve this equation for velocity: $\tfrac{1}{2}mv^2 = \tfrac{1}{2}kA^2 - \tfrac{1}{8}kA^2 = \tfrac{3}{8}kA^2 \quad \therefore$

$$v = \sqrt{\frac{3k}{4m}}A = \sqrt{\frac{3(142\text{ N/m})}{4(1.25\text{ kg})}}(0.0837\text{ m}) = 0.773\text{ m/s}$$

**Insight** It was energy conservation that required the energy at $x = A$ to be the same as at $x = A/2$.

---

## Practice Quiz

**8.** When the kinetic energy of an object undergoing SHM is maximum, its potential energy is

**(a)** maximum.

**(b)** equal to the kinetic energy.

**(c)** half its maximum.

**(d)** twice the kinetic energy.

**(e)** None of the above.

**9.** An object of mass 3.3 kg oscillates with SHM according to the equation $x = (8.7\text{ cm})\cos[(1.4\text{ rad/s})t]$. What is the mechanical energy of this system?

**(a)** 29 J **(b)** 40 J **(c)** 3.6 J **(d)** 0.024 J **(e)** 0.0059 J

## 13–6 The Pendulum

A **simple pendulum** is a mass (the *bob*), *m*, suspended by a cord or rod (of negligible mass) of length, *L*. When a simple pendulum is displaced from equilibrium by small amounts, its oscillation about equilibrium is very nearly simple harmonic motion. If we take the lowest point as the reference level for zero gravitational potential energy, then the potential energy of the simple pendulum is given by

$$U = mgL\left(1 - \cos\theta\right)$$

where $\theta$ is the angle the cord makes with the vertical. By analyzing the restoring force (due to gravity) for small-angle oscillations about the vertical line, we can see that the "force constant" for a simple pendulum works out to be $k = mg/L$. We can use this result for *k* to characterize the SHM of the simple pendulum in precisely the same way we used *k* previously. Therefore, substituting *mg*/*L* for *k*, we find the period of the simple pendulum (the primary useful quantity for a simple pendulum) to be

$$T = 2\pi\sqrt{\frac{L}{g}}$$

Notice that the period does not depend on the mass of the pendulum, just its length and the local acceleration of gravity.

For a simple pendulum, we ignore the mass of the rod that holds the bob. When more precision is needed, this approximation may not be good enough. A pendulum for which the mass of the rod cannot be ignored is called a **physical pendulum.** For a physical pendulum, we must consider it to be a rotational system and take into account its moment of inertia, *I*. When we do this analysis, we find that the period of a physical pendulum is

$$T = 2\pi\sqrt{\frac{I}{mg\ell}}$$

where $\ell$ is the distance from the axis about which the pendulum swings to the center of mass of the rod-bob system.

---

**Exercise 13–5 On Top of Mount Everest** Pendulums are sometimes used to measure the local acceleration of gravity. Mount Everest in Nepal is approximately 8850 m tall at its peak. If a simple pendulum of length 1.75 m set in small oscillations at the top of Mount Everest has a measured period of 2.655 s, what is the acceleration of gravity at the top of Mount Everest?

**Solution:** We are given the following information.

**Given:** $h = 8850$ m, $L = 1.75$ m, $T = 2.655$ s; **Find:** $g$

Because we seek to determine $g$, we first solve for $g$:

$$T = 2\pi\sqrt{\frac{L}{g}} \quad \Rightarrow \quad g = \frac{4\pi^2 L}{T^2}$$

Now we get the final numerical result:

$$g = \frac{4\pi^2 (1.75\ \text{m})}{(2.655\ \text{s})^2} = 9.80\ \text{m/s}^2$$

---

## Practice Quiz

**10.** A meter-stick is fixed to a horizontal rod at one end, while the other end is allowed to swing freely. A small Styrofoam ball is glued to the free-swinging end of the stick. This system should be treated as

**(a)** a simple pendulum.

**(b)** a physical pendulum.

**(c)** None of the above.

## 13–7 – 13–8 Damped and Driven Oscillations, Resonance

In realistic oscillating systems, energy is lost during the motion. Because of this energy loss, the amplitude of the oscillation decreases. We call this type of motion **damped oscillation.** Frequently, damping forces (like air resistance) are proportional to the velocity of the object on which the force is applied, so that

$$\vec{\mathbf{F}} = -b\vec{\mathbf{v}}$$

where the coefficient $b$ is called the *damping constant*.

The situation in which a system oscillates with decreasing amplitude occurs when the damping constant is small and is referred to as **underdamped** motion. A larger damping constant can lead to the case in which the system will no longer oscillate; this is called **critically damped** motion. Critically damped motion is on the borderline between a system that oscillates and one that just moves back to its equilibrium configuration. An even larger damping constant leads to **overdamped** motion. With overdamping, the system relaxes to equilibrium more slowly than with critical damping, and there is room to make the value of $b$ smaller without causing oscillations.

The effects of damping forces can be overcome by putting energy into the system in order to maintain the oscillation. When this energy is input we say that the motion is a **driven oscillation**. With driven oscillations, it is important to recognize that systems tend to have **natural frequencies** of oscillation, that is, frequencies at which they will oscillate if no driving force is present (consider the pendulum discussed above). When a system is being driven at a natural frequency of oscillation, the result is oscillations of larger amplitude than occur when the system is driven at a different frequency. This phenomenon is called **resonance** (for this reason, natural frequencies are sometimes called *resonant frequencies*). Resonance plays an important role in many applications and in the generation and detection of sound, which will be discussed in a later chapter.

---

**Example 13–6 Swinging** A 45-lb child enjoys swinging very high. If the length of the chain that holds the swing is 2.5 m, approximately how often should the child be pushed to get a good high swing?

**Picture the Problem** The sketch shows the child on a swing.

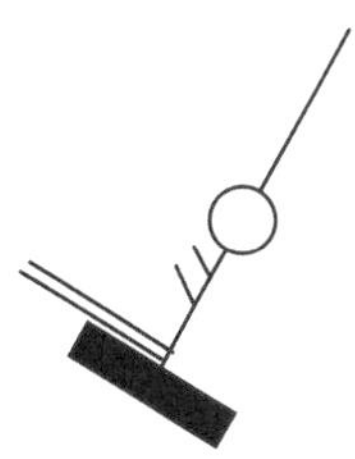

**Strategy** We seek only an approximate answer, so we will make two approximations. We will treat the child and swing as a simple pendulum with the child as the bob, and we will use the SHM results and ignore any slight inaccuracies that result from the fact that the child swings through large angles.

Since we want the child's motion to have large amplitude, we should push the child at her resonant frequency.

**Solution**

**1.** Determine the natural frequency for the child: $f = \frac{1}{T} = \frac{1}{2\pi}\sqrt{\frac{g}{L}} = \frac{1}{2\pi}\sqrt{\frac{9.81\ \text{m/s}^2}{2.5\ \text{m}}} = 0.32\ \text{Hz}$

**2.** Interpret the result: $f \approx \frac{1}{3\text{s}}$ $\therefore$ approximately 1 push every 3 sec.

**Insight** Recognize how the concepts of natural frequency and resonance fit together.

## Reference Tools and Resources

### I. Key Terms and Phrases

**periodic motion** motion that repeats itself

**complete cycle** when the position, velocity, and acceleration of an oscillatory motion repeat themselves between two successive passes

**period** (for oscillations) the amount of time for one complete cycle

**frequency** the number of cycles per unit of time

**simple harmonic motion** the oscillatory motion that results from a force that obeys Hooke's law

**amplitude** the maximum displacement from equilibrium

**angular frequency** $2\pi$ times the frequency

**simple pendulum** a mass suspended by a cord or rod of negligible mass

**physical pendulum** a mass distribution that is suspended and free to oscillate

**damped oscillation** when an oscillating system looses energy

**underdamped oscillation** a small damping constant causes oscillatory motion of decreasing amplitude

**critically damped oscillation** when the damping constant is just large enough to prevent oscillations

**overdamped oscillation** the damping constant is more than just large enough to prevent oscillations

**driven oscillation** when an external agent forces a system to oscillate

**natural frequency** a frequency at which a system will oscillate if no driving force is applied

**resonance** that large-amplitude oscillations occur when a system is driven at a natural frequency

## II. Important Equations

| Name/Topic | Equation | Explanation |
|---|---|---|
| frequency | $f = \dfrac{1}{T}$ | The frequency equals the inverse of the period |
| simple harmonic motion | $x = A\cos\left(\dfrac{2\pi}{T}t\right)$ | Position as a function of time in SHM assuming the motion begins at the maximum displacement |
| angular frequency | $\omega = 2\pi f$ | The relationship between angular frequency and frequency |
| period | $T = 2\pi\sqrt{\dfrac{m}{k}}$ | The period of a mass, *m,* oscillating on a spring with force constant *k* |
| mechanical energy | $E = \frac{1}{2}kA^2 = \frac{1}{2}mA^2\omega^2$ | The mechanical energy in SHM |
| period | $T = 2\pi\sqrt{\dfrac{L}{g}}$ | The period of a simple pendulum of length *L* |

## III. Know Your Units

| Quantity | Dimension | SI Unit |
|---|---|---|
| period ( $T$ ) | [T] | s |
| frequency ( $f$ ) | [$\mathrm{T}^{-1}$] | Hz |
| angular frequency ( $\omega$ ) | [$\mathrm{T}^{-1}$] | rad/s |
| amplitude ( $A$ ) | [L] | m |

# Puzzle

### THE BUNGEE BUCKET

Little Danny stands next to a large amusement park bungee bucket. The empty bucket is hanging on two large, identical springs. It oscillates with an amplitude of 2 meters and a period of 2 seconds. It is shown at the equilibrium position in the sketch, 2.5 meters above the floor. Danny waits until the bucket comes to a momentary stop, ready to start its journey upward. He quickly takes a seat in the bucket. Danny's mass is equal to the mass of the empty bucket. What happens next? Does the bucket take Danny for a ride?

(You can treat the two springs as a single spring with the spring constant $k$ equal to twice the $k$ of each single spring. In this estimate, ignore the mass of the springs.)

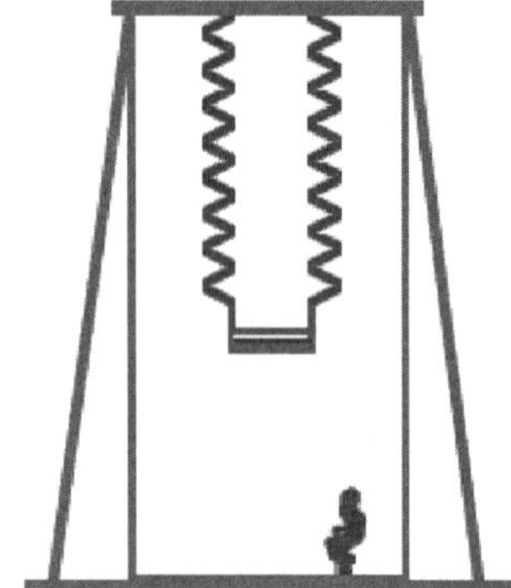

## Answers to Selected Conceptual Questions

**4.** Recall that the maximum speed of a mass on a spring is $v_{\text{max}} = \omega A$, where $\omega = \sqrt{k/m}$. It follows that the maximum kinetic energy is $K_{\text{max}} = \frac{1}{2}mv_{\text{max}}^2 = \frac{1}{2}m\left(kA^2/m\right) = \frac{1}{2}kA^2$. Note that the mass cancels in our final expression for the maximum kinetic energy. Therefore, the larger mass moves more slowly by just the right amount so that the kinetic energy is unchanged.

**8.** The period of a pendulum is independent of the mass of its bob. Therefore, the period should be unaffected.

## Solutions to Selected End-of-Chapter Problems and Conceptual Exercises

7. **Picture the Problem**: As gasoline burns inside the engine of a car, it causes the pistons to expand, which in turn causes the crankshaft to rotate. Increasing the gas in the engine (revving) causes the crankshaft to rotate faster. The frequency of the car's engine is measured as the number of times the crankshaft rotates per minute.

**Strategy:** The frequency is given in units of rev/min which can be converted to hertz. The period can then be found by inverting the frequency. Reverse the process to convert a period back into a frequency in hertz.

**Solution: 1. (a)** Convert $f$ to hertz: $f = \left(2700\ \frac{\text{rev}}{\text{min}}\right)\left(\frac{1\ \text{min}}{60\ \text{s}}\right) = \boxed{45\ \text{Hz}}$

**2.** Invert the frequency to obtain the period: $T = \frac{1}{f} = \frac{1}{45\ \text{Hz}} = \boxed{0.022\ \text{s}}$

**3. (b)** Invert the new period and convert seconds to minutes to obtain the rpm: $f = \frac{1}{T} = \frac{1\ \text{rev}}{0.044\ \text{s}} \times \frac{60\ \text{s}}{\text{min}} = \boxed{1400\ \text{rpm}}$

**Insight:** Since period and frequency are inverses of each other, a longer period resulted in a lower frequency.

15. **Picture the Problem**: When two or more atoms are bound in a molecule, they are separated by an equilibrium distance. If the atoms get too close to each other, the binding force is repulsive. When the atoms are too far apart, the binding force is attractive. The nature of the binding force, therefore, is to cause the atoms to oscillate about the equilibrium distance.

**Strategy:** Since the mass starts at $x = A$ at time $t = 0$, this is a cosine function given by $x = A\cos(\omega t)$. From the data given we need to identify the constants $A$ and $\omega$. A cosine function is at its maximum at $t = 0$, but a sine function equals zero at $t = 0$.

**Solution: 1. (a)** Identify the amplitude as $A$: $A = 3.50\text{ nm}$

**2.** Calculate the angular frequency from the frequency: $\omega = 2\pi f = 4.00\pi \times 10^{14}\text{ s}^{-1}$

**3.** Substitute the amplitude and angular frequency into the cosine equation: $x = \boxed{(3.50\text{ nm})\cos\left[\left(4.00\pi \times 10^{14}\text{ s}^{-1}\right)t\right]}$

**4. (b)** It will be a $\boxed{\text{sine function}}$, $x = A\sin(\omega t)$, because sine satisfies the initial condition of $x = 0$ at $t = 0$.

**Insight:** A cosine function has a maximum amplitude at $t = 0$. A sine function has zero amplitude at $t = 0$.

31. **Picture the Problem**: The figure shows the rider on the mechanical horse.

**Strategy: (a)** The rider will separate from the mechanical horse if $a_{max}$ at the top of the motion exceeds the acceleration of gravity, $g$. Therefore, because $a_{max}$ is related to the angular speed and the amplitude, we can set $a_{max}$ equal to $g$ and solve for the amplitude.

**Solution: 1. (b)** Write $a_{max}$ in terms of $A$ and $\omega$ and set it equal to $g$: $a_{max} = A\omega^2 = g$

**2.** Write angular speed in terms of period and solve for amplitude: $A\left(\dfrac{2\pi}{T}\right)^2 = g \Rightarrow A = \left(\dfrac{T}{2\pi}\right)^2 g$

**3.** Insert the numeric values for the period and gravity: $A = \left(\dfrac{0.74\text{ s}}{2\pi}\right)^2 (9.81\text{ m/s}^2) = \boxed{0.14\text{ m}}$

**Insight:** As long as the maximum acceleration is less than gravity, the rider will remain in contact with the horse, because an upward normal force is required to support the rider. When the maximum acceleration is greater than gravity, the normal force is no longer necessary and the rider will accelerate up from the horse unless an additional downward force is applied to the rider.

39. **Picture the Problem**: The picture shows the unstretched spring and the spring with a 0.50-kg mass attached to it.

**Strategy:** We can use the displacement of the spring to calculate the spring force constant. The spring force constant and period can then be inserted into the period equation to solve for the necessary mass.

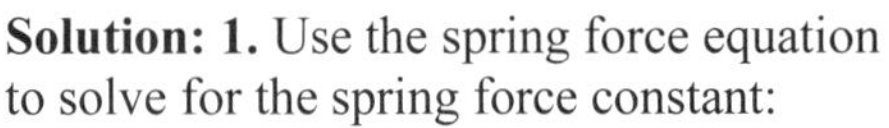

**Solution: 1.** Use the spring force equation to solve for the spring force constant: $F = ky \Rightarrow k = \dfrac{F}{y} = \dfrac{mg}{y}$

**2.** Insert the numeric values to obtain $k$: $k = \dfrac{0.50\text{ kg}(9.81\text{ m/s}^2)}{15\times 10^{-2}\text{ m}} = 32.7\text{ N/m}$

**3.** Solve the period equation for the mass: $T = 2\pi\sqrt{\dfrac{m}{k}} \Rightarrow m = \left(\dfrac{T}{2\pi}\right)^2 k$

**4.** Insert the numeric values to obtain the mass: $m = \left(\dfrac{0.75\text{ s}}{2\pi}\right)^2 (32.7\text{ N/m}) = \boxed{0.47\text{ kg}}$

**Insight:** Since the period is proportional to the square root of the mass, increasing the mass will increase the period.

55. **Picture the Problem**: A block and spring are initially at rest as a bullet is fired at high speed directly toward them. The bullet then embeds in the block and compresses the spring.

**Strategy:** The bullet and block first undergo an inelastic collision. Then they jointly compress the spring, converting their kinetic energy into potential energy of the spring. Use conservation of energy to relate the speed $v$ of the block and bullet to the compression distance $x$. Finally, use conservation of momentum to find the initial speed of the bullet $v_0$ from the combined speed of bullet and block. The time elapsed from impact to rest is one-quarter of a period.

**Solution: 1. (a)** Set the initial kinetic energy of the block and bullet to the final potential energy of the spring:

$$K_{\text{i}} = U_{\text{f}}$$
$$\frac{1}{2}(M+m)v^2 = \frac{1}{2}kA^2$$

**2.** Solve for the speed of the bullet and block:

$$v = \sqrt{\frac{kA^2}{M+m}} = \sqrt{\frac{(785\text{ N/m})(0.0588\text{ m})^2}{1.500\text{ kg} + 0.00225\text{ kg}}} = \underline{\underline{1.344\text{ m/s}}}$$

**3.** Using conservation of momentum write the initial speed of the bullet in terms of the final speed of bullet and block:

$$mv_0 = (M+m)v$$
$$v_0 = \left(\frac{M+m}{m}\right)v$$

**4.** Calculate initial speed of bullet:

$$v_0 = \left(\frac{1.500\text{ kg} + 0.00225\text{ kg}}{0.00225\text{ kg}}\right)(1.344\text{ m/s}) = \boxed{897\text{ m/s}}$$

**5. (b)** Calculate one-quarter period:

$$\frac{T}{4} = \frac{\pi}{2}\sqrt{\frac{M+m}{k}} = \frac{\pi}{2}\sqrt{\frac{1.50225\text{ kg}}{785\text{ N/m}}} = \boxed{0.0687\text{ s}}$$

**Insight:** The initial kinetic energy of the bullet does not equal the final energy of the compressed spring. Some of the initial kinetic energy is lost due to the inelastic collision with the block.

56. **Picture the Problem**: The picture shows a metronome that oscillates with a mass (the bow tie) attached to a thin metal rod that pivots about a point near the belly of the penguin.

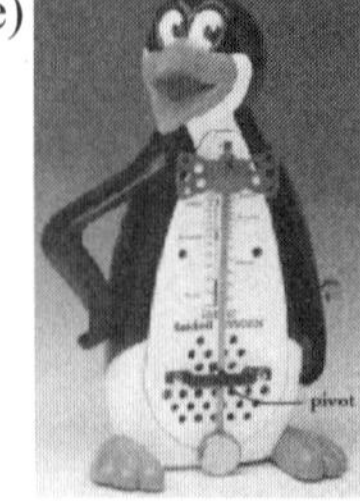


**Strategy:** The device can be considered a physical pendulum whose moment of inertia about the pivot point can be adjusted by moving the bow tie up and down the thin metal rod. Use equation 13-21 to answer the conceptual question.

**Solution:** Equation 13-21 indicates that the period of a physical pendulum is proportional to the square root of the moment of inertia. In order to reduce the period and increase the frequency of oscillation the moment of inertia should be decreased. We conclude that the penguin's bow tie should be moved downward in order to increase the frequency.

**Insight:** The reverse would be true if this were a standard pendulum with the pivot point at the top and the mass down below. In such a case the mass would have to be moved upward in order to decrease the moment of inertia.

62. **Picture the Problem**: A simple pendulum is a mass attached to a string. The mass is displaced so the string is slightly away from the vertical and released. The mass then oscillates about the vertical with a period determined by the length of the string and gravity.

**Strategy:** Calculate the length of the pendulum from its period.

**Solution: 1.** Solve the period equation for length:

$$T = 2\pi\sqrt{\frac{L}{g}} \Rightarrow L = \left(\frac{T}{2\pi}\right)^2 g$$

**2.** Insert the numeric values:

$$L = \left(\frac{1.00\text{ s}}{2\pi}\right)^2 \left(9.81\text{ m/s}^2\right) = \boxed{24.8\text{ cm}}$$

**Insight:** This is the length of the pendulum in many older clocks. Larger clocks, such as a grandfather clock, have pendulums about a meter long with a period of 2 seconds.

## Answers to Practice Quiz

**1**. (d) **2**. (b) **3**. (a) **4**. (c) **5**. (e) **6**. (b) 7. (b) **8**. (e) **9**. (d) **10**. (b)

# CHAPTER 14

# WAVES AND SOUND

## Chapter Objectives

After studying this chapter, you should

1. know the two main types of waves.
2. know the main characteristics of waves and wave motion.
3. be able to determine the speed of a wave on a string.
4. understand the nature of sound waves.
5. understand the relationship between sound intensity and the human perception of sound.
6. understand the Doppler effect for sound.
7. understand the superposition and interference of waves.
8. know how standing waves are generated.
9. understand the modes of vibration of standing waves on strings and in air columns.
10. know what beats are and what causes them.

## Warm-Ups

1. A certain transverse traveling wave on a string is represented by $y(x,t) = A\sin(kx + \omega t)$, where $y$ is the displacement of the string, $x$ is the position of a point on the string, and $t$ is time. What are the units of the coefficients of the $x$ and $t$ terms in the above expression? What do we call those coefficients?
2. Consider the wave represented by the expression in question 1. A wave can also be represented by a similar equation in which the two terms in the argument of the sine function have opposite signs. Is this wave different from the original one? If the answer is yes, what is the difference?
3. Suppose you suspend a 3-meter nylon rope from a hook in the ceiling and tie a 10-kg ball to the end of the rope. The mass of the rope is 0.5 kg. Now, you pluck the rope. That sends a wave up and down the rope. Estimate the speed of propagation of that wave.
4. How much faster does sound travel in steel than in water? What properties of the two materials are responsible for this difference?

## Chapter Review

In this chapter, we study **waves**. You can view a wave as resulting from the connection of a series of oscillators (oscillations were studied in the previous chapter) or as a propagating oscillation. Generally, any disturbance that propagates can be called a wave. In this chapter we focus on **harmonic waves** in which the oscillation that gives rise to the wave is a simple harmonic oscillation. The study of waves is important in almost every branch of physics and has many applications.

### 14–1 Types of Waves

There are two main types of waves. These types are distinguished by the relationship between the direction of the oscillation of the medium in which the wave is traveling and the direction of propagation of the wave. In a **transverse wave**, the direction of oscillation is perpendicular (or transverse) to the direction of propagation. A wave on a string is a good example of a transverse wave. The other main type of wave is a **longitudinal wave**, in which the direction of oscillation is along the same line as the direction of propagation. A compression wave traveling along a spring (such as a slinky) is a good example of a longitudinal wave.

Because a harmonic wave results from simple harmonic motion within a medium, the main characteristics of waves are related to the cycle of this repeating motion. One of these characteristics relates to the minimum time it takes for a wave to repeat itself, the **period**, $T$. As with any simple harmonic motion, the inverse of the period is called the **frequency**, $f$.

Waves also repeat themselves spatially; the minimum repeat length of a wave is called its **wavelength**, $\lambda$. If you consider the following diagram of a transverse wave, the wavelength equals the distance between successive crests, or troughs, of the wave (other corresponding points may also be used).

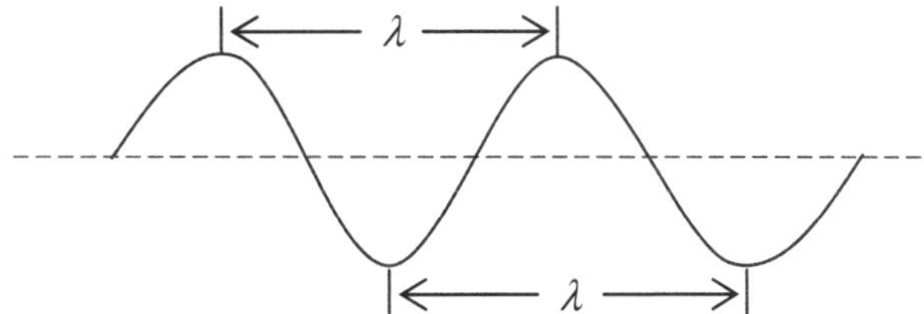


A third important characteristic of a wave is its speed of travel. This speed equals the distance the wave travels before it repeats (the wavelength) divided by the time it takes the wave to repeat (the period); therefore, the speed of a wave is given by

$$v = \frac{\lambda}{T} = \lambda f$$

## Practice Quiz

1. Two troughs of a wave of frequency 200 Hz are 1.7 m apart. What is the speed of this wave?

   **(a)** 5.0 mm/s **(b)** 340 m/s **(c)** 8.5 mm/s **(d)** 120 m/s **(e)** none of the above

2. If two waves traveling through the same medium have the same period, but wave 1 has half the wavelength of wave 2, how do the speeds of these waves compare?

   **(a)** $v_1 = 2v_2$ **(b)** $v_1 = 4v_2$ **(c)** $v_2 = 4v_1$ **(d)** $v_2 = 2v_1$ **(e)** $v_2 = v_1$

## 14–2 Waves on a String

Waves traveling on a string (or any linear cord) is a common way to generate sounds, so it is important to understand this behavior. The speed at which a wave on a string travels is determined by two properties of the string: its tension and *linear density,* $\mu$ (mass per unit length). The more tightly pulled the string is, the more rapidly it will oscillate, and the faster the wave will travel. The heavier a segment of the string is, the more sluggishly (or slowly) it will move under a given tension, and the slower the wave will travel. Detailed analysis of this situation shows that the relationship between these three quantities is

$$v = \sqrt{\frac{F}{\mu}}$$

where $F$ is the tension in the string (force transmitted through it), and $\mu$ is the mass per unit length, which is given by $\mu = m/L$, where $m$ and $L$ are the mass and length of the string respectively.

---

**Example 14–1 Sending a Wave** A string of length 2.59 m has a mass of 5.11 g and is fixed at one end. A person takes the other end and oscillates it up and down with a frequency of 3.47 Hz. If it takes the resulting wave 0.862 s to travel the length of the string, **(a)** determine the tension in the string, and **(b)** determine the wavelength of the wave.

**Picture the Problem** The sketch shows the person oscillating one end of the string while the other end remains fixed.

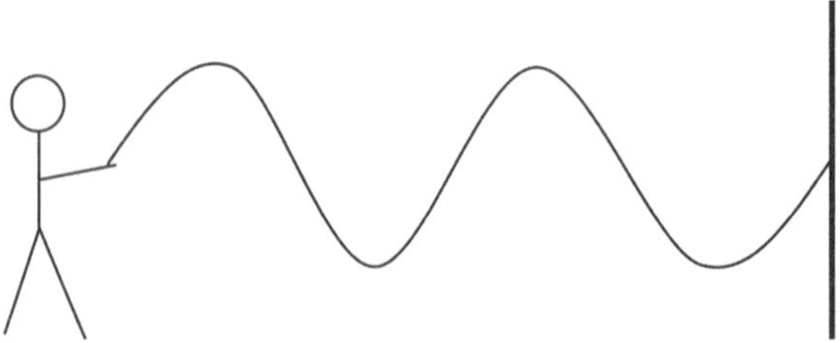

**Strategy** The given information allows us to find both $v$ and $\mu$ so we can use the relationship between $v$, $\mu$, and $F$ in part (a). Knowing $v$ and frequency, we can then calculate the wavelength.

**Solution**

**Part (a)**

1. Determine mass per unit length: $\mu = \frac{m}{L} = \frac{0.00511 \text{ kg}}{2.59 \text{ m}} = 1.973 \times 10^{-3} \text{ kg/m}$

2. Determine the speed of the wave: $v = \frac{L}{t} = \frac{2.59 \text{ m}}{0.862 \text{ s}} = 3.005 \text{ m/s}$

3. Solve for tension, $F$: $v = \sqrt{\frac{F}{\mu}} \Rightarrow F = \mu v^2$

4. Obtain the numerical result: $F = \mu v^2 = (1.973 \times 10^{-3} \text{ kg/m})(3.005 \text{ m/s})^2 = 0.0178 \text{ N}$

**Part (b)**

5. Use the relationship between $v$ and $\lambda$ to solve for $\lambda$: $v = \lambda f \Rightarrow \lambda = \frac{v}{f} = \frac{3.005 \text{ m/s}}{3.47 \text{ Hz}} = 0.866 \text{ m}$

**Insight** Notice that the wave speed connects the wave properties ($f$, $\lambda$) with those of the medium ($F$, $\mu$).

---

In the above example, a wave is sent along a string with a fixed end. When the wave reaches that end, it is reflected back in the opposite direction along the string. Because the end is fixed, it inverts each wave pulse upon reflection as a result of applying a force in a direction opposite that of the force applied by the string on the fixed connection. If the end of the string opposite the person is free to move, the reflected wave is not inverted relative to the initial wave because the end oscillates along with the rest of the string. This issue of how a wave is reflected will become important in later discussions.

## Practice Quiz

3. A certain string sustains waves of speed, $v$, when under a tension $F$. If the tension in the string is reduced to $F' = \frac{1}{3}F$, is the speed of the wave reduced or enhanced, and by what factor is it reduced or enhanced?

   **(a)** reduced by a factor of 0.58
   **(b)** reduced by a factor of 3
   **(c)** enhanced by a factor of 1.7
   **(d)** enhanced by a factor of 3
   **(e)** the speed stays the same

## *14–3 Harmonic Wave Functions

A traveling harmonic wave can be described by a reasonably simple functional form. For clarity we will consider a transverse wave for which the direction of oscillation is the $y$ direction and the direction of propagation is the $x$ direction. The position of a point on the wave (i.e., $y$) depends both on where you look in space (i.e., on $x$) and when you look in time (i.e., on $t$), so the position $y$ is a function of both $x$ and $t$. A harmonic wave traveling in the $+x$ direction can be described by the equation

$$y(x,t) = A\cos\left(\frac{2\pi}{\lambda}x - \frac{2\pi}{T}t\right)$$

where $A$ is the amplitude of the wave (its maximum displacement from equilibrium). Examination of this expression shows that for fixed $t$ (a snapshot of the wave), the wave repeats wherever $x$ increases by a distance $\lambda$. Similarly, for fixed $x$ (at a given location) the wave repeats whenever $t$ increases by a time $T$.

---

**Exercise 14–2 A Harmonic Wave** A transverse harmonic wave is described by the function

$$y = (3.2\text{ m})\cos\left[\left(0.25\text{ m}^{-1}\right)x - \left(1.7\text{ s}^{-1}\right)t\right].$$

What is the frequency of this wave?

**Solution**

Comparing the given function with the general form shows us that $\frac{2\pi}{T} = 1.7\text{ s}^{-1}$. This means that

$$\frac{1}{T} = \frac{1.7\text{ s}^{-1}}{2\pi}$$

Since frequency is given by $f = 1/T$, we can conclude that

$$f = \frac{1.7\text{ s}^{-1}}{2\pi} = 0.27\text{ Hz}$$

---

### Practice Quiz

4. What is the speed of a wave described by the expression $y = (0.21\text{ m})\cos\left[\left(0.13\text{ m}^{-1}\right)x - \left(2.4\text{ s}^{-1}\right)t\right]$?

   **(a)** 48 m/s **(b)** 0.38 m/s **(c)** 18 m/s **(d)** 0.21 m/s **(e)** 2.4 m/s

## 14–4 – 14–5 Sound Waves and Sound Intensity

Sound is one of the most important everyday applications of waves. Sound waves are longitudal waves in the air (often, longitudinal waves in *any* medium are called sound waves). The precise speed of sound in air depends on the atmospheric conditions at the time and location of the wave; however, under normal conditions this speed approximately equals 343 m/s (or 770 mi/h). This is the value we will use unless otherwise specified.

The pitch of a sound is determined by the frequency of the wave carrying the sound. A high-pitched sound has a relatively large (or high) frequency and a low-pitched sound has a relatively small (or low) frequency. Just as the frequency of sound can take a wide range of values, so can the wavelength. As mentioned previously, in air the speed of sound is roughly constant. Through the relation $v = \lambda f$, we can see that the wavelength takes on a range of values in association with the range of frequencies such that the product remains constant.

As just discussed, frequency is the physical property of a sound wave that determines the human perception of a sound's pitch. The human perception of the loudness of a sound is determined by the **intensity**, $I$, of the wave carrying the sound. The average intensity of a wave is the amount of energy that passes through a given area per unit time divided by the area, that is, energy per unit time per unit area. Since energy per unit time is power (in watts) delivered by the wave, on average the intensity is given by

$$I = \frac{P}{A}$$

where $P$ is the power, and $A$ is the area over which this power is spread. As can be seen from this relation, the SI unit of intensity is W/m$^2$. For a (point) source of sound whose power output $P$ is dispersed equally in all directions (an isotropic source), the intensity of the sound decreases with the inverse square of the distance $r$ from the source:

$$I = \frac{P}{4\pi r^2}$$

**Example 14–3 Inverse Square Law** The speakers of a small radio emit 55 W of power. **(a)** What is the approximate intensity of its sound 3.5 m away? **(b)** How far from the radio would you have to be to receive one-quarter of this intensity?

**Picture the Problem** The diagram shows a radio (modeled as a point source) giving off sound waves in all directions. The circles represent outward-moving crests in the direction of the dashed arrows.

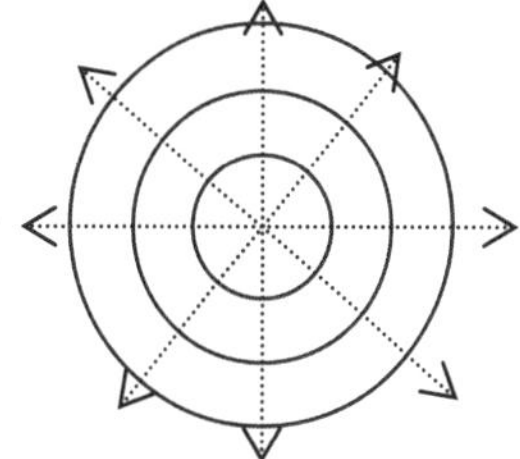

**Strategy** We make use of the variation of intensity with distance.

**Solution**

**Part (a)**

1. Use the expression for the intensity from an isotropic point source:

$$I = \frac{P}{4\pi r^2} = \frac{55\text{ W}}{4\pi(3.5\text{ m})^2} = 0.36\text{ W/m}^2$$

**Part (b)**

2. Relate the two intensities:

$$\frac{I}{I_x} = \frac{P/4\pi r^2}{P/4\pi x^2} = \frac{x^2}{r^2}$$

3. Use the fact that $I_x = \frac{1}{4}I$ to solve for $x$:

$$4 = \frac{x^2}{r^2} \quad \Rightarrow \quad x = 2r = 2(3.5\text{ m}) = 7.0\text{ m}$$

**Insight** Try to think of two other ways to solve part (b), one of which requires little or no calculations.

---

As mentioned earlier, human perception of the loudness of a sound is determined by the intensity of the sound. However, because the range of intensities that humans hear is very large, we use a more convenient measure called the **intensity level**, $\beta$. The intensity level measures loudness by comparing the intensity of a sound to a standard reference intensity, $I_0 = 10^{-12}\text{ W/m}^2$; this ratio is then rescaled by taking its logarithm. This quantity $\beta$ is dimensionless, but we refer to the values of $\beta$ as being measured in a "unit" called the decibel (dB). This situation is similar to quoting angular measures in radians even though angles are dimensionless. Values of $\beta$ can be obtained from the intensity $I$ by the expression

$$\beta = 10\log\left(\frac{I}{I_0}\right)$$

When more than one source of sound contributes, the intensities of the individual sources add. The intensity level is determined from the resulting sum.

---

**Exercise 14–4 The Desirable Range of Sound** Given that the lower threshold intensity for human hearing is about $1.0 \times 10^{-12}$ W/m$^2$ and the pain threshold is about 1.0 W/m$^2$, determine the desirable range of sound in decibels.

**Solution**

The lower threshold in decibels corresponds to the lower intensity; therefore, since $I_0 = 1.0\times10^{-12}\text{ W/m}^2$,

$$\beta_{lower} = 10\log\left(\frac{I}{I_0}\right) = 10\log\left(\frac{I_0}{I_0}\right) = 10\log(1) = 0\text{ dB}$$

The upper part of the desirable range is

$$\beta_{upper} = 10\log\left(\frac{I}{I_0}\right) = 10\log\left(\frac{1.0\ \text{W/m}^2}{1.0\times10^{-12}\ \text{W/m}^2}\right) = 10\log(10^{12}) = 120\log(10) = 120\ \text{dB}$$

Notice here that by using intensity level $\beta$ instead of intensities $I$, the range of values you have to work with is much smaller without any loss of information. This is one of the nice uses of the logarithm.

---

## Practice Quiz

**5.** The intensity of a sound 2.66 m from an isotropic source is $8.48 \times 10^{-8}$ W/m$^2$. What will the intensity be 5.32 m from the source (in W/m$^2$)?

**(a)** $4.24 \times 10^{-8}$ **(b)** $1.70 \times 10^{-7}$ **(c)** $3.39 \times 10^{-7}$ **(d)** $2.12 \times 10^{-8}$ **(e)** none of the above

**6.** If the intensity of a sound becomes a factor of 10 greater, describe what happens to the intensity level.

**(a)** The intensity level increases by 1 dB.

**(b)** The intensity level decreases by 1 dB.

**(c)** The intensity level increases by a factor of 10.

**(d)** The intensity level increases by a factor of log(10).

**(e)** None of the above.

## 14–6 The Doppler Effect

When there is relative motion between the source of a sound and the observer (or receiver) of the sound, the pitch of the sound changes. The pitch gets higher if the source and the observer are moving closer to each other, and it gets lower if they are moving further apart. The phenomenon just described is called the **Doppler effect**. This effect has analogs for all other types of waves in addition to sound.

Because the pitch of a sound is directly associated with its frequency, we treat the Doppler effect by finding the frequency perceived by the observer, $f'$, as compared with the frequency emitted by the source, $f$. We take the speed of sound in air to be constant, $\upsilon$ = 343 m/s, and call $u_o$ the speed of the observer and $u_s$ the speed of the source. With these definitions, and the provision that both $u_o$ and $u_s$ be less than $v$, we represent the Doppler effect for sound by the equation

$$f' = \left(\frac{1 \pm u_o/\upsilon}{1 \mp u_s/\upsilon}\right) f$$

In the numerator, the upper sign is used if the motion of the observer is toward the source, and the lower sign is used if it is away from the source. Similarly, in the denominator, the upper sign is used if the motion of the source is toward the observer, and the lower sign is used if it is away. Hence, upper signs are for "toward" and lower signs are for "away." There are four possibilities, as indicated below.

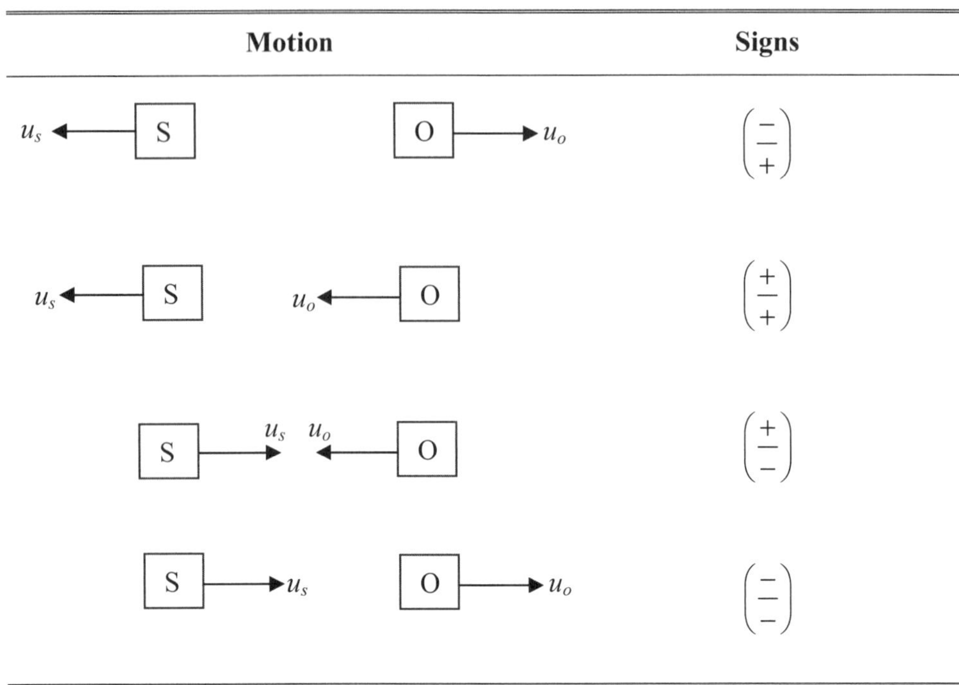


**Example 14–5 Doppler Effect** **(a)** A car moves with a speed of 45.0 mi/h toward a stationary observer as its horn is blown, emitting a frequency of 445 Hz. What frequency is heard by the observer? **(b)** Two cars move toward each other, each with a speed of 22.5 mi/h. If one car's horn is blown with a frequency of 445 Hz, what frequency is heard by observers in the other car?

**Picture the Problem** The sketch shows (a) the source car moving toward the stationary observer car and (b) source and observer cars moving toward each other.

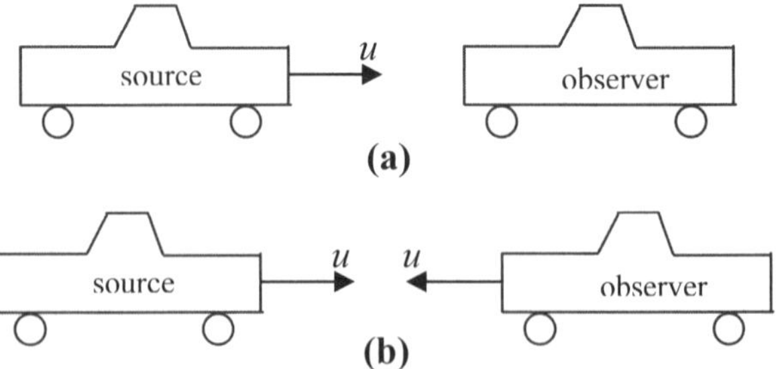


**Strategy** We apply the expression for the Doppler effect in both parts (a) and (b), making sure to use the proper signs to describe the given situations.

**Solution**

**Part (a)**

1. Since the source is moving toward the observer, we use the minus sign in the denominator, with $u_o = 0$:

$$f' = \left(\frac{1 \pm u_o / v}{1 \mp u_s / v}\right) f \;\Rightarrow\; f' = \left(\frac{1}{1 - u_s / v}\right) f$$

2. Convert the speed to m/s:

$$u_s = 45.0 \text{ mi/h}\left(\frac{0.447 \text{ m/s}}{\text{mi/h}}\right) = 20.115 \text{ m/s}$$

3. Solve for the numerical answer:

$$f' = \left(\frac{1}{1 - u_s / v}\right) f = \left(1 - \frac{20.115 \frac{\text{m}}{\text{s}}}{343 \frac{\text{m}}{\text{s}}}\right)^{-1} (445 \text{ Hz}) = 473 \text{ Hz}$$

**Part (b)**

4. Convert the speeds to m/s:

$$u_s = u_o = 22.5 \text{ mi/h}\left(\frac{0.447 \text{ m/s}}{\text{mi/h}}\right) = 10.058 \text{ m/s}$$

5. For this case, we have a + in the numerator and a – in the denominator:

$$f = \left(\frac{1 + u_o / v}{1 - u_s / v}\right) f$$

$$\therefore\; f' = \left(\frac{1 + (10.058 \text{ m/s})/(343 \text{ m/s})}{1 - (10.058 \text{ m/s})/(343 \text{ m/s})}\right)(445 \text{ Hz})$$

$$= 472 \text{ Hz}$$

**Insight** Even though the relative speeds between the two cars are the same in both parts (a) and (b), the frequencies heard are slightly different. Why?

## Practice Quiz

7. A source of sound is moving away from a stationary observer at 15 m/s. If the frequency emitted by the source increases, the frequency heard by the observer will

   **(a)** increase. **(b)** decrease. **(c)** stay the same. **(d)** None of these.

8. A source of sound is moving west at 15 m/s behind an observer who is also moving west at 12 m/s. The frequency heard by the observer, compared with the frequency emitted by the source, will be

   **(a)** lower. **(b)** higher. **(c)** the same. **(d)** None of these.

## 14–7 – 14–8 Superposition, Interference, and Standing Waves

When more than one wave passes through the same medium simultaneously, the resulting wave is called a **superposition** of the individual waves. The superposition of two waves, $y_1(x,t)$ and $y_2(x,t)$, results from the addition of the individual displacements to form a single wave, $y(x,t) = y_1(x,t) + y_2(x,t)$. After the interaction, the individual waves continue on their way unaltered. The wave that results from the superposition of other waves is said to produce an **interference pattern** of the individual waves.

Characteristic of wave interference patterns are regions where two special cases can be identified. One of these special cases occurs when the individual waves add in such a way that their maxima and/or their minima are at the same place at the same time. The result of this occurrence is that the amplitude of the resultant wave equals the sum of the amplitudes of the individual waves (the resultant wave is, in a sense, enhanced relative to the individual waves). This effect is called **constructive interference**. In short, constructive interference between two waves occurs at places where the waves are **in phase**. The other special case, **destructive interference**, occurs when the maximum of one wave meets the minimum of another. In this latter case, the amplitude of the resultant wave equals the difference of the amplitudes of the individual waves (the resultant wave is therefore diminished relative to the individual waves). When destructive interference occurs, the waves are said to have **opposite phase**. In general, constructive (destructive) interference occurs whenever and wherever waves add in phase (with opposite phase), not only at crests and troughs.

---

**Example 14–6 Two Sources with Opposite Phases** Each of two point sources a distance $s$ = 3.0 m apart emits sound waves of equal wavelength with opposite phase. If a detector placed 5.2 m from the line joining the sources and one-quarter of the way from one to the other, detects constructive interference between the two waves, what is the maximum possible wavelength of the sound wave?

**Picture the Problem** The diagram shows the two sources (top and bottom). The detector is indicated as being a distance, $L$, from the line joining the sources and 1/4 of the way between them from the top source.

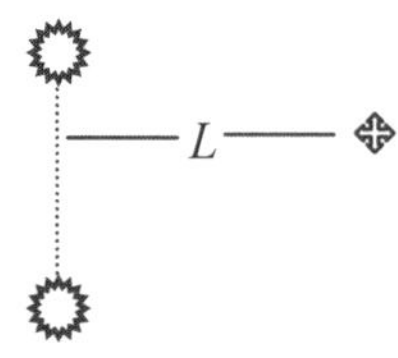


**Strategy** We need to determine the path difference between waves from each source in terms of the distance $L$. We must then correctly identify how this path difference relates to the wavelength to produce constructive interference.

**Solution**

1. The path length of the top source is the hypotenuse of the triangle with legs $L$ and $s/4$:

$$d_T = \sqrt{L^2 + (s/4)^2} = \sqrt{(5.2\text{ m})^2 + \left(3.0\text{ m}/4\right)^2} = 5.254\text{ m}$$

2. The path length of the bottom source is the hypotenuse of the triangle with legs $L$ and $\frac{3}{4}s$:

$$d_B = \sqrt{L^2 + \left(\frac{3s}{4}\right)^2} = \sqrt{(5.2\ \text{m})^2 + \left(\frac{9.0\ \text{m}}{4}\right)^2} = 5.666\ \text{m}$$

3. The path difference between the two sources is:

$$\Delta d = d_B - d_T = 5.67\ \text{m} - 5.25\ \text{m} = 0.412\ \text{m}$$

4. For constructive interference, the path difference must be an integral multiple of $\frac{1}{2}\lambda$ because the waves start with opposite phases. Thus, the maximum possible wavelength is:

$$\Delta d = \lambda/2 \Rightarrow \lambda = 2(\Delta d) = 2(0.412\ \text{m}) = 0.82\ \text{m}$$

**Insight** The answer gives only one possible wavelength. What are some others?

---

One of the most important consequences of the superposition of waves occurs when a wave and its reflection travel through a medium. Under the right conditions, these two waves, traveling in opposite directions, combine to produce what is called a **standing wave**. A standing wave is a wave that oscillates in time but is fixed in its spatial location. Standing waves contain positions (or regions) called **nodes**,

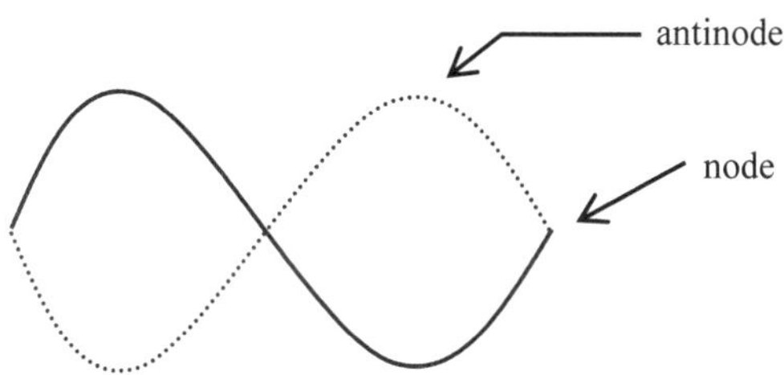


where no oscillation occurs, and other positions where the oscillation of the medium has maximum amplitude, called **antinodes**.

### Waves on a String

To establish a standing wave on a string with fixed ends, such as a guitar string, each fixed end must be a node. Therefore, the only standing waves that can be established on such a string are those that meet the condition of having nodes at the fixed ends. The longest wavelength of a standing wave that will satisfy these conditions is called the **fundamental mode** (in terms of wavelength) or the **first harmonic** (in terms of frequency). Try to convince yourself by drawing a picture that for a string of length $L$, the wavelength of the fundamental mode must be $\lambda = 2L$. The corresponding frequency of the first harmonic is found from $v = \lambda f_1$ to be $f_1 = v/2L$. If we continue to examine the different modes possible on strings with fixed ends, we see that the wavelengths and frequencies can be determined by

$$\lambda_n = \frac{2L}{n} \quad \text{and} \quad f_n = nf_1$$

where $n = 1, 2, 3, \ldots$ .

**Vibrating Columns of Air**

Standing waves in air columns are analyzed similarly to standing waves on strings; however, it is important to recognize that in air columns we are talking about longitudinal sound waves rather than transverse vibrations. Two types of air columns are in common use for generating sounds: one type has one end of the column closed off (with the other end open) and the other type has both ends open.

Standing waves in air columns with both ends open display the same characteristics as waves on a string with both ends fixed:

$$f_n = nf_1 \quad \text{and} \quad \lambda_n = \frac{2L}{n}$$

for $n = 1, 2, 3, \ldots$ The fundamental frequency is given by

$$f_1 = \frac{v}{2L}$$

where $v$ is the speed of sound, and $L$ is the length of the column.

In air columns with one closed end, standing waves exhibit somewhat different characteristics. In this case the closed end is a node, and the open end is an antinode. Thus, unlike in the two previous cases, the ends of the standing wave are different. The longest wavelength of a standing wave (the fundamental mode) that can be set up in a column of length $L$ is $\lambda_1 = 4L$. The other wavelengths are

$$\lambda_n = \frac{4L}{n} \quad \text{with} \quad n = 1,3,5,\ldots$$

For the frequencies, only odd harmonics (odd multiples of the fundamental frequency) are present:

$$f_n = nf_1 \quad \text{with} \quad n = 1,3,5,\ldots$$

where the fundamental frequency is

$$f_1 = \frac{v}{4L}$$

All the above information can be used to explain the generation of sound, especially music.

---

**Example 14–7 Guitar String** A certain guitar string of length 0.90 m has a linear density of 0.0075 kg/m. When properly tuned, the string has a $4^{\text{th}}$ harmonic of 1024 Hz. What is the tension for properly tuning this string?

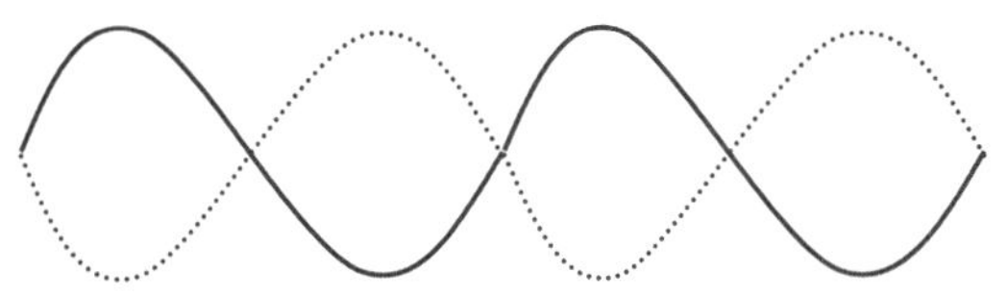

**Picture the Problem** The sketch shows the 4th harmonic of a guitar string with fixed ends.

**Strategy** Since we know how tension relates to the speed of the wave, we can express the speed in terms of the wavelength and frequency and relate these to the tension.

**Solution**

1. From the diagram, we see that we have two wavelengths for the fourth harmonic: $L = 2\lambda \quad \therefore \quad \lambda = L/2 = (0.90 \text{ m})/2 = 0.45 \text{ m}$

2. The relationship between $\lambda$, $f$, and $F$ is: $\lambda f = \sqrt{\dfrac{F}{\mu}} \quad \therefore \quad F = \mu \lambda^2 f^2$

3. Obtain the numerical result: $F = \mu\lambda^2 f^2 = (0.0075 \text{ kg/m})(0.45 \text{ m})^2 (1024 \text{ Hz})^2 = 1600 \text{ N}$

**Insight** Practice drawing various modes and harmonics of standing waves on strings and in air columns.

## Practice Quiz

9. Suppose two sources of identical waves emit them in phase. When these waves superimpose in the surrounding space, which of the following is a condition on the path difference, $\Delta d$, between the waves that will produce destructive interference?

   **(a)** $\Delta d = \lambda$ **(b)** $\Delta d = \lambda/2$ **(c)** $\Delta d = \lambda/3$ **(d)** $\Delta d = \lambda/4$ **(e)** none of the above

10. What is the wavelength of the third harmonic for standing waves on a string of length 1.35 m with fixed ends?

    **(a)** 1.5 m **(b)** 1.35 m **(c)** 0.900 m **(d)** 4.05 m **(e)** 0.45 m

11. What is the wavelength of the third harmonic for standing waves in an air column of length 1.35 m with only one open end?

    **(a)** 5.4 m **(b)** 1.35 m **(c)** 1.80 m **(d)** 1.08 m **(e)** 0.45 m

### 14–9 Beats

The superposition of waves of different frequencies gives rise to the phenomenon called **beats**. These beats appear as a regular fluctuation in the intensity of the wave that results from the superposition. This variation in intensity results from a variation in the amplitude of the resultant wave. The frequency of the successive intensity maxima is called the **beat frequency**. Two waves of frequencies $f_1$ and $f_2$ will produce a beat frequency equal to the difference between them:

$$f_{beat} = |f_1 - f_2|$$

Beats are often heard when two guitar strings are played at the same time and are often used to tune musical instruments to the desired frequency.

## Reference Tools and Resources

### I. Key Terms and Phrases

**wave** results from the connection of a series of oscillators

**harmonic waves** waves for which the oscillators move with simple harmonic motion

**transverse waves** waves for which the oscillation is perpendicular to the direction of propagation

**longitudinal waves** waves for which the oscillation is along the direction of propagation

**period** (for waves) the minimum amount of time it takes for a wave to repeat

**frequency** the number of cycles of a wave's oscillations per unit of time

**wavelength** the minimum repeat length of a wave

**intensity** the amount of energy per unit area per unit time

**intensity level** a measure of a sound's loudness relative to a standard reference

**Doppler effect** the shift in frequency due to relative motion between the source and the observer

**superposition** the addition of two or more waves

**interference pattern** the wave pattern that results from the superposition of two or more waves

**in phase** when the crests and/or troughs of different waves occur at the same time

**opposite phase** when the crests of a wave occur at the same time as the troughs of another wave

**constructive interference** when waves superimpose in phase resulting in a wave of larger amplitude

**destructive interference** when waves superimpose out of phase resulting in a wave of smaller amplitude

**standing wave** a stationary wave from the superposition of two waves traveling in opposite directions

**node** positions on a standing wave that do not oscillate

**antinode** positions on a standing wave that oscillate with maximum amplitude

**fundamental mode** (first harmonic) the longest wavelength standing wave

**beats** variation in the intensity of a wave resulting from the superposition of waves of different frequency

**beat frequency** the frequency of successive intensity maxima of a wave that exhibits beats

## II. Important Equations

| Name/Topic | Equation | Explanation |
|---|---|---|
| wave speed | $v = \frac{\lambda}{T} = \lambda f$ | The speed of a traveling wave |
| wave speed | $v = \sqrt{\frac{F}{\mu}}$ | The speed of a wave on a string of tension $F$ and linear density $\mu$ |
| harmonic waves | $y(x,t) = A\cos\left(\frac{2\pi}{\lambda}x - \frac{2\pi}{T}t\right)$ | The functional form for a harmonic traveling wave |
| sound intensity | $I = \frac{P}{4\pi r^2}$ | The intensity of a sound from a point source that emits isotropically falls off as the inverse square of the distance from the source |
| intensity level | $\beta = 10\log\left(\frac{I}{I_0}\right)$ | The intensity level (in decibels) for the loudness of a sound |
| Doppler effect | $f' = \left(\frac{1 \pm u_o/v}{1 \mp u_s/v}\right)f$ | The 4 possible relationships between the frequency heard $f'$ and the frequency emitted $f$ due to the Doppler effect |
| standing waves | $\lambda_n = 2L/n$ and $f_n = nf_1$ | The modes and harmonics for standing waves on strings with fixed ends and in air columns with both ends open. The possible values of $n$ are 1, 2, 3,… |

| | | |
|---|---|---|
| standing waves | $\lambda_n = \frac{4L}{n}$ and $f_n = nf_1$ | The modes and harmonics for standing waves in air columns with only one open end. The possible values of $n$ are 1, 3, 5,… |
| beats | $f_{beat} = \lvert f_1 - f_2 \rvert$ | The frequency of beats produced by the superposition of two waves |

## III. Know Your Units

| Quantity | Dimension | SI Unit |
|---|---|---|
| period ( $T$ ) | [T] | s |
| frequency ( $f$ ) | $[T^{-1}]$ | Hz |
| wavelength ( $\lambda$ ) | [L] | m |
| linear density ( $\mu$ ) | [M]/[L] | kg/m |
| intensity ( $I$ ) | $[M]/[T^3]$ | $W/m^2$ |
| intensity level ( $\beta$ ) | dimensionless | dB |

# Puzzle

(This puzzle by Professor Evelyn Patterson, *United States Air Force Academy*, is used by permission.)

Little Sally is floating peacefully in her small plastic boat, at the position marked A in the figure below. The lake is calm, and the day is sunny. A motorboat speeds by, about 30-m offshore, and the waves reach the shore about half a minute later. Suddenly, little Sally's older brother and sister decide to give Sally some fun. Each sibling is in a rowboat firmly tied to moorings out in the lake, as shown in the figure. They both jump up and down in their respective rowboats, bobbing up and down in synch together, about 15 times every half a minute. Soon, large waves are emanating from each of the bobbing boats, and the older brother and sister are giggling fiendishly.

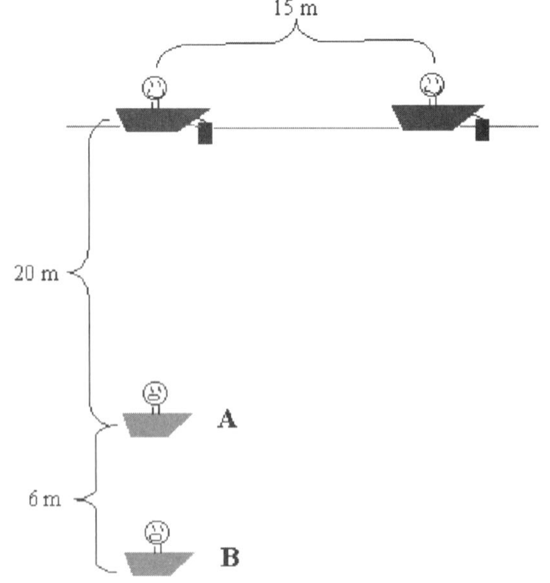

Sally's father, seeing the situation and having taken the equivalent of honors physics, does some quick thinking and moves Sally to the position marked B in the figure. Sally likes to float in calm water. Does Sally's father do Sally a favor by moving her?

## Answers to Selected Conceptual Questions

**6.** The Doppler effect applies to radar waves as well as to sound waves. In particular, the ball sees a Doppler-shifted radar frequency for the waves coming from the radar gun. Then, the ball acts as a moving source for waves of this frequency, giving a second Doppler shift to the echoes that are picked up by the radar gun. This provides a one-to-one correspondence between the final observed frequency and the speed of the ball.

**8.** The sliding part of a trombone varies the length of the vibrating air column that produces the trombone's sound. By adjusting this length, the player controls the resonant frequencies of the instrument. This, in turn, varies the frequency of sound produced by the trombone.

**10.** The thicker string is used to produce the low-frequency notes. This follows because the fundamental frequency depends directly on the speed of waves on the string. Therefore, for a given tension, a string with a greater mass per length has a smaller wave speed and hence a lower frequency.

## Solutions to Selected End-of-Chapter Problems and Conceptual Exercises

5. **Picture the Problem**: A wave of known amplitude, frequency, and wavelength travels along a string.

**Strategy:** Multiply the time by the wave speed, where the wave speed is given by equation 14-1, to calculate the horizontal distance traveled by the wave. To find the distance traveled by a knot on the string, note that a point on the string travels up and down a distance four times the amplitude during each period. Determine the fraction of a period that is spanned by the elapsed time and multiply it by 4*A* to find the distance traveled by the knot.

**Solution: 1. (a)** Calculate the distance traveled by the wave:

$$d_{\text{w}} = vt = (\lambda f)t = (27\times10^{-2}\text{ m})(4.5\text{ Hz})(0.50\text{ s}) = \boxed{0.61\text{ m}}$$

**2. (b)** Multiply 4*A* by the fraction of a period, noting that $f = 1/T$::

$$d_{\text{k}} = (4A)\left(\frac{t}{T}\right) = 4A\,f\,t = 4(12\times10^{-2}\text{ m})(4.5\text{ Hz})(0.50\text{ s}) = \boxed{1.1\text{ m}}$$

**3. (c)** The distance traveled by a wave peak is independent of the amplitude, so the answer in part (a) is unchanged. The distance traveled by the knot varies directly with the amplitude, so the answer in part (b) is halved.

**Insight:** A point on the string travels four times the wave amplitude in the same time that the crest travels one wavelength.

16. **Picture the Problem**: A wave takes 0.094 seconds to travel across a wire of known length and density.

**Strategy:** First find the linear mass density of the wire by dividing the mass by the length. Solve equation 14-2 for the tension in the wire. The velocity is given by the length of the wire divided by the time for the sound to travel across it.

**Solution: 1. (a)** Solve equation 14-2 for the tension:

$$v = \sqrt{\frac{F}{\mu}}$$

$$F = \mu v^2 = \frac{m}{L}\left(\frac{L}{t}\right)^2 = \frac{mL}{t^2}$$

**2.** Insert the given mass, length, and time:

$$F = \frac{(0.087 \text{ kg})(5.2 \text{ m})}{(0.094 \text{ s})^2} = \boxed{51 \text{ N}}$$

**3. (b)** The tension is proportional to the mass (if $L$ and $t$ remain constant), so the tension found in part (a) would be $\boxed{\text{larger}}$ if the mass of the wire were greater than 87 g.

**4. (c)** Solve with a mass of 0.097 kg.

$$F = \frac{(0.097 \text{ kg})(5.2 \text{ m})}{(0.094 \text{ s})^2} = \boxed{57 \text{ N}}$$

**Insight:** A heavier string requires greater tension for a wave to travel across it in the same time.

24. **Picture the Problem**: We are asked to sketch the equation $y = (15 \text{ cm})\cos\left(\frac{\pi}{5.0 \text{ cm}}x - \frac{\pi}{12 \text{ s}}t\right)$ for specific values of $t$.

**Strategy:** Insert the given times into the equation and plot the equation as a function of position. When a point on the graph moves from $y = 15$ cm to $y = 0$ cm, it has traveled from the maximum amplitude to equilibrium. This is one-quarter of a period. Calculate this time from the period given in the equation.

**Solution: 1. (a)** For $t = 0$, sketch the function $y = (15 \text{ cm})\cos\left(\frac{\pi}{5.0 \text{ cm}}x\right)$:

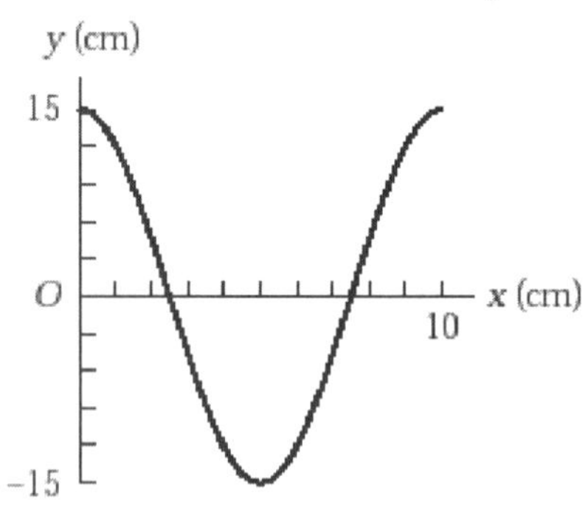


**2. (b)** For $t = 3.0$ s, sketch the function

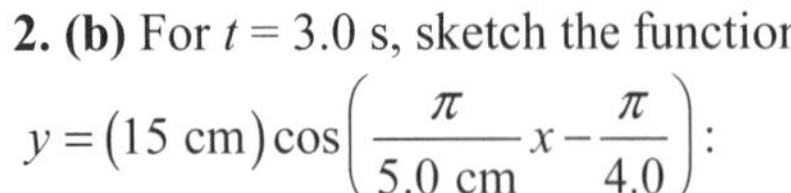

$y = (15 \text{ cm})\cos\left(\frac{\pi}{5.0 \text{ cm}}x - \frac{\pi}{4.0}\right)$:

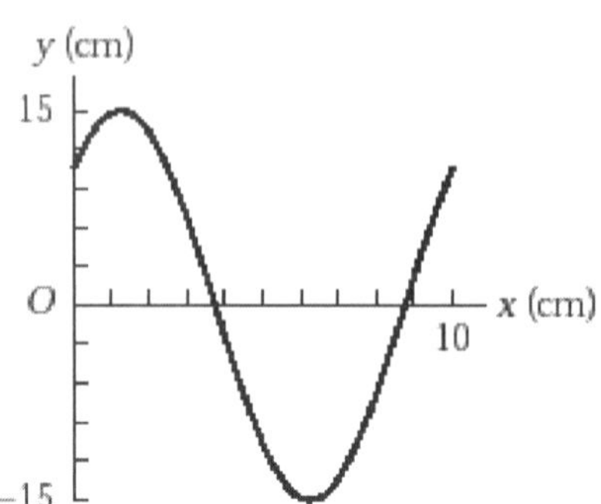

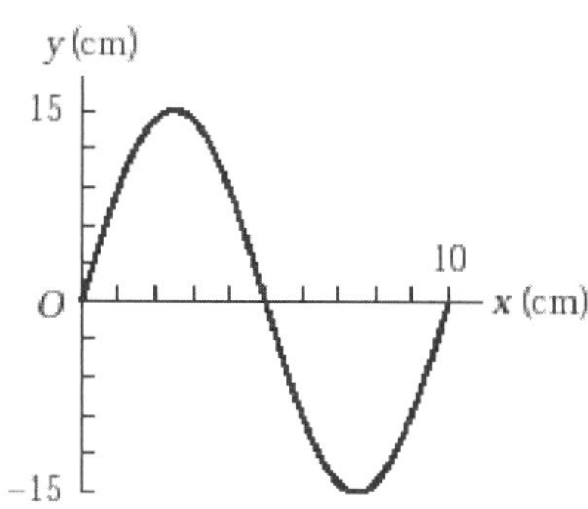


**3. (c)** For $t = 6.0$ s, sketch the function $y = (15\text{ cm})\cos\left(\frac{\pi}{5.0\text{ cm}}x - \frac{\pi}{2.0}\right)$:

**4. (d)** Calculate one-quarter of a period: $t = \frac{1}{4}T = \frac{1}{4}(24\text{ s}) = \boxed{6.0\text{ s}}$

**Insight:** Note from the initial graph, at $t = 0$, the wave is at a maximum at $x = 0$. At $t = 6$ sec the wave at $x = 0$, is at $y = 0$, as predicted in part (d).

30. **Picture the Problem**: The figure represents you dropping a rock down a well and listening for the splash.

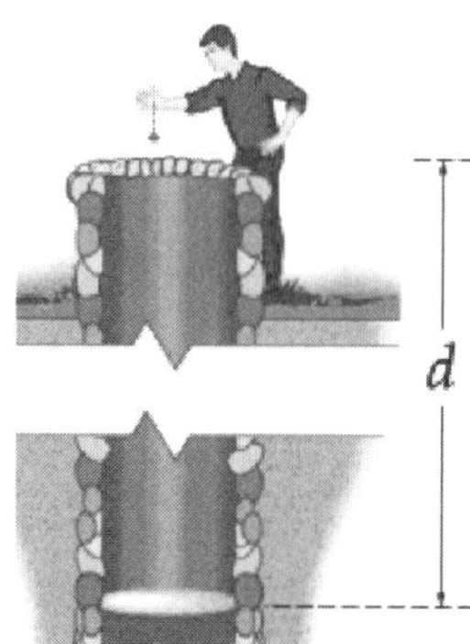


**Strategy:** The time to hear the splash, $t = 1.5$ s, is the sum of the time for the rock to fall to the water, $t_1$, and the time for the sound of the splash to reach you, $t_2$. Solve the free-fall equation (equation 2-13) for the time to fall and add it to the time required for the sound to travel the same distance. Set the sum of these times equal to the time to hear the splash and solve for the distance.

**Solution: 1. (a)** Solve equation 2-13 for the falling time: $t_1 = \sqrt{\frac{2d}{g}}$

**2.** Solve for the time for the sound to travel up the well: $t_2 = \frac{d}{v_s}$

**3.** Sum the two times to equal the total time: $t = t_1 + t_2 = \sqrt{\frac{2d}{g}} + \frac{d}{v_s}$

**4.** Rewrite as a quadratic equation in terms of the variable $\sqrt{d}$ :

$$0 = \frac{1}{v_s}\left(\sqrt{d}\right)^2 + \sqrt{\frac{2}{g}}\sqrt{d} - t$$

$$0 = \left(\frac{1}{343\text{ m/s}}\right)\left(\sqrt{d}\right)^2 + \left(\sqrt{\frac{2}{9.81\text{ m/s}^2}}\right)\sqrt{d} - 1.2\text{s}$$

**5.** Solve for $\sqrt{d}$ using the quadratic formula and square the result: $\sqrt{d} = \sqrt{3.2537\text{ m}} \Rightarrow d = 10.587\text{ m} = \boxed{11\text{ m}}$

**6. (b)** The time to hear the sound would be [less then 3.0 seconds] because, although the sound travel time would double, the fall time would less than double.

**Insight:** The time to hear the sound for a 21-meter-deep well is 2.1 s, which is indeed less than 3.0 s.

37. **Picture the Problem**: The intensity level at a distance of 2.0 meters is given. We want to find the intensity levels at 12 m and 21 meters. We also want to find the distance for which the intensity level is 0, the farthest point at which the siren can be heard.

**Strategy:** Insert equation 14-6 into equation 14-8 to create an equation for intensity level in relation to distance. Use this relationship to calculate the intensity levels at 12 m and 21 m. To calculate the farthest distance at which the siren can be heard, set the intensity level to zero and solve the relation for distance.

**Solution: 1.** Insert equation 14-6 into equation 14-8:

$$\beta = 10 \log\left(\frac{I_2}{I_0}\right) = 10 \log\left[\left(\frac{r_1}{r_2}\right)^2\left(\frac{I_1}{I_0}\right)\right]$$

$$= 10 \log\left(\frac{I_1}{I_0}\right) + 10 \log\left(\frac{r_1}{r_2}\right)^2$$

**2. (a)** Insert $\beta$ at $r_1 = 2.0$ m and set $r_2 = 12$ m:

$$\beta = 120 + 10 \log\left(\frac{2.0 \text{ m}}{12 \text{ m}}\right)^2 = \boxed{104 \text{ dB}}$$

**3. (b)** Repeat for $r_2 = 21$ m:

$$\beta = 120 + 10 \log\left(\frac{2.0 \text{ m}}{21 \text{ m}}\right)^2 = \boxed{99.6 \text{ dB}}$$

**4. (c)** Set the intensity level equal to zero:

$$\beta = 10 \log\left(\frac{I_1}{I_0}\right) + 10 \log\left(\frac{r_1}{r_2}\right)^2$$

$$0 = 120 + 10 \log\left(\frac{2.0 \text{ m}}{r}\right)^2$$

**5.** Solve for $r$:

$$-10 \log\left(\frac{2.0 \text{ m}}{r}\right)^2 = 120$$

$$\left(\frac{2.0 \text{ m}}{r}\right)^2 = 10^{-12} \Rightarrow r = \frac{2.0 \text{ m}}{10^{-6}} = \boxed{2.0\times10^6 \text{ m}}$$

**Insight:** This is a theoretical limit that could be realized in an ideal case. In a more realistic scenario, ambient noise, as well as energy losses when the sound waves are reflected or absorbed by surfaces, would prevent us from hearing the sound 2000 km away. Sometimes the real-world factors we ignore make a huge difference!

53. **Picture the Problem**: The figure shows two bicyclists approaching each other at the same speed.

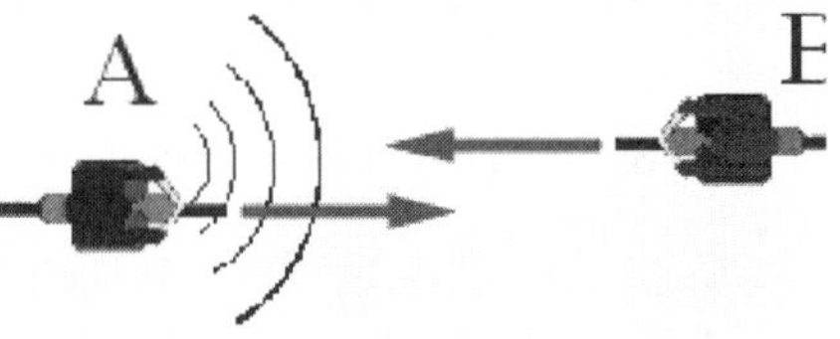


**Strategy:** We want to calculate the frequency at which Cyclist B hears Cyclist A's horn. Because both the source and the observer are moving, we should use equation 14-11. Since they are approaching each other, the plus sign is used in the numerator and the minus sign in the denominator.

**Solution: 1. (a)** Solve equation 14-11 for the observed frequency:

$$f' = \left(\frac{1+u_o/v}{1-u_s/v}\right) f = \left[\frac{1+(8.50 \text{ m/s}/343 \text{ m/s})}{1-(8.50 \text{ m/s}/343 \text{ m/s})}\right](315 \text{ Hz}) = \boxed{0.33 \text{ kHz}}$$

**2. (b)** The greater increase in the frequency heard by bicyclist B would occur when $\boxed{\text{(i) bicyclist A speeds up by 1.50 m/s}}$. For equal changes in speed, the Doppler shift due to a moving source is larger than that due to a moving observer.

**Insight:** Increasing the Cyclist B's speed by 1.5 m/s results in an observed frequency of 332 Hz, while increasing Cyclist A's speed by 1.5 m/s results in an observed frequency of 333 Hz.

65. **Picture the Problem**: The image shows two out-of-phase speakers separated by 3.5 meters and an observer standing by a wall 5.0 meters away. When the observer moves 0.84 meters along the wall he goes from the central destructive interference to the first constructive interference. We want to calculate the frequency of sound emitted by the speaker.

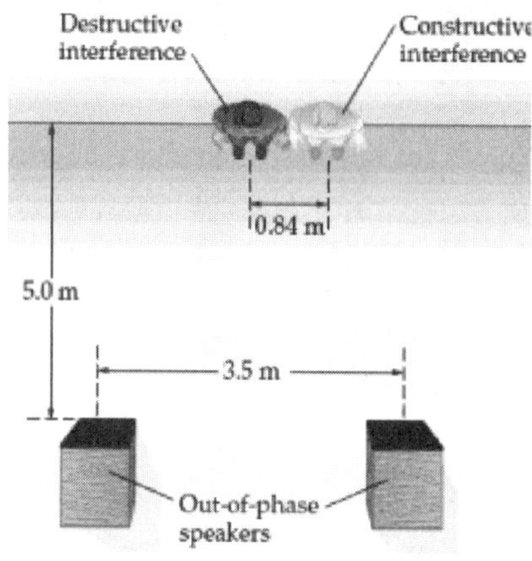


**Strategy:** Use the Pythagorean Theorem to calculate the distance of the observer from each speaker. Calculate the difference in distances to each speaker. Because the speakers are out of phase, constructive interference occurs when the difference in distances is equal to a half wavelength. Set the difference in distances equal to a half wavelength and use equation 14-1 to calculate the resulting frequency.

**Solution: 1.** Calculate $d_1$:

$$d_1 = \sqrt{(5.0\text{ m})^2 + \left[\tfrac{1}{2}(3.5\text{ m}) + 0.84\text{ m}\right]^2} = 5.631\text{ m}$$

**2.** Calculate $d_2$:

$$d_2 = \sqrt{(5.0\text{ m})^2 + \left[\tfrac{1}{2}(3.5\text{ m}) - 0.84\text{ m}\right]^2} = 5.082\text{ m}$$

**3.** Set $d_1 - d_2 = \frac{1}{2}\lambda$:

$$\lambda = 2(d_1 - d_2) = 2(5.631\text{ m} - 5.082\text{ m}) = 1.098\text{ m}$$

**4.** Use equation 14-1 to calculate the frequency:

$$f = \frac{v}{\lambda} = \frac{343\text{ m/s}}{1.098\text{ m})} = \boxed{0.31\text{ kHz}}$$

**Insight:** Constructive interference will also occur at the observer's position for other frequencies for which $d_1 - d_2$ is an integer multiple of half the wavelength, corresponding to $f$ = 0.94 kHz, 1.6 kHz, etc.

75. **Picture the Problem**: The figure shows a standing wave inside an organ pipe 2.75 m long. We want to calculate the frequency of the wave and the fundamental frequency of the pipe.

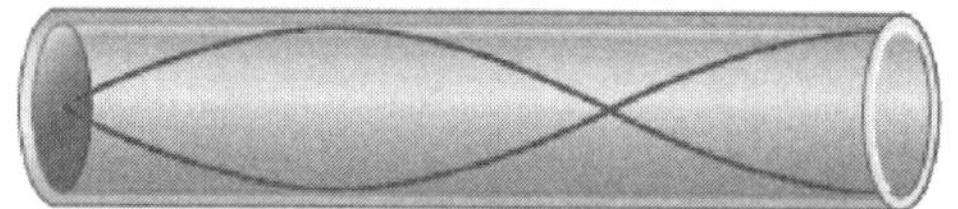

**Strategy:** The pipe has a node at one end and an antinode at the other, so it is closed at one end. The wave corresponds to the third harmonic because there is one node inside the pipe. Use equation 14-14 to calculate the frequency of this harmonic, setting $n$ = 3. Then calculate the fundamental frequency using $n$ = 1.

**Solution: 1. (a)** Solve equation 14-14 with $n$ = 3:

$$f_3 = \frac{3v}{4L} = \frac{3(343\text{ m/s})}{4(2.75\text{ m})} = \boxed{93.5\text{ Hz}}$$

**2. (b)** Solve equation 14-14 with $n$ = 1:

$$f_1 = \frac{v}{4L} = \frac{343\text{ m/s}}{4(2.75\text{ m})} = \boxed{31.2\text{ Hz}}$$

**Insight:** The next harmonic possible on this pipe is the fifth harmonic, whose frequency is 156 Hz.

85. **Picture the Problem**: We are given the length and frequency of one cello string. A second string, which is identical to the first, is shortened. When sounding together, the two strings produce a beat frequency of 4.33 Hz.

**Strategy:** We want to calculate the length of the shortened string. Because the two strings are identical, the speed of the waves on each string is the same. Use equation 14-12 to write a relation between the frequencies and lengths, given the same velocities. Solve this relation for the length of the second string. Since the second string has a shorter length, it has the higher frequency. Write the frequency of the second string as the sum of the first frequency and the beat frequency.

**Solution: 1.** Solve equation 14-12 for the velocity in terms of frequency and length: $f = \dfrac{v}{2L} \Rightarrow v = 2f_1 L_1 = 2f_2 L_2$

**2.** Solve for the length of the second string: $L_2 = \dfrac{f_1}{f_2} L_1$

**3.** Set the second frequency equal to the sum of the first frequency and the beat frequency: $L_2 = \dfrac{f_1}{f_1 + f_{beat}} L_1 = \left( \dfrac{130.9 \text{ Hz}}{130.9 \text{ Hz} + 4.33 \text{ Hz}} \right)(1.25 \text{ m}) = \boxed{1.21 \text{ m}}$

**Insight:** The string has been shortened by a distance of 1.25 m – 1.21 m = 0.04 m or 4.0 cm.

## Answers to Practice Quiz

**1**. (b) **2**. (d) **3**. (a) **4**. (c) **5**. (d) **6**. (e) 7. (a) **8**. (b) **9**. (b) **10**. (c) **11**. (d)

# CHAPTER 15
# FLUIDS

## Chapter Objectives

After studying this chapter, you should

1. understand the essential characteristics of fluid behavior.
2. know the difference between atmospheric and gauge pressure.
3. understand how pressure depends on depth in a static fluid.
4. understand, and be able to apply, Pascal's principle.
5. understand, and be able to apply, Archimedes' principle.
6. understand the behavior described by the equation of continuity.
7. understand, and be able to apply, Bernoulli's equation.
8. understand the basic reasons for viscosity and surface tension.

## Warm-Ups

1. The atmospheric pressure at sea level is $1.013 \times 10^5$ N/m$^2$. The unit of pressure, N/m$^2$, is also called a pascal (Pa). The surface area of an average adult body is about 2 m$^2$. How much crushing force does the atmosphere exert on people? Why don't we get crushed? Calculate this value in pounds using the fact that the weight of a 1-kilogram mass is 2.2 pounds.
2. The atmospheric pressure decreases with altitude exponentially according to the relation $p = p_o e^{-(0.00012)h}$, where $h$ is the altitude above sea level in meters, and $p_o = 1.013 \times 10^5$ N/m$^2$. The average atmospheric pressure in Florida is close to $p_o$. Estimate the average atmospheric pressure in Denver (The elevation at Denver is about 1500 m).
3. Water is pumped up a pipeline as shown. The water pours out at the top and to the ground. The pump is running at constant speed. Compare the water speed at the three points A, B, and C in the pipeline. *State your reasons*.

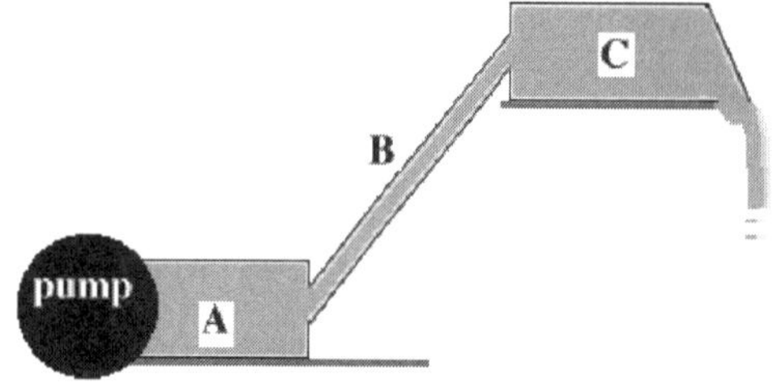

4. Estimate the force on a 1.5-m$^2$ windshield as a 60 mi/h (30 m/s) wind passes near its surface. The air in the car is stationary, and the car windows are shut tight.

## Chapter Review

In this chapter, we study **fluids**. Fluids are characterized by their ability to flow; both liquids and gases are fluids. An understanding of fluid behavior is essential to life, and applications of this understanding are essential to many of the conveniences of modern living.

### 15–1 – 15–2 Density and Pressure

One of the most convenient properties used to describe a fluid is its **density**, $\rho$. The density of a substance is a measure of how compact the substance is, that is, how much mass is packed into a volume of the substance. On average, the density of a substance is the amount of mass $M$ divided by the volume $V$ taken up by that mass

$$\rho = \frac{M}{V}$$

Another important quantity in the study of fluids is **pressure**, $P$. On average, the pressure that is applied to an object is the amount of force, $F$, (normal to the surface) divided by the area, $A$, over which the force spreads

$$P = \frac{F}{A}$$

As the preceding expressions indicate, the unit of pressure is that of force divided by area; in SI units this is called the pascal (Pa): 1 Pa = 1 N/m$^2$. An important property of the pressure in a fluid is that it is equally applied in all directions (at a given depth) and applies forces that are perpendicular to any surface in the fluid.

When considering the pressure on or within a fluid, it is important to recognize that for many applications we must account for a constant atmospheric pressure ($P_{at} = 1.01 \times 10^5$ Pa). Because of the constant presence of the atmosphere, we are often interested only in the pressure above and beyond that applied by the atmosphere. This additional pressure is called the **gauge pressure**, $P_g$

$$P_g = P - P_{at}$$

where P is the total pressure applied (sometimes called the *absolute pressure*).

**Exercise 15–1 The Density of a Fluid** What is the density of a 1.00 gallon, 3.22-lb fluid?

**Solution:** We are given the following information:

**Given:** $V = 1.00$ gal, $W = 3.22$ lb; **Find:** $\rho$

Since the density is given by $\rho = m/V$, we need to find the mass from the weight and convert everything to SI units. To get the mass we use

$$m = \frac{W}{g} = \frac{3.22 \text{ lb}\left(\frac{4.45 \text{ N}}{\text{lb}}\right)}{9.81 \text{ m/s}^2} = 1.461 \text{ kg}$$

We can immediately convert the volume to give

$$V = 1.00 \text{ gal}\left(\frac{3.785\times10^{-3}\text{m}^3}{\text{gal}}\right) = 3.785\times10^{-3}\text{m}^3$$

We are now ready to calculate the density:

$$\rho = \frac{m}{V} = \frac{1.461 \text{ kg}}{3.785\times10^{-3}\text{m}^3} = 386 \text{ kg/m}^3$$

**Example 15–2 Gauge Pressure of Water** A uniform cylindrical container has a radius of 7.8 cm and a height of 13.2 cm. If this container is completely filled with water, what gauge pressure does the water apply to the bottom of the container?

**Picture the Problem** The sketch shows a cylindrical container filled with water.

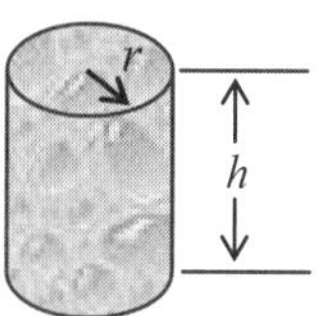


**Strategy** The pressure applied by the water is the force that the water applies (equal to its weight) divided by the bottom area of the cylinder.

**Solution**

1. The volume of water equals the volume of the cylinder: $V = Ah = \pi r^2 h = \pi(0.078\text{m})^2(0.132\text{m}) = 2.523\times10^{-3}\text{m}^3$

2. Determine the mass of the water: $m = \rho V = \left(1000 \frac{\text{kg}}{\text{m}^3}\right)\left(2.523\times10^{-3}\text{ m}^3\right) = 2.523 \text{ kg}$

3. Determine the weight of the water: $W = mg = (2.523 \text{ kg})(9.81 \text{ m/s}^2) = 24.75 \text{ N}$

**4.** Obtain the gauge pressure: $P = \frac{F}{A} = \frac{mg}{\pi r^2} = \frac{24.75\text{ N}}{\pi(0.078\text{ m})^2} = 1.3\times10^3\text{ Pa}$

**Insight** The answer is the gauge pressure because we ignored atmospheric pressure.

## Practice Quiz

1. If twice as much volume of fluid 1 weighs the same as fluid 2, how do their densities compare?

   **(a)** Fluid 1 is twice as dense as fluid 2.

   **(b)** Fluid 1 is half as dense as fluid 2.

   **(c)** Fluid 1 and fluid 2 have equal densities.

   **(d)** Fluid 1 is four times less dense than fluid 2.

   **(e)** None of the above.

2. If a uniform cylinder of height 0.850 m and radius 0.250 m is completely filled with water, what is the total pressure on the bottom of the cylinder?

   **(a)** 8.34 kPa **(b)** 133 kPa **(c)** 2.08 kPa **(d)** 109 kPa **(e)** 46.2 kPa

## 15–3 Static Equilibrium in Fluids: Pressure and Depth

In this and the next several sections, we shall consider the properties of static fluids, that is, fluids that do not flow. In static fluids, every part of the fluid and every object within it is in static equilibrium. One of the basic properties of fluids is **Pascal's principle**:

*External pressure applied to an enclosed fluid is transmitted unchanged throughout the fluid.*

Pascal's principle is important in determining the dependence of pressure on the depth within a fluid. Without any external pressure applied to the outer surface of a fluid, the pressure measured at a depth, *h,* beneath the surface arises from the weight of the fluid above the given level. The amount of this increase in pressure is given by $\rho gh$. If there is external pressure on the fluid, such as from the atmosphere and/or any other source, this pressure is transmitted undiminished to every point in the fluid and it must be added to the pressure due to the weight of the fluid. Thus, for the dependence of pressure on depth we have

$$P_2 = P_1 + \rho gh$$

where $P_2$ is the pressure a distance, *h,* below the level at which it is $P_1$.

**Example 15–3 Blood Pressure with Depth** Human blood has a density of approximately $1.05 \times 10^3$ kg/m$^3$. Use this information to estimate the difference in blood pressure between the brain and the feet in a person who is approximately 6 feet tall.

**Picture the Problem** The sketch shows a person approximately 6 feet in height.

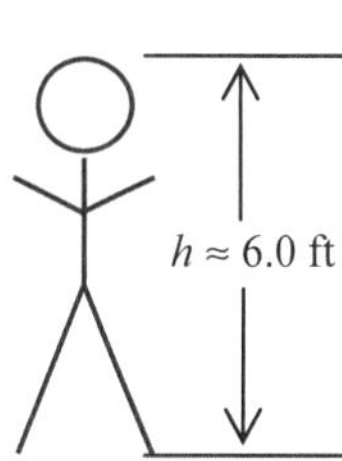


**Strategy** We attempt this approximation to two significant figures by applying the above result for the dependence of pressure on depth.

**Solution**

1. Convert the height to meters: $h = 6.0\text{ ft}\left(\dfrac{\text{m}}{3.28\text{ ft}}\right) = 1.83\text{ m}$

2. The difference in pressure is given by: $P_2 - P_1 = \rho g h$

3. Obtain the numerical result: $P_2 - P_1 = \left(1.05\times10^3\ \frac{\text{kg}}{\text{m}^3}\right)\left(9.81\ \frac{\text{m}}{\text{s}^2}\right)(1.83\text{ m}) = 19\text{ kPa}$

**Insight** This is only an estimate for many reasons. The blood in the body is not a static fluid; it flows, and between the head and the feet is a pump (the heart) that will affect the result. If you plan to study medicine, see if you can find out the difference in blood pressure between the head and the feet.

Pascal's principle is also key to understanding the hydraulic lift. This device uses fluid pressure to convert a small input force into a large output force. The input force $F_1$ is applied to a fluid over a comparatively small area, $A_1$, giving rise to a pressure change of $F_1/A_1$. Since this pressure change is transmitted undiminished throughout the fluid, $F_1/A_1 = F_2/A_2$, we must get a larger output force $F_2 > F_1$ if it is spread over a larger area $A_2 > A_1$; therefore,

$$F_2 = F_1\left(\frac{A_2}{A_1}\right)$$

The consequence of getting this larger output force is that the distance through which this force can move an object at the output, $d_2$, is smaller than the distance at the input, $d_1$. Because the same volume of fluid moves at the input and output, $A_1 d_1 = A_2 d_2$, the output distance is given by

$$d_2 = d_1\left(\frac{A_1}{A_2}\right)$$

**Exercise 15–4 Hydraulic Lift** Your job at Dave's Manufacturing Company is to design a hydraulic lift for a client. This client typically needs to raise objects through a height of 0.500 meters. The system should easily be used by people of average height, so the input distance should not exceed 5.00 ft. What would be a good ratio of output area to input area to consider?

**Solution:** We are given the following information.

**Given:** $d_1 = 5.00$ ft , $d_2 = 0.500$ m; **Find:** $A_2/A_1$

From the information given, we know that the ratio of the input distance to the output distance is directly proportional to the ratio we seek:

$$\frac{d_1}{d_2} = \frac{A_2}{A_1}$$

Therefore, the ratio is

$$\frac{A_2}{A_1} = \frac{5.00\ \text{ft}\left(\frac{\text{m}}{3.28\ \text{ft}}\right)}{0.500\ \text{m}} = 3.05$$

The output area should be at least 3.05 times greater than the input area. Of course, in a more realistic situation you'd want more information from the client. What additional information might you want?

## Practice Quiz

3. What is the gauge pressure 3.4 m below the surface of a 550-kg/m$^3$ density fluid?
   **(a)** 33 kPa **(b)** 1900 Pa **(c)** 18 kPa **(d)** 100 kPa **(e)** 550 Pa

4. The gauge pressure at a particular location in a fluid of density 870 kg/m$^3$ is 120 kPa. What is the gauge pressure in the fluid 5.9 m above this location?
   **(a)** 70 kPa **(b)** 50 kPa **(c)** 62 kPa **(d)** 58 kPa **(e)** 20 kPa

5. In the figure, $h_1 = 0.16$ cm and $h_2 = 4.34$ cm; $h_2$ is the height of an unknown liquid. The remaining liquid in the U-tube is water (1000 kg/m$^3$). What is the density of the unknown liquid (in kg/m$^3$)?
   **(a)** 36.9 **(b)** 996 **(c)** 1012 **(d)** 1000 **(e)** 963

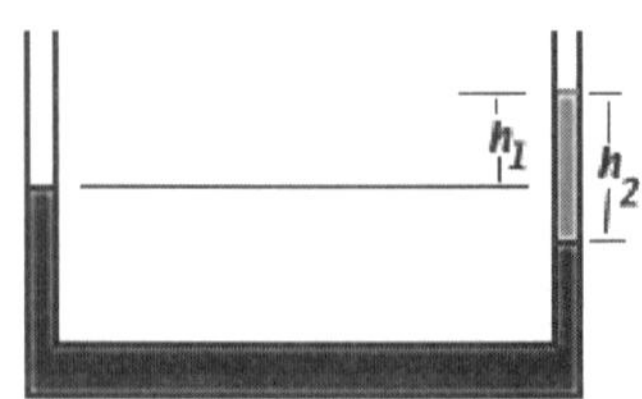

## 15–4 – 15–5 Archimedes' Principle and Buoyancy

When an object is submerged in a fluid, the volume taken up by the object displaces an equal volume of the fluid. The pressure applied by the fluid onto the object results in an upward force on the object; this phenomenon is called **buoyancy** and is governed by **Archimedes' principle**:

> *An object immersed in a fluid experiences an upward force equal to the weight of the fluid displaced by the object.*

The weight of the fluid displaced by the object equals the mass of this fluid times the acceleration due to gravity, $mg$. When dealing with buoyancy, it is usually more convenient to express the mass in terms of the density, $m = \rho V$; therefore, for an object submerged in a fluid, the buoyant force on it is

$$F_b = \rho g V$$

Archimedes' principle explains the phenomenon of **flotation**, which occurs when the buoyant force acting on an object equals its weight. Often, floating objects are not completely submerged in the fluid. The amount of volume submerged, $V_{sub}$, for a solid object of volume, $V_s$, floating in a fluid of density $\rho_f$ is

$$V_{sub} = V_s\left(\frac{\rho_s}{\rho_f}\right)$$

where $\rho_s$ is the density of the solid object.

---

**Exercise 15–5 The Secret of Magic** Many magic tricks are based on physical principles. In order to fool her audience, a magician uses an object that sinks in the freshwater made available to the audience, but floats in the seawater that she uses on stage. What is the maximum percentage of the object's volume that will float above the seawater?

**Picture the Problem** The sketch shows a floating object partially submerged in seawater.

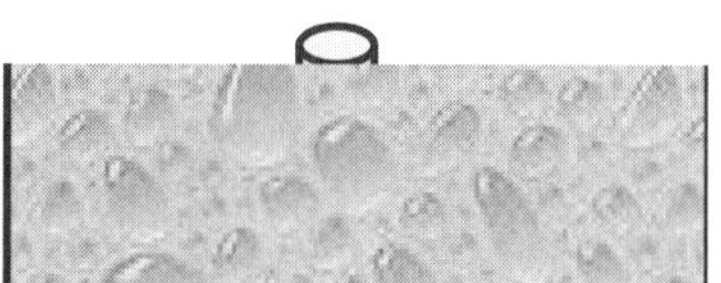

**Strategy** According to the above discussion, more of an object will be submerged if its density approaches that of the fluid, so we get the maximum above-surface float for the smallest possible object density. Since it must sink in fresh water, the smallest object density is 1000 kg/m$^3$.

**Solution**

1. The minimum fraction of volume submerged is: $\dfrac{V_{sub}}{V} = \dfrac{\rho_{obj}}{\rho_{sea}} = \dfrac{1000 \text{ kg/m}^3}{1025 \text{ kg/m}^3} = 0.976$

2. Find the maximum amount of volume floating above the surface: $V_{above} = (1 - 0.976)V = 0.024V$

3. Calculate the percentage: $\dfrac{V_{above}}{V} = 0.024 = 2.4\%$

**Insight** What could the magician do to the seawater to make a larger percentage of the object float?

## Practice Quiz

6. An object of density 750 kg/m$^3$ is half submerged in a fluid. What is the density of this fluid?
   **(a)** 1500 kg/m$^3$ **(b)** 188 kg/m$^3$ **(c)** 375 kg/m$^3$ **(d)** 2250 kg/m$^3$ **(e)** 750 kg/m$^3$

7. If the volume of the object in question 5 is 0.33 m$^3$, what is the buoyant force on this object?
   **(a)** 9810 N **(b)** 1210 N **(c)** 3240 N **(d)** 1000 N **(e)** 2430 N

## 15–6 Fluid Flow and Continuity

In this section, we begin to discuss properties of fluid flow. During the smooth flow of a constrained fluid (e.g., through a pipe), we can assume that the same amount of mass passes through each cross section of pipe in a given amount of time. This smooth-flow condition leads to what is known as the **equation of continuity**, which says that the mass, $m_1$, flowing through an area, $A_1$, in a given time equals the mass, $m_2$, flowing through area, $A_2$, in that same amount of time. The amount of mass per unit time of a fluid of density, $\rho$, flowing through an area, $A$, at speed, $v$, is $\rho A v$. Therefore, the equation of continuity is

$$\rho_1 A_1 v_1 = \rho_2 A_2 v_2$$

Usually, liquids are considered to be incompressible because the density of the liquid hardly changes as it flows from one place to another. In such cases, the densities in the equation of continuity are equal, $\rho_1 = \rho_2$, and we can write the equation as

$$A_1 v_1 = A_2 v_2$$

The quantity $Av$ equals the *volume flow rate* of the fluid. Thus, the preceding equation says that the volume flow rate is constant for an incompressible fluid.

**Example 15–6 Continuity** Plastic bottles that are used to hold water for athletes often have a long slender nozzle out of which the water emerges. If the end of the nozzle has a diameter of 1.0 cm, and you determine the water to emerge at 25 cm/s for a typical squeeze of the bottle, what is the initial speed of the water in the neck of the bottle if its diameter is 6.0 cm?

**Picture the Problem** The sketch shows a squeeze water bottle with a thin nozzle at the end.

**Strategy** Because we don't expect the density of the water to change when flowing from inside the bottle to outside, we need only to use the fact that the volume flow rate is constant.

**Solution**

1. Use the equation of continuity: $A_1 v_1 = A_2 v_2 \Rightarrow \frac{\pi d_1^2}{4} v_1 = \frac{\pi d_2^2}{4} v_2 \Rightarrow d_1^2 v_1 = d_2^2 v_2$

2. Solve for $v_1$: $v_1 = v_2 \left( \frac{d_2}{d_1} \right)^2$

3. Obtain the numerical result: $v_1 = (25 \text{ cm/s}) \left( \frac{1.0 \text{ cm}}{6.0 \text{ cm}} \right)^2 = 0.69 \text{ cm/s}$

**Insight** The fact that a fluid flows more rapidly when squeezed is used in many different applications. Can you think of some others?

## Practice Quiz

8. If the area through which an incompressible fluid flows decreases, the speed of flow will
   **(a)** decrease. **(b)** increase. **(c)** stay the same.

9. If the area through which an incompressible fluid flows is cut in half, the speed of flow will
   **(a)** also be cut in half.
   **(b)** double.
   **(c)** decrease to ¼ its speed.
   **(d)** triple.
   **(e)** None of the above.

## 15–7 – 15–8 Bernoulli's Equation

In general, the same concepts that we use to describe the dynamics of particles also apply to fluid dynamics. With fluids it is often more convenient to express these concepts in terms of density and pressure rather than mass and force. One such example comes from the application of the work-energy theorem to fluids. The result is a mathematical relation known as **Bernoulli's equation**.

With a fluid we can replace the concept of a particle with a small region of the fluid called a *fluid element* of density, $\rho$, moving at speed, $\upsilon$, while sweeping out a volume, $\Delta V$, under the action of a differential fluid pressure $\Delta P$. Bernoulli's equation can be obtained by applying the work-energy theorem ($W_{net} = \Delta K$) to this fluid element. The work done on a fluid element due to a change in the fluid pressure is $\Delta W_{pressure} = (P_1 - P_2)(\Delta V)$. The work done by gravity as the fluid element changes vertical level from a height $y_1$ to $y_2$ is $\Delta W_{gravity} = -\rho\Delta V(y_2 - y_1)$. The change in the kinetic energy of the fluid element is $\Delta K = \left(\frac{1}{2}\rho\upsilon_2^2 - \frac{1}{2}\rho\upsilon_1^2\right)\Delta V$. These three quantities combine to give Bernoulli's equation

$$P_1 + \tfrac{1}{2}\rho\upsilon_1^2 + \rho g y_1 = P_2 + \tfrac{1}{2}\rho\upsilon_2^2 + \rho g y_2$$

This expression holds in the absence of frictional losses.

Another way of stating Bernoulli's equation is

$$P + \tfrac{1}{2}\rho\upsilon^2 + \rho g y = \text{constant}$$

This form of the equation helps make the physical consequences a little more clear because you can see, for example, that for a fluid flowing at a constant vertical level, an increase in the speed of flow must be accompanied by a decrease in pressure (and vice versa). This effect, often called *Bernoulli's principle*, is important in understanding the consequence of airflow in many applications.

You should also notice that Bernoulli's equation is consistent with the dependence of pressure on depth that was discussed above. Examining this dependence leads to **Torricelli's law** for the speed of a fluid flowing from an aperture in a container placed at depth, $h$, below the surface of the fluid. If, for example, both the surface of the fluid and the aperture are open to the air, then the pressure at both locations is atmospheric pressure, so that $P_1 = P_2$ in Bernoulli's equation. Assuming that the fluid is essentially static at the surface, ($\upsilon_1 = 0$) we find

$$\upsilon_2 = \sqrt{2gh}$$

where $h = y_1 - y_2$.

**Example 15–7 Water Flow at Constant Pressure** Water flows in a horizontal segment of pipe with a speed of 2.6 m/s. The pipe widens, so that its area becomes 35% larger. If the flow is to be at constant pressure, how far above the initial horizontal level should the pipe divert the water?

**Picture the Problem** The diagram shows a pipe carrying water first horizontally, then uphill, and horizontally again.

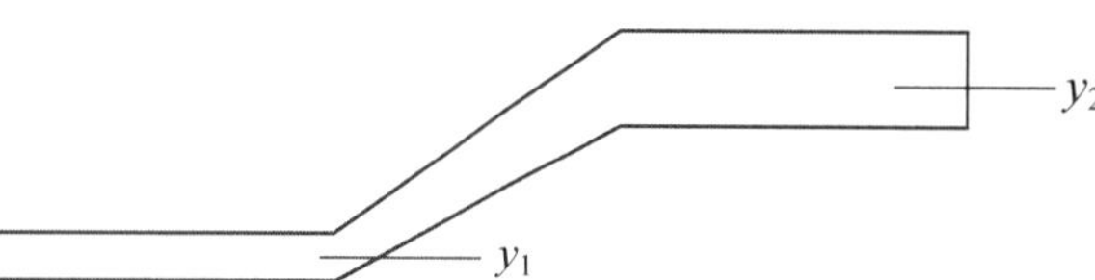


**Strategy** Bernoulli's equation relates the pressure to the speed of flow and the change in vertical level. However, the changing width of the pipe can be handled by the equation of continuity, so both expressions should be used.

**Solution**

1. Using the fact that the pressure is constant, $P_1 = P_2$, Bernoulli's equation becomes:

$$\frac{1}{2}\rho v_1^2 + \rho g y_1 = \frac{1}{2}\rho v_2^2 + \rho g y_2$$

2. Solve for the difference in horizontal level:

$$(y_2 - y_1) = \frac{1}{2g}\left(v_1^2 - v_2^2\right)$$

3. Use the continuity equation to solve for $v_2$ in terms of $v_1$:

$$v_2 = \frac{A_1 v_1}{A_2} = \frac{A_1 v_1}{A_1 + 0.35A_1} = \frac{v_1}{1.35}$$

4. Substitute $v_2$ into the expression for the change in vertical level:

$$(y_2 - y_1) = \frac{1}{2g}\left[v_1^2 - (v_1/1.35)^2\right] = (0.451)\frac{v_1^2}{2g}$$

5. Obtain the numerical result:

$$(y_2 - y_1) = (0.451)\frac{(2.6\text{ m/s})^2}{2\left(9.81\text{ m/s}^2\right)} = 0.16\text{ m}$$

**Insight** As this example illustrates, you should keep in mind that it is often useful to use both Bernoulli's equation and the equation of continuity to perform a complete analysis.

## Practice Quiz

**10.** If the speed of a horizontally flowing fluid decreases, the pressure in the fluid will

**(a)** increase. **(b)** decrease. **(c)** stay the same.

**11.** If, for a constant speed of flow, a fluid begins to flow downhill, the pressure in the fluid begins to

**(a)** increase. **(b)** decrease. **(c)** stay the same.

## *15–9 Viscosity and Surface Tension

### Viscosity

When a particle moves, it usually experiences some sort of frictional resistance to its motion. The same is true for fluid flow. With fluids, this resistance is called **viscosity**. All of the previous discussion assumed an ideal fluid, which, in part, means that we assumed that there was no viscosity. In this section, we will take a glimpse at the effect of including this unavoidable phenomenon.

For a fluid flowing through a tube of length, $L$, and cross-sectional area, $A$, the flow results from a pressure differential across the length of the tube, $P_1 - P_2$. What are some of the physical quantities related to this pressure difference? It works out that the pressure difference is directly proportional to the speed, $v$, at which the fluid flows (it requires a greater pressure difference to get faster flow). We also find that $P_1 - P_2$ is directly proportional to the length of the tube (it requires a greater pressure difference to sustain the flow of the larger amount of fluid contained in a longer tube). Finally, the pressure difference is inversely proportional to the cross-sectional area of the tube (a wider tube provides more room for the fluid to flow more freely requiring less pressure). Thus, we have that $(P_1 - P_2) \propto vL/A$. The constant of proportionality turns out to be $8\pi\eta$, where $\eta$ is called the **coefficient of viscosity**. Therefore, we have the following relation

$$P_1 - P_2 = 8\pi\eta\frac{vL}{A}$$

We can see from this equation that larger values of $\eta$ mean the fluid is more viscous because more pressure is needed for the fluid to flow at a certain speed. The SI unit of $\eta$ is $\text{N}\cdot\text{s/m}^2$; however, a commonly used unit of $\eta$ is the *poise*, which is defined as

$$1 \text{ poise} = 1 \text{ dyne}\cdot\text{s/cm}^2 = 0.1 \text{ N}\cdot\text{s/m}^2$$

As mentioned previously, fluid flow is often characterized by the volume flow rate, $\Delta V/\Delta t = Av$. When the above expression for viscous flow is rewritten in terms of the volume flow rate for a tube of circular cross section ($A = \pi r^2$), we get Poiseuille's equation:

$$\frac{\Delta V}{\Delta t} = \frac{(P_1 - P_2)\pi r^4}{8\eta L}$$

which shows that the volume flow rate increases as the fourth power of the radius.

**Surface Tension**

The surface of a fluid, especially a liquid, is observed to behave in a way similar to that of an elastic membrane when objects are placed on the surface. This effect is due to the **surface tension** of the fluid. Surface tension results from internal forces between the molecules of a fluid that collectively resist the deformation of the fluid's surface from its equilibrium configuration. This effect is analogous to what happens in a spring and is responsible for the ability of insects to walk on water.

---

**Exercise 15–8 Viscosity** Using the result of Example 15–3, estimate the volume flow rate of blood from the head to the feet of the six-foot-tall person. (Assume an effective radius of 23 cm.)

**Solution:** We know the following information:

**Given:** $P_1 - P_2 = 18.8$ kPa, $L = 1.83$ m, $r = 0.23$ m, $\eta = 0.00272$ N·s/m$^2$; **Find:** $\Delta V/\Delta t$

Using the given information, together with the viscosity of blood (from Table 15-3 in the text), we can directly use Poiseuille's equation:

$$\frac{\Delta V}{\Delta t} = \frac{(P_1 - P_2)\pi r^4}{8\eta L} = \frac{\pi\left(18.8\times10^3 \text{ Pa}\right)\left(0.23\text{ m}\right)^4}{8\left(0.00272 \text{ N}\cdot\text{s/m}^2\right)\left(1.83\text{ m}\right)} = 4.2\times10^3 \text{ m}^3/\text{s}$$

Notice that we used three significant figures for the pressure difference. We did this because the final result of Example 15–3 is now just an intermediate result for this example.

---

### Practice Quiz

**12.** In general, we expect the volume flow rate of a fluid with a large coefficient of viscosity to be

**(a)** large. **(b)** small. **(c)** $\eta$ has no relevance to the volume flow rate.

## Reference Tools and Resources

### I. Key Terms and Phrases

**fluid** a liquid or a gas

**density** a measure of the compactness of an object or substance, given by its mass per unit volume

**pressure** the normal force per unit area acting on an object or within a fluid

**gauge pressure** the measure of pressure that excludes the atmospheric pressure

**Pascal's principle** the principle that an external pressure is transmitted undiminished throughout a fluid

**buoyancy** the phenomenon that fluid pressure applies an upward force on immersed objects

**Archimedes' principle** the buoyant force equals the weight of the fluid displaced by an immersed object

**flotation** occurs when the buoyant force equals an object's weight

**equation of continuity** the equation that expresses the constant mass flow rate within a fluid

**Bernoulli's equation** the application of the work-energy theorem to fluid flow

**Torricelli's law** the equation that determines the speed with which a fluid will flow from an aperture below the surface of a fluid in an open container

**viscosity** resistance to fluid flow

**surface tension** the surfaces of fluids often behave in a way similar to an elastic membrane

## II. Important Equations

| Name/Topic | Equation | Explanation |
|---|---|---|
| density | $\rho = \dfrac{M}{V}$ | The density of a fluid is the mass of the fluid divided by its volume |
| pressure | $P = \dfrac{F}{A}$ | The pressure in a fluid is the normal force per unit area |
| pressure with depth | $P_2 = P_1 + \rho gh$ | The pressure in a fluid increases with depth, $h$ |
| buoyancy | $F_b = \rho gV$ | The buoyant force on an object when a volume $V$ is submerged in a fluid of density $\rho$ |
| equation of continuity | $A_1 v_1 = A_2 v_2$ | The equation of continuity for an incompressible fluid |
| Bernoulli's equation | $P_1 + \frac{1}{2}\rho v_1^2 + \rho g y_1 = P_2 + \frac{1}{2}\rho v_2^2 + \rho g y_2$ | The work-energy theorem applied to fluid flow |
| viscosity | $P_1 - P_2 = 8\pi\eta \dfrac{vL}{A}$ | The pressure difference required to cause a viscous fluid to flow with speed, $v$, through a pipe of length, $L$, and cross-sectional area $A$ |

## III. Know Your Units

| Quantity | Dimension | SI Unit |
|---|---|---|
| density ( $\rho$ ) | $[M]/[L^3]$ | $kg/m^3$ |
| pressure ( $P$ ) | $[M][L^{-1}][T^{-2}]$ | Pa |
| coefficient of viscosity ( $\eta$ ) | $[M][L^{-1}][T^{-1}]$ | $N{\cdot}s/m^2$ |

# Puzzle

**SINK OR SWIM**

A rectangular block of wood floats submerged, 70% in water 30% in oil, as shown in the sketch. What happens if you add some more oil? What happens if you add some more water? What happens if you siphon off the oil? Do the percentages change? Which way?

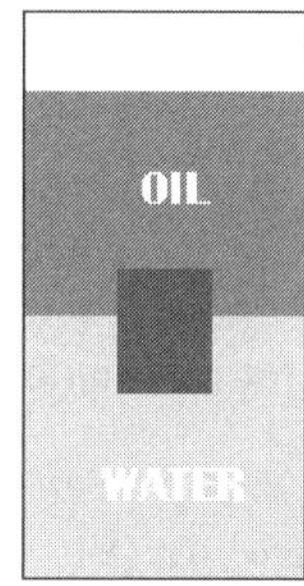


# Answers to Selected Conceptual Questions

**4.** A suction cup is held in place by atmospheric pressure. When the cup is applied, you push it flat against the surface you want to stick it to. This expels most of the air in the cup, and leads to a larger pressure on the outside of the cup. Thus, atmospheric pressure pushes the outside of the cup against the surface.

**8.** In a hot-air balloon, vertical motion is controlled by adding heat to the air in the balloon, or by letting it cool off. As the temperature of the air in the balloon changes, so too does its density. By controlling the overall density of the balloon, one can control whether it rises, falls, or is neutrally buoyant.

**10.** The physics in this case is pretty "ugly". Ice floats in water, whether it is a house-sized iceberg, a car-sized chunk, or a thimble-sized ice cube. If the Earth is warming and icebergs are breaking up into smaller pieces, each of the smaller pieces will be just as buoyant as the original berg.

# Solutions to Selected End-of-Chapter Problems and Conceptual Exercises

11. **Picture the Problem**: When you ride a bicycle, your weight and the weight of the bicycle are supported by the air pressure in both tires spread out over the area of contact between the tires and the road.

**Strategy:** To calculate your weight, first solve equation 15-2 for supporting force of the air pressure on the tires. Set this force equal to the sum of your weight and the weight of the bicycle. Subtract the weight of the bicycle to determine your weight.

**Solution: 1.** Multiply the tire pressure by the contact area to calculate the supporting force on the bicycle:

$$F = PA = \left(70.5\ \text{lb/in}^2\right)\left(\frac{1.01\times10^5\ \text{Pa}}{14.7\ \text{lb/in}^2}\right)\left(2\times7.13\ \text{cm}^2\right)\left(\frac{1\ \text{m}}{100\ \text{cm}}\right)^2 = \underline{\underline{690.7\ \text{N}}}$$

**2.** Set the supporting force equal to the sum of your weight and the weight of the bicycle:

$$F = W_{\text{you}} + W_{\text{bicycle}} = W_{\text{you}} + m_{\text{bicycle}}g$$

**3.** Solve for your weight:

$$W_{\text{you}} = F - m_{\text{bicycle}}g = 690.7\ N - 7.7\text{kg}\left(9.8\text{m/s}^2\right) = \boxed{615\ \text{N}}$$

**Insight:** When "popping a wheelie" on the bicycle, such that only one wheel is touching the ground, that wheel must support the entire weight of the bicycle and rider. Therefore, because the tire pressure has not changed, the area of contact for the single tire would double. In this problem the area would increase to 14.26 cm$^2$.

27. **Picture the Problem**: An IV solution is elevated above the injection point. The pressure in the bag is atmospheric pressure, while the pressure at the injection point is 109 kPa.

**Strategy:** Solve equation 15-6 for the height of the solution above the injection point.

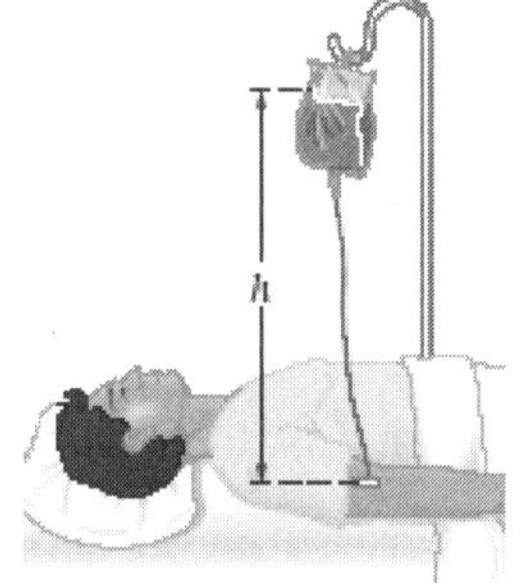


**Solution: 1. (a)** Solve equation 15-6 for $h$:

$$P = P_{\text{at}} + \rho gh \Rightarrow h = \frac{P - P_{\text{at}}}{\rho g}$$

**2.** Insert the given data:

$$h = \frac{109\ \text{kPa} - 101.3\ \text{kPa}}{\left(1020\ \text{kg/m}^3\right)\left(9.81\ \text{m/s}^2\right)} = \boxed{0.770\ \text{m}}$$

**3. (b)** From the equation in step 1, we see that the height is inversely proportional to the density of the fluid. Therefore if a less dense fluid is used, the height must be $\boxed{\text{increased}}$.

**Insight:** If the density of the fluid were reduced to 920 kg/m$^3$, the bag would need to be suspended at a height of 0.853 m, which is, as predicted, higher than the 0.770 meters.

34. **Picture the Problem**: A helium balloon displaces heavier air, causing it to rise. We want to calculate the maximum additional weight that the balloon can lift.

**Strategy:** Calculate the buoyant force on the balloon using equation 15-9 with the density of air and the volume of a sphere. Subtract the weight of the balloon and the weight of the helium ( $W = \rho_{He}Vg$ ) to calculate the additional weight the balloon can lift.

**Solution: 1.** Calculate the buoy-ant force on the balloon:

$$F_{\text{b}} = \rho_{\text{air}}Vg = \rho_{\text{air}}\left[\frac{4}{3}\pi r^3\right]g = \left(1.29\ \text{kg/m}^3\right)\left[\frac{4}{3}\pi\left(4.9\text{m}\right)^3\right]\left(9.81\ \text{m/s}^2\right) = \underline{\underline{6236\ \text{N}}}$$

**2.** Subtract the weight of the balloon and of the helium:

$$W = F_b - m_{balloon}g - \rho_{He}V_{He}g = 6236\text{ N} - (3.2\text{ kg})(9.81\text{ m/s}^2)$$
$$-(0.179\text{ kg/m}^3)\left[\frac{4}{3}\pi(4.9\text{ m})^3\right](9.81\text{ m/s}^2) = \boxed{5.3\text{ kN}}$$

**Insight:** In this problem the weight of the balloon (31 N) is negligible. The additional mass that the balloon can lift is equal to the difference in densities of the air and helium times the volume of the balloon.

47. **Picture the Problem**: A person is floating in water. Because the person is not accelerating, the buoyant force must equal her weight. When an upward force is applied to the swimmer, her body rises out of the water such that the new buoyant force and the applied force are equal to her weight.

**Strategy:** Set the buoyant force (equation 15-9) equal to the person's weight and solve for the submerged volume. Subtract the submerged volume from the total volume to calculate the volume above the surface. Set the sum of the buoyant force and the applied force equal to the weight and solve for the applied force.

**Solution: 1. (a)** Set the buoyant force equal to the weight and solve for volume submerged:

$$F_b = W$$
$$\rho_w V_{sub} g = mg \quad \Rightarrow \quad V_{sub} = \frac{m}{\rho_w}$$

**2.** Subtract submerged volume from total volume:

$$V_{above} = V_{total} - V_{sub} = V_{total} - \frac{m}{\rho_w}$$
$$= 0.089\text{ m}^3 - \frac{81\text{ kg}}{1000\text{ kg/m}^3} = \boxed{0.008\text{ m}^3}$$

**3. (b)** Set the applied force and buoyant force equal to the weight and solve for the applied force:

$$F_{applied} + F_b = W$$
$$F_{applied} = W - F_b = mg - \rho_w V_{sub2} g$$

**4.** Set the submerged volume equal to the initially submerged volume minus the given change in volume:

$$F_{applied} = mg - \rho_w\left(V_{sub} - 0.0018\text{ m}^3\right)g$$
$$= mg - \rho_w\left(\frac{m}{\rho_w} - 0.0018\text{ m}^3\right)g$$
$$= \rho_w\left(0.0018\text{ m}^3\right)g$$
$$F_{applied} = (1000\text{ kg/m}^3)(0.0018\text{ m}^3)(9.81\text{ m/s}^2) = \boxed{18\text{ N}}$$

**Insight:** Note that the applied force equals the change in the buoyant force on the person, that is the density of the water times the change in volume submerged times gravity.

55. **Picture the Problem**: This solution contains an art image.

**Strategy**: The volume flow rate is equal to the cross-sectional area of the blood vessel times the velocity of the blood through the blood vessel. Divide the flow rate by the cross-sectional area to calculate the speed. In the capillaries the flow rate through each capillary will be equal to the total flow rate divided by the number of capillaries.

**Solution: 1. (a)** Divide the volume flow rate in the arteriole branch by the branch's cross-sectional area:

$$\frac{\Delta V}{\Delta t} = Av \Rightarrow v = \frac{\frac{\Delta V}{\Delta t}}{A} = \frac{5.5\times10^{-6}\ \text{cm}^3/\text{s}}{\pi\left[\frac{1}{2}(0.0030\ \text{cm})\right]^2} = \boxed{0.78\ \text{cm/s}}$$

**2. (b)** Divide the volume flow rate in the capillaries (1/340th of the arteriole branch flow rate) by the cross-sectional area of the capillary:

$$v = \frac{\frac{\Delta V}{\Delta t}}{A} = \frac{\left(\frac{5.5\times10^{-6}\ \text{cm}^3/\text{s}}{340}\right)}{\pi\left[\frac{1}{2}(4.0\times10^{-4}\ \text{cm})\right]^2} = \boxed{0.13\ \text{cm/s}}$$

**Insight:** The blood speed in the capillaries is much slower than in the other blood vessels.

63. **Picture the Problem**: Water sprays out of a leak in a stationary garden hose and rises up to a height of 0.68 meters.

**Strategy:** Solve equation 15-15 for the pressure $P_1$, where point 1 is inside the garden hose with $y_1 = 0$ and point 2 is the highest point that the water sprays.

**Solution:** Solve equation 15-15 for $P_1$:

$$P_1 + \rho g y_1 = P_2 + \rho g y_2$$
$$P_1 = P_{\text{atm}} + \rho g y_2$$
$$= 1.01\times10^5\ \text{Pa} + (1000\ \text{kg/m}^3)(9.81\ \text{m/s}^2)(0.68\ \text{m})$$
$$P_1 = \boxed{1.08\times10^5\ \text{Pa}}$$

**Insight:** When the water is flowing through the garden hose, the spray does not reach as high as when the water is stationary. This is because the pressure inside the hose is less when the speed is not zero.

67. **Picture the Problem**: The pressure difference between the inside of the house, where the wind speed is zero, and the top of the roof, where the wind speed is $v = 47.7$ m/s, creates an upward force on the roof, as shown in the figure.

**Strategy:** Solve Bernoulli's equation (equation 15-14) for the change in pressure across the roof. Multiply the change in pressure by the area of the roof to calculate the net force on the roof.

**Solution: 1. (a)** Solve equation 15-14 for the change in pressure:

$$P_{\text{top}} + \tfrac{1}{2}\rho v_{\text{top}}^2 = P_{\text{in}} + \tfrac{1}{2}\rho v_{\text{in}}^2$$
$$P_{\text{in}} - P_{\text{top}} = \tfrac{1}{2}\rho\left(v_{\text{top}}^2 - v_{\text{in}}^2\right)$$

**2.** Multiply the change in pressure by the area of the roof:

$$\Delta F = \Delta P A = \left[\tfrac{1}{2}\rho\left(v_{\text{top}}^2 - v_{\text{in}}^2\right)\right]A$$
$$= \tfrac{1}{2}(1.29\ \text{kg/m}^3)\left[(47.7\ \text{m/s})^2 - 0\right](668\ \text{m}^2) = \boxed{980\ \text{kN}}$$

**3. (b)** The force is directed $\boxed{\text{upward}}$. Stationary air exerts a larger pressure than air moving rapidly past the outside of the roof.

**Insight:** This upward force is over 100 tons of force. It is no wonder that many roofs are blown off during a high wind storm!

73. **Picture the Problem**: Water flows through a garden hose of known diameter at a given volume flow rate. We wish to calculate the water speed and pressure drop across the hose.

**Strategy:** Divide the volume flow rate by the cross-sectional area of the hose to calculate the water velocity. Solve Poiseuille's equation (15-19) for the pressure change across the hose. To calculate the effect of cutting the cross-sectional area in half on the velocity and flow rate, find the ratio of the initial to final radii and solve Poiseuille's equation for the ratio of the flow rates. Calculate the ratio of the velocities by dividing the flow rates by their corresponding cross-sectional areas.

**Solution: 1. (a)** Divide flow rate by cross-sectional area to calculate velocity:

$$v = \frac{\left(\frac{\Delta V}{\Delta t}\right)}{A} = \frac{5.0\times10^{-4}\ \text{m}^3/\text{s}}{\pi\left(1.25\times10^{-2}\ \text{m}\right)^2} = \boxed{1.0\ \text{m/s}}$$

**2. (b)** Solve Poiseuille's equation for $\Delta P$:

$$\Delta P = \frac{8\eta L(\Delta V/\Delta t)}{\pi r^4}$$

$$= \frac{8\left(1.0055\times10^{-3}\ \text{N}\cdot\text{s/m}^2\right)\left(15\ \text{m}\right)\left(5.0\times10^{-4}\ \text{m}^3/\text{s}\right)}{\pi\left(1.25\times10^{-2}\ \text{m}\right)^4}$$

$$\Delta P = \boxed{0.79\ \text{kPa}}$$

**3. (c)** Set the new area equal to half the initial area and solve for the ratio of radii:

$$A_2 = \tfrac{1}{2}A_1$$

$$\pi r_2^2 = \tfrac{1}{2}\pi r_1^2 \Rightarrow r_2 = r_1/\sqrt{2}$$

**4.** Solve Poiseuille's equation for the ratio of flow rates:

$$\left(\frac{\Delta V}{\Delta t}\right)_2 \Big/ \left(\frac{\Delta V}{\Delta t}\right)_1 = \frac{(P_1 - P_2)\pi r_2^4}{8\eta L} \Big/ \frac{(P_1 - P_2)\pi r_1^4}{8\eta L}$$

$$= \left(\frac{r_2}{r_1}\right)^4 = \left(\frac{r_1/\sqrt{2}}{r_1}\right)^4 = \frac{1}{4}$$

**5.** Divide the flow rates by the area to calculate the ratio of velocities:

$$\frac{v_2}{v_1} = \frac{(\Delta V/\Delta t)_2/A_2}{(\Delta V/\Delta t)_1/A_1} = \left(\frac{1}{4}\right)\left(\frac{A_1}{\frac{1}{2}A_1}\right) = \frac{1}{2}$$

The water speed is multiplied by a factor of $\boxed{1/2}$.

**6. (d)** Use the ratio of flow rates from step 4:

The volume flow rate is multiplied by a factor of $\boxed{1/4}$.

**Insight:** Because the volume flow rate is proportional to the fourth power of the radius, and the cross-sectional area is proportional to the square of the radius, the volume flow rate is proportional to the square of the cross-sectional area. Equivalently the water speed is proportional to the cross-sectional area.

## Answers to Practice Quiz

**1.** (b) **2.** (d) **3.** (c) **4.** (a) **5.** (e) **6.** (a) **7.** (e) **8.** (b) **9.** (b) **10.** (a) **11.** (a) **12.** (b)

# CHAPTER 16
# TEMPERATURE AND HEAT

## Chapter Objectives

After studying this chapter, you should

1. understand the meaning of thermal equilibrium and the role of temperature.
2. know how to convert among the Celsius, Fahrenheit, and Kelvin temperature scales.
3. understand the basic phenomenology of thermal expansion.
4. understand the relationship between heat and work.
5. understand the difference between specific heat and heat capacity.
6. be able to perform basic calorimetry calculations.
7. know and describe the 3 main processes for heat transfer.
8. understand the basic phenomenology of heat transfer by conduction and radiation.

## Warm-Ups

1. A piece of wood at 130 degrees C can be picked up comfortably, but a piece of aluminum at the same temperature will give a painful burn. Why is this?
2. Estimate the amount of heat necessary to raise your body temperature by 1 degree Fahrenheit.
3. A metal plate has a circular hole cut in it. If the temperature of the plate increases, will the diameter of the hole increase, decrease, or remain unchanged?

4. In an attempt to open a new jar of peanut butter, you run very hot tap water over the (steel) lid. Estimate the change in the diameter of the lid.

## Chapter Review

This is the first of three chapters on thermodynamics, which can loosely be described as the study of heat and the physical processes associated with heat transfer. In this chapter, we focus on the concepts of temperature and heat as well as a few physical consequences of heat transfer.

## 16–1 Temperature and the Zeroth Law of Thermodynamics

In this section, we seek to develop working definitions of **temperature** and **heat**. The two concepts are intimately related and must be defined together. Heat is a form of energy; it is energy that flows between two systems. Systems are said to be in **thermal contact** when it is possible for heat to flow between them; however, heat does not always flow between systems that are in thermal contact. The property of systems that determines whether heat flow will occur is called the *temperature*. If there is a temperature difference between two systems that are in thermal contact, heat will flow from the system with higher temperature to the system with lower temperature. Two systems are said to be in **thermal equilibrium** if, when they are brought into thermal contact, no heat transfer occurs. In this latter case, the systems must have equal temperatures. These statements are embodied in the **zeroth law of thermodynamics**:

> *If system A is in thermal equilibrium with system B, and system C is also in thermal equilibrium with system B, then systems A and C will be in thermal equilibrium if brought into thermal contact.*

We thus obtain the following working definitions:

- *Heat* is the energy that is transferred between systems because of a temperature difference.
- *Temperature* is the property of systems that determines the existence and direction of the heat flow between systems when they are in thermal contact.

Temperature is a new fundamental quantity, not defined in terms of length, mass, and time. It is assigned the dimensional symbol [K].

### Practice Quiz

1. Which of the following statements is most accurate?
   **(a)** If systems A and B are at different temperatures, heat will flow from system A to system B.
   **(b)** If systems A and B are at different temperatures, heat will flow from system B to system A.
   **(c)** If two systems are at different temperatures, heat will flow from one system to the other.
   **(d)** If two systems are at different temperatures, heat will flow from the system with higher temperature to the system with lower temperature.
   **(e)** If two systems are at different temperatures, heat will flow from the system with higher temperature to the system with lower temperature if they are brought into thermal contact.

## 16–2 Temperature Scales

The primary temperature scales in common use are the Celsius, Fahrenheit, and Kelvin scales. Each scale is based on different choices for setting values for two convenient fixed points. The Celsius scale takes the freezing temperature of water to be 0 °C and the boiling temperature of water to be 100 °C. On the Fahrenheit scale, water freezes at 32 °F and boils at 212 °F. In addition to having different settings for the freezing and boiling points of water, the scales of these two systems are different in that a temperature change of one degree on the Fahrenheit scale corresponds to a change of only 5/9 degrees on the Celsius scale. Conversions between the Celsius and Fahrenheit scales can be accomplished using the following formulas:

$$T_F = \left(\frac{9}{5}\,^\circ\text{F}/^\circ\text{C}\right)T_C + 32\,^\circ\text{F}$$

$$T_C = \left(\frac{5}{9}\,^\circ\text{C}/^\circ\text{F}\right)\left(T_F - 32\,^\circ\text{F}\right)$$

---

**Exercise 16–1 Change in Temperature** One day in Michigan, the temperature rose from a morning low of –8.0 °F to an afternoon high of 22 °F. By how many Celsius degrees did the temperature rise?

**Solution:** We are given the following information:

**Given:** $T_i = -8.0$ °F, $T_f = 22$ °F; **Find:** $\Delta T$ in C°

Since we know how to convert from Fahrenheit to Celsius, let's find an expression for $\Delta T_C$:

$$\Delta T_C = T_{C,f} - T_{C,i} = \left(\frac{5}{9}\,^\circ\text{C}/^\circ\text{F}\right)\left(T_{F,f} - 32\,^\circ\text{F}\right) - \left(\frac{5}{9}\,^\circ\text{C}/^\circ\text{F}\right)\left(T_{F,i} - 32\,^\circ\text{F}\right) = \left(\frac{5}{9}\,^\circ\text{C}/^\circ\text{F}\right)\left(T_{F,f} - T_{F,i}\right)$$

Therefore, we have

$$\Delta T_C = \left(\frac{5}{9}\,^\circ\text{C}/^\circ\text{F}\right)\left[22\,^\circ\text{F} - \left(-8.0\,^\circ\text{F}\right)\right] = 17\,\text{C}^\circ$$

---

The Kelvin temperature scale is based on the existence of a temperature below which (even *to* which) it is impossible to cool any system. This temperature is called **absolute zero**. The Kelvin scale is identical to the Celsius scale except that the zero mark is shifted to correspond to absolute zero (–273.15 °C). The conversion between temperatures on the Kelvin scale and the Celsius scale is as follows:

$$T = T_C + 273.15$$

### Quoting Temperatures

You should be aware of the following with regard to quoting temperatures and temperature differences:

- When quoting a temperature in Celsius or Fahrenheit, the degree symbol is both written and spoken first. That is, 10 °C ("10 degrees Celsius"), for example.
- When quoting temperature differences in Celsius or Fahrenheit, the degree symbol is both written and spoken last. That is, 10 C° ("10 Celsius degrees"), for example.
- The Celsius scale is no longer alternatively referred to as the centigrade scale.
- No degree symbol is written or spoken with the Kelvin scale. Therefore, both temperature values and temperature differences are identically quoted. For example, the temperature of 10 kelvins and a temperature difference of 10 kelvins are both written as 10 K.

---

**Exercise 16–2 A Temperature Conversion** The normal body temperature of 98.6 °F corresponds to what temperature on the Kelvin scale?

**Solution:** We are given the following information:

**Given:** $T_F = 98.6$ °F; **Find:** $T$

Because we know how to convert from Fahrenheit to Celsius and from Celsius to Kelvin, we can combine the two to get a conversion from Fahrenheit to Kelvin. To get $T_C$ we use

$$T_C = \left(\frac{5}{9}\ {}^\circ\text{C}/{}^\circ\text{F}\right)\left(T_F - 32\ {}^\circ\text{F}\right)$$

and substitute this expression into the conversion for $T$ to get

$$T = T_C + 273.15 = \left(\frac{5}{9}\ {}^\circ\text{C}/{}^\circ\text{F}\right)\left(T_F - 32\ {}^\circ\text{F}\right) + 273.15$$

Therefore,

$$T = \left(\frac{5}{9}\ {}^\circ\text{C}/{}^\circ\text{F}\right)\left(98.6\ {}^\circ\text{F} - 32\ {}^\circ\text{F}\right) + 273.15 = 310\ \text{K}$$

---

## Practice Quiz

**2.** A temperature change of 1 kelvin corresponds to how much of a change in Fahrenheit degrees?

**(a)** 0.55 **(b)** 1.8 **(c)** 5/9 **(d)** 32 **(e)** 273.15

**3.** What temperature corresponds to absolute zero on the Fahrenheit scale?

**(a)** 0 °F **(b)** –273.15 °F **(c)** – 459.67 °F **(d)** – 212 °F **(e)** – 100 °F

## 16–3 Thermal Expansion

Most substances expand when heated and contract when cooled. The amount of expansion is different for different substances. We can identify some aspects of this behavior that are the same for nearly all substances. It is found that the amount by which a substance will expand, that is, its change in length ($\Delta L$), area ($\Delta A$), or volume ($\Delta V$), is directly proportional to the temperature change ($\Delta T$) that drives the expansion. Furthermore, these changes in size are also directly proportional to the original size ($L_0$, $A$, or $V$) of the object being heated or cooled.

Thus, for the linear expansion of an object we have $\Delta L \propto L_0 \Delta T$. The constant of proportionality depends on the substance and is called the **coefficient of linear expansion**, $\alpha$. An expression that describes the linear expansion of an object due to a change in its temperature is

$$\Delta L = \alpha L_0 \Delta T$$

As you can determine by dimensional analysis, the SI unit of $\alpha$ is $K^{-1}$ (the non SI unit $C^{\circ -1}$ is also in common use). Values of the coefficient of linear expansion for different substances can be found in Table 16–1 of the text.

For the expansion of areas, the description is very much like that of linear expansion. To a very high approximation, the coefficient of area expansion works out to be twice that of the coefficient of linear expansion, leading to the expression

$$\Delta A = (2\alpha) A (\Delta T)$$

Therefore, the same table of values (Table 16–1) can be used to determine area expansions. For volume expansions, we have $\Delta V \propto V \Delta T$, with a proportionality constant $\beta$ called the **coefficient of volume expansion**. Values of $\beta$ for different substances are listed in Table 16–1. For substances not listed in the table under volume expansion, a good approximation is to take $\beta \approx 3\alpha$; therefore, we have

$$\Delta V = \beta V (\Delta T)$$

**Example 16–3 The Expansion of Iron** An iron cube has an edge length of 2.300 cm. If its temperature is raised from –12.00 °C to 22.00 °C, what is its volume at the higher temperature?

**Picture the Problem** The sketch shows exaggerated views of an iron cube **(a)** before and **(b)** after the temperature has been raised.

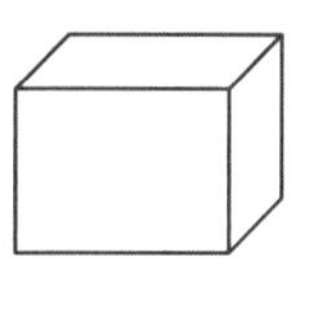


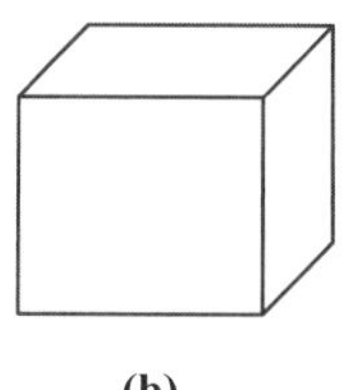


**Strategy** We apply the relation that describes volume expansion to determine $\Delta V$, then add this change to the original volume to get the final result.

**Solution**

1. Calling the length of one edge, $L$, determine the original volume of the cube: $V = L^3 = (2.300\text{ cm})^3 = 12.167\text{ cm}^3$

2. Since no value of $\beta$ is given for iron in Table 16–1, make use of the result for $\alpha$: $\beta = 3\alpha = 3\,(12 \times 10^{-6}\text{ K}^{-1}) = 36 \times 10^{-6}\text{ K}^{-1}$

3. Find the temperature difference: $\Delta T = 22.00\text{ °C} - (-12.00\text{ °C}) = 34.00\text{ C°}$

4. Find the change in volume:
$$\Delta V = \beta V(\Delta T) = (36\times10^{-6}\text{ K}^{-1})(12.167\text{ cm}^3)(34.00\text{ K}) = 0.014892\text{ cm}^3$$

5. Then, calculate the final volume:
$$V_f = V + \Delta V = 12.167\text{ cm}^3 + 0.014892\text{ cm}^3 = 12.18\text{ cm}^3$$

**Insight** In order to show a meaningful change in the volume, we kept four significant figures despite the fact that we were given only two figures for the value of $\alpha$. Also, notice that we calculated the temperature change in Celsius degrees but used the change in Kelvin. This latter practice is permissible only when dealing with temperature *differences* on these scales.

## Practice Quiz

4. What is the coefficient of volume expansion for lead?

   **(a)** $29\times10^{-6}\text{ K}^{-1}$ **(b)** $58\times10^{-6}\text{ K}^{-1}$ **(c)** $87\times10^{-6}\text{ K}^{-1}$ **(d)** $44\times10^{-6}\text{ K}^{-1}$ **(e)** $99\times10^{-6}\text{ K}^{-1}$

5. If a 12.6-cm copper rod is cooled from 53.2 °C to –10.8 °C, by how much will its length change?

   **(a)** $1.4\times10^{-2}$ cm **(b)** $9.1\times10^{-3}$ cm **(c)** $1.1\times10^{-2}$ cm **(d)** $2.3\times10^{-3}$ cm **(e)** $1.7\times10^{-5}$ cm

6. A steel ($\alpha = 12 \times 10^{-6}$ C°) container with a volume of 525 cm$^3$ is filled with oil ($\beta = 0.7 \times 10^{-3}$ C°). If the temperature is increased by 30.8 C°, how many cubic centimeters of oil will overflow?
   **(a)** 7 **(b)** 31 **(c)** 0 **(d)** 25 **(e)** 11

## 16–4 Heat and Mechanical Work

We have already noted that heat, $Q$, is a transfer of energy. From previous chapters we also know that there is a close relationship between the transfer of energy and mechanical work. It follows then that heat can at least sometimes be converted into mechanical work. Two specialized units have been adopted for dealing with the mechanical work associated with heat flow.

One unit of heat in common use is the **calorie** (cal). One calorie is the amount of heat needed to change the temperature of 1 gram of water by 1 Celsius degree. The equivalent amount of energy in joules is called the **mechanical equivalent of heat**:

$$1 \text{ cal} = 4.186 \text{ J}$$

Also in common use is the kilocalorie (kcal), sometimes called a "food calorie" because it is used when quoting the energy content of foods. It is important to realize that the most common notation for the kilocalorie is 'Cal' with a capital $C$, so whenever energy in calories is being discussed you must note whether or not the $C$ is capitalized.

Another unit of heat in common use is the **British Thermal Unit** (Btu). One Btu is the amount of energy needed to change the temperature of one pound of water by 1 Fahrenheit degree. In terms of the other units of heat we have

$$1 \text{ Btu} = 0.252 \text{ kcal} = 1055 \text{ J}$$

## Practice Quiz

7. If you eat 2200 Calories of food per day, what is your energy intake in units of Btu?
   **(a)** 8.7 **(b)** 4786 **(c)** 2200 **(d)** 8700 **(e)** 1055

## 16–5 Specific Heats

We have already pointed out that heat is associated with a temperature difference between two systems. It is also true that heat can result in the change in temperature of a given system. The amount of heat needed to change the temperature of an *object* is directly proportional to the change in temperature required; that is, $Q \propto \Delta T$. The proportionality constant between the two quantities depends on the type (and mass) of substance of which the object is made. The proportionality constant $C$ is called the **heat capacity** of the object, so we have

$$Q = C\,\Delta T$$

The SI unit of heat capacity is J/K (J/C° is also in common use).

The concept of heat capacity refers to specific objects because you need to know how much of the substance you are heating up or cooling down (and the heat capacity contains that information). It is often more useful to have a quantity that is independent of the amount of substance and depends only on the nature of the substance. Such a quantity is called the **specific heat** ($c$), which is basically the heat capacity per unit mass: $c = C/m$. Therefore, in terms of specific heat, the amount of heat needed to change the temperature of a substance is given by

$$Q = mc\,\Delta T$$

where $m$ is the mass of the substance into or out of which the energy flows. The SI unit of specific heat is J/(kg·K). Table 16–2 in the text contains values of the specific heats of several substances.

---

**Example 16–4 Calorimetry** An insulated container of negligible heat capacity contains 2.11 kg of water at 22.0 °C. A hot aluminum ball of mass 0.435 kg and temperature 90.0 °C is placed into the water. What will be the final equilibrium temperature of the ball and the water?

**Picture the Problem** The sketch shows an aluminum ball sitting in a container of water.

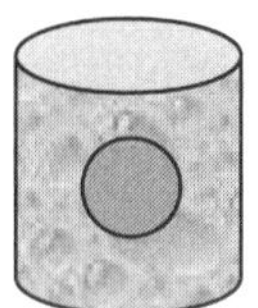

**Strategy** Since the container has negligible heat capacity, we can ignore any heat transfer to or from it. Heat will transfer from the hot ball to the cooler water. Energy conservation requires that the net heat flow, into or out of the system, be zero.

**Solution**

1. By conservation of energy: $Q_w + Q_a = 0$

2. The heat gained by the water is: $Q_w = m_w c_w \left(T - T_{0w}\right)$

3. Heat is lost by the aluminum: $Q_a = m_a c_a \left(T - T_{0a}\right)$

4. The conservation of energy equation becomes: $m_w c_w \left(T - T_{0w}\right) + m_a c_a \left(T - T_{0a}\right) = 0$

5. Solve for the equilibrium temperature, $T$:

$$T = \frac{m_w c_w T_{0w} + m_a c_a T_{0a}}{m_w c_w + m_a c_a}$$

6. Obtain the numerical result using Table 16–2 in the text to get the specific heats:

$$T = \left[(2.11\text{ kg})\left(4186\,\tfrac{\text{J}}{\text{kg}\cdot\text{C}^\circ}\right) + (0.435\text{ kg})\left(900\,\tfrac{\text{J}}{\text{kg}\cdot\text{C}^\circ}\right)\right]^{-1} \times \left\{\begin{array}{l}(2.11\text{ kg})\left(4186\,\tfrac{\text{J}}{\text{kg}\cdot\text{C}^\circ}\right)(22.0\ ^\circ\text{C}) \\ +(0.435\text{ kg})\left(900\,\tfrac{\text{J}}{\text{kg}\cdot\text{C}^\circ}\right)(90.0\ ^\circ\text{C})\end{array}\right\}$$

$$= 24.9\ ^\circ\text{C}$$

**Insight** Notice that in step 3 the heat flow is negative (because the equilibrium temperature is less than the initial temperature of the aluminum), which is indicative of heat lost as opposed to heat gained.

## Practice Quiz

8. If an object made out of a certain substance has a heat capacity $C$, a similar object made of the same substance but having twice the mass will have a heat capacity of

**(a)** $C/2$ **(b)** $C^2$ **(c)** $C$ **(d)** $2C$ **(e)** $\sqrt{C}$

9. An object is made out of a certain substance that has a specific heat $c$. To do calorimetry with a similar object made of the same substance but having twice the mass, we must use a specific heat of

**(a)** $c/2$ **(b)** $c^2$ **(c)** $c$ **(d)** $2c$ **(e)** $\sqrt{c}$

## 16–6 Conduction, Convection, and Radiation

### Conduction

There are three main processes by which heat is transferred. One of these processes is called **conduction**. Heat conduction occurs when the energy flows directly through a material because of a temperature difference, $\Delta T$, across the material. Consider a cylindrical object of circular cross-sectional area, $A$, and length, $L$. If the ends of this cylinder have different temperatures, $T_2$ and $T_1$ (with $T_2 > T_1$), then heat will conduct through the cylinder until the ends are at the same temperature. The average rate at which energy will flow, $Q/t$ (where $t$ is time), is found to depend on three clearly identifiable quantities:

- The heat flow rate is directly proportional to the temperature difference: $Q/t \propto \Delta T$ ($\Delta T = T_2 - T_1$).
- The heat flow rate is directly proportional to the area through which it flows: $Q/t \propto A$.
- The heat flow rate is inversely proportional to the distance through which it flows: $Q/t \propto L^{-1}$.

Combining these considerations, we get $Q/t \propto A(\Delta T)/L$. The constant of proportionality, $k$, is called the **thermal conductivity** of the substance. The SI unit of thermal conductivity is W/(m·K). Table 16–3 in your text contains values of thermal conductivity for several substances. The final result is an equation that is commonly written in two ways:

$$\frac{Q}{t} = kA\frac{\Delta T}{L} \quad \text{or} \quad Q = kA\left(\frac{\Delta T}{L}\right)t$$

**Exercise 16–5 Heat Loss through a Window** If a 3.0-ft × 4.0-ft glass window, 3.0-mm thick, has an inner surface temperature of 30.0 °F and an outer surface temperature 20.0 °F, at what rate is energy lost through the window?

**Picture the Problem** The sketch shows a rectangular window and indicates the inside and outside surfaces.

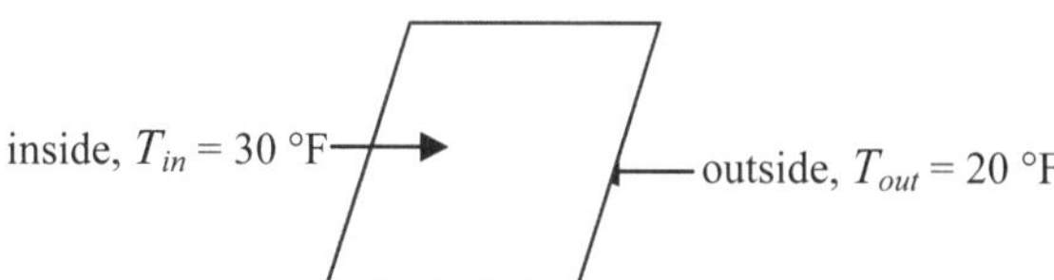


**Strategy** We need to apply the expression for the rate of thermal conduction. Before we use this expression we should convert quantities to SI units.

**Solution**

**1**. Using the result of Exercise 16–1, convert the temperature difference to Celsius degrees:

$$\Delta T_C = \left(\frac{5}{9}\ {}^\circ\text{C}/{}^\circ\text{F}\right)\Delta T_F = \left(\frac{5}{9}\ {}^\circ\text{C}/{}^\circ\text{F}\right)\left(10\ \text{F}^\circ\right) = 5.56\ \text{C}^\circ$$

**2**. Calculate the area in SI units:

$$A = 3.0\ \text{ft} \times 4.0\ \text{ft} = 12.0\ \text{ft}^2\left(\frac{\text{m}}{3.28\,\text{ft}}\right)^2 = 1.12\ \text{m}^2$$

**3**. Use Table 16–3 for the thermal conductivity of glass to solve for the rate of heat conduction:

$$\frac{Q}{t} = kA\frac{\Delta T}{L} = \left(0.84\ \tfrac{\text{W}}{\text{m}\cdot\text{C}^\circ}\right)\left(1.12\ \text{m}^2\right)\left(\frac{5.56\ \text{C}^\circ}{3.0\times10^{-3}\ \text{m}}\right)$$
$$= 1700\ \text{W}$$

**Insight** This is a pretty high rate of energy loss. The prospect of such high energy loss is why the energy efficiency of windows is a large and profitable part of the home improvement business.

## Practice Quiz

**10**. Connecting two identical rods end to end has what effect on the rate of heat flow across the two rods, compared with the rate at which heat would conduct across just one of them?

**(a)** Heat will conduct at twice the rate for the connected rods.

**(b)** Heat will conduct at half the rate for the connected rods.

**(c)** Heat will conduct at the same rate for both the connected rods and the single rod.

**(d)** There will be no heat conduction for the connected rods.

**(e)** None of the above.

### Convection

Heat transfer also occurs by the bulk movement of matter from one place to another, such as the rising of hot air in a room. This type of heat transfer is called **convection**. Convection is an important process in fluids where material is free to move.

### Radiation

All objects emit and absorb heat in the form of **radiation**. By radiation we mean electromagnetic waves, which will be studied in a later chapter. Visible light, infrared, microwaves, X-rays, gamma rays, radio and television waves, and ultraviolet light are all forms of electromagnetic radiation. Infrared and visible light are the two that are most applicable to heat.

It has been determined that the radiant power of an object is directly proportional to both the surface area of the object from which the radiation flows and the fourth power of the Kelvin temperature: $P \propto AT^4$. The proportionality factor for this expression is written as $e\sigma$, where $\sigma$ is a fundamental constant called the Stefan-Boltzmann constant

$$\sigma = 5.67\times10^{-8}\ \mathrm{W/(m^2\cdot K^4)}$$

and $e$ is the *emissivity*, which depends on the nature of the surface. The emissivity is a dimensionless quantity whose value lies between 0 and 1; it is a measure of how effectively an object radiates heat. If $e = 1$, the object is said to be a perfect radiator. The power radiated by an object is therefore given by

$$P = e\sigma AT^4$$

Objects also absorb radiant energy according to this same relation, except that the temperature is the temperature of the environment instead of the temperature of the object. Keep in mind that when using the above relation the temperature should be in Kelvin. Because we are not dealing with a temperature difference or temperature change, the Kelvin and the Celsius scales are not interchangeable.

---

**Example 16–6 Net Radiation** A metal ball of emissivity 0.65 and surface area 0.66 m$^2$ is heated to a temperature of 92 °C and placed in a room in which the air temperature is 22 °C. **(a)** Does heat flow into or out of the metal ball? **(b)** At what net rate does radiant heat flow between the ball and the air?

**Solution** We are given the following information:

**Given:** $e = 0.65$, $T_{ball} = 92\ °\text{C}$, $T_{air} = 22\ °\text{C}$; **Find:** **(a)** the direction of heat flow, **(b)** $P_{net}$

**(a)** Since the ball is at a higher temperature than the air and heat flows from higher temperatures to lower temperatures, the heat must flow out of the metal ball.

**(b)** The rate at which the metal ball radiates energy away is given by

$$P_{out} = e\sigma A T_{ball}^4$$

The rate at which the metal ball absorbs energy from the air is given by

$$P_{in} = e\sigma A T_{air}^4$$

Therefore, the net rate at which the ball loses energy is

$$P_{net} = e\sigma A\left(T_{ball}^4 - T_{air}^4\right)$$

Before using this result for $P_{net}$, let's first convert the temperatures to Kelvin:

$$T_{ball,K} = T_{ball,C} + 273.15 = 92 + 273.15 = 365.15\text{ K}$$
$$T_{air,K} = T_{air,C} + 273.15 = 22 + 273.15 = 295.15\text{ K}$$

We can now calculate the net power radiated by the ball as

$$P_{net} = (0.65)\left(5.67\times10^{-8}\ \frac{\text{W}}{\text{m}^2\cdot\text{K}^4}\right)\left(0.66\text{ m}^2\right)\left[(365.15\text{ K})^4 - (295.15\text{ K})^4\right] = 250\text{ W}$$

**Insight** The net power is the difference between what is emitted and absorbed. Whether this net energy is emitted or absorbed is determined by the temperature difference between the object and its environment.

## Practice Quiz

**11.** Which process of heat transfer is least important when cooking on an electric range?

**(a)** conduction **(b)** convection **(c)** radiation **(d)** they are all equally important

**12.** At what rate does an object of emissivity 0.55 absorb radiation if it has a surface area of 0.33 $\text{m}^2$ and sits in an environment of ambient temperature 290 K?

**(a)** 73 J/s **(b)** $3.0 \times 10^{-6}$ J/s **(c)** 290 J/s **(d)** insufficient information to determine

# Reference Tools and Resources

## I. Key Terms and Phrases

**temperature** the property of systems that determines the existence and direction of the heat flow between them when they are in thermal contact

**heat** the energy that is transferred between systems because of a temperature difference

**thermal contact** exists between systems when it is possible for heat to flow between them

**thermal equilibrium** exists when systems are brought into thermal contact and no heat transfer occurs

**zeroth law of thermodynamics** the fundamental law that allows a working definition of temperature

**absolute zero** the lower limit on physically attainable temperatures (zero on the Kelvin scale)

**coefficient of expansion** the substance-dependent proportionality factor that determines how much an object will expand as the result of a temperature change

**calorie** the heat needed to change the temperature of 1 gram of water by 1 Celsius degree

**British Thermal Unit** the heat needed to change the temperature of 1 lb of water by 1 Fahrenheit degree

**heat capacity** a quantity that determines how much heat is needed to change the temperature of an object by a certain amount

**specific heat** a quantity that determines how much heat is needed to change the temperature of a unit mass of an object by a certain amount

**conduction** heat transfer by direct flow through an object due to a temperature difference across it

**thermal conductivity** a quantity that determines the rate at which heat conducts through a given object

**convection** heat transfer by direct movement of matter from one place to another

**radiation** heat transfer by emission or absorption of electromagnetic waves

**Stefan-Boltzmann constant** determines the rate at which an object emits or absorbs radiation

**emissivity** a number between 0 and 1 that measures how effectively an object radiates heat

## II. Important Equations

| Name/Topic | Equation | Explanation |
|---|---|---|
| temperature scales | $T_C = \left(\frac{5}{9}\,^\circ\text{C}/^\circ\text{F}\right)\left(T_F - 32\,^\circ\text{F}\right)$<br>$T = T_C + 273.15$ | The conversions between Celsius and Fahrenheit and Kelvin and Celsius |
| thermal expansion | $\Delta L = \alpha L_0 \Delta T$<br>$\Delta V = \beta V(\Delta T)$ | The change in size of an object due to a temperature change |
| specific heat | $Q = mc\,\Delta T$ | The heat needed to change the temperature of an amount of mass, *m*, of a substance by $\Delta T$ |
| conduction | $\frac{Q}{t} = kA\left(\frac{\Delta T}{L}\right)t$ | The heat that conducts through an object of length, *L*, and cross-sectional area, *A*, in time *t* |
| radiation | $P = e\sigma AT^4$ | The rate of heat transfer by electromagnetic radiation |

## III. Know Your Units

| Quantity | Dimension | SI Unit |
|---|---|---|
| temperature ( $T$ ) | [K] | K |
| heat ( $Q$ ) | $[\text{M}][\text{L}^2][\text{T}^{-2}]$ | J |
| coefficient of expansion ($\alpha, \beta$) | $[\text{K}^{-1}]$ | $\text{K}^{-1}$ |
| heat capacity ( $C$ ) | $[\text{M}][\text{L}^2][\text{T}^{-2}][\text{K}^{-1}]$ | J/K |
| specific heat ( $c$ ) | $[\text{L}^2][\text{T}^{-2}][\text{K}^{-1}]$ | J/(kg·K) |
| thermal conductivity ( $k$ ) | $[\text{M}][\text{L}][\text{T}^{-3}][\text{K}^{-1}]$ | W/(m·K) |
| Stefan-Boltzmann constant ($\sigma$) | $[\text{M}][\text{T}^{-3}][\text{K}^{-4}]$ | $\text{W/(m}^2\cdot\text{K}^4)$ |
| emissivity ( $e$ ) | dimensionless | — |

## Puzzle

**POT PHYSICS**

Advertisements for some expensive cookware have diagrams that look much like the figure. The diagram is labeled to indicate that the inner and outer surfaces of the pot are made of stainless steel and the inner layer is made of another metal such as aluminum or copper. Please explain why pots are made this way. Discuss why the different metals are chosen and why using a layered structure is better than using one metal and the same total thickness.

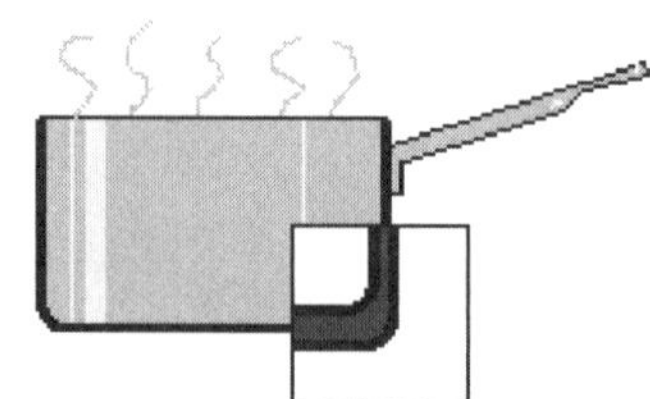

## Answers to Selected Conceptual Questions

**4.** No. Heat is not a quantity that one object has more of than another. Heat is the energy that is transferred between objects of different temperatures.

**8.** Heating the glass jar and its metal lid to the same higher temperature results in a greater expansion in the lid than in the glass. As a result, the lid can become loose enough to turn.

**14.** Both the metal and the wood are at a lower temperature than your skin. Therefore, heat will flow from your skin to both the metal and the wood. The metal feels cooler, however, because it has a greater thermal conductivity. This allows the heat from your skin to flow to a larger effective volume than is the case with the wood.

## Solutions to Selected End-of-Chapter Problems and Conceptual Exercises

11. **Picture the Problem:** Our sketch shows the pressure in the gas thermometer as a function of the temperature. Note that at $T = 100°\text{C}$ the pressure is 227 mmHg and the pressure extrapolates to zero at the temperature $-273.15°\text{C}$. Assuming ideal behavior we want to find the temperature associated with a pressure of 162 mmHg.

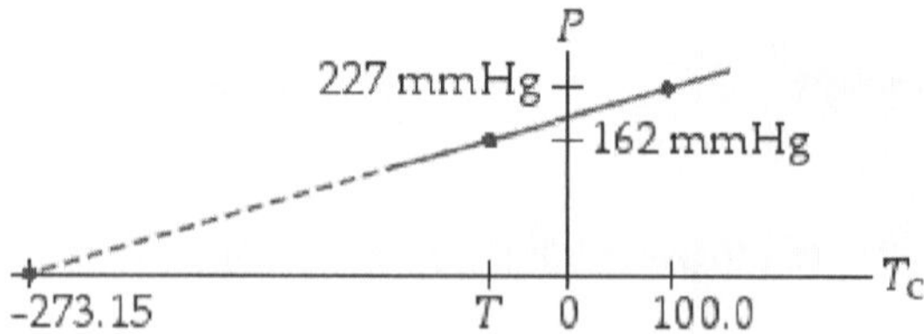


**Strategy:** We assume that the pressure lies on a straight line. Using the known pressures at 100°C and −273.15°C, calculate the rate at which pressure increases as a function of temperature. Calculate the temperature at 162 mmHg using this rate.

**Solution: 1.** Divide the change in pressure by the change in temperature to obtain the rate:

$$\text{rate} = \frac{227\text{ mmHg} - 0}{100°\text{C} - (-273.15°\text{C})} = 0.60833\text{ mmHg/C°}$$

**2.** Solve the rate equation for the temperature:

$$\text{rate} = \frac{P - P_0}{T - T_0} \Rightarrow T = T_0 + \frac{P - P_0}{\text{rate}}$$

**3.** Insert given data:

$$T = 100°\text{C} + \frac{162 \text{ mmHg} - 227 \text{ mmHg}}{0.60833 \text{ mmHg/C°}} = \boxed{-6.85°\text{C}}$$

**Insight:** The gas thermometer is first placed in the boiling water to calibrate it, because water boils at a known temperature. Once the thermometer is calibrated, the pressure variation can accurately give temperatures over a wide range.

25. **Picture the Problem:** The volume of an aluminum saucepan expands as the temperature of the pan increases. Water, which initially fills the saucepan to the brim, also increases in temperature and expands. If the water expands more than the saucepan, the water will spill over the top. If the saucepan expands more than the water, the water level will drop below the brim of the pan.

**Strategy:** Use equation 16-6 to calculate the change in volumes of the saucepan and of the water. Subtract the change in volume of the saucepan from the change in volume of the water to determine the volume of water that overflows the saucepan. The coefficient of volume expansion for water is given in Table 16-1. The coefficient of volume of expansion of aluminum is three times its coefficient of linear expansion, which is also given in Table 16-1.

**Solution: 1. (a)** Because water has a larger coefficient of volumetric expansion, its volume will increase more than the volume of the aluminum saucepan. Therefore, water will overflow from the pan.

**2. (b)** Calculate the initial volumes of the saucepan and water:

$$V_0 = \pi r^2 h = \pi \left(\frac{23 \text{ cm}}{2}\right)^2 (6.0 \text{ cm}) = 2493\text{c m}^3$$

**3.** Use equation 16-6 to write the changes in volume:

$$\Delta V_\text{w} = \beta_\text{w} V_0 \Delta T$$

$$\Delta V_\text{Al} = 3\alpha_\text{Al} V_0 \Delta T$$

**4.** Subtract the change on volume of the pan from the water to calculate the volume of water spilled:

$$V_\text{spill} = \Delta V_\text{w} - \Delta V_\text{Al} = (\beta_\text{w} - 3\alpha_\text{Al}) V_0 \Delta T$$

$$= (21 - 3\times 2.4)\times 10^{-5} (\text{C°})^{-1} (2493 \text{ cm}^3)\left[(88 - 19)°\text{C}\right]$$

$$V_\text{spill} = \boxed{24 \text{ cm}^3}$$

**Insight:** This is the same principle that enables a mercury thermometer to work. The mercury expands faster than the surrounding glass, causing the mercury column to rise.

29. **Picture the Problem:** The figure shows Joule's apparatus. As the weights fall at constant speed, the work done is converted into thermal energy, raising the temperature of the water.

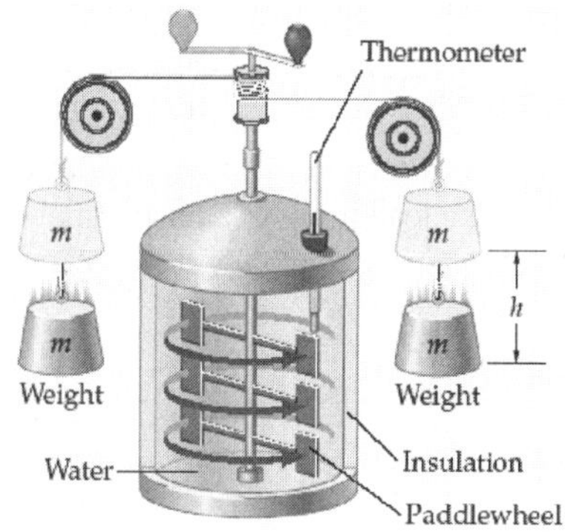


**Strategy:** We need to calculate the increase in temperature for a given amount of work. Calculate the work done by gravity on the weights as they fall through the height *h*. Set this work equal to the increase in thermal energy of the water. Multiply the result by $(1.0 \text{ C°}/6200 \text{ J})$ to determine the increase in temperature.

**Solution: 1. (a)** Calculate the work done by gravity on the two weights:

$$W = mgh = \left[2(0.95 \text{ kg})\right](9.81 \text{ m/s}^2)(0.48 \text{ m}) = 8.95$$

**2.** Multiply the work by the change in temperature per work:

$$\Delta T_C = \left(\frac{1.0\ \text{C}^\circ}{6200\ \text{J}}\right)(8.95\ \text{J}) = \boxed{1.4\times10^{-3}\ \text{C}^\circ}$$

**3. (b)** Fahrenheit degrees are smaller than Celsius degrees, so the temperature rise in Fahrenheit degrees would be greater than the result in part (a).

**4. (c)** Convert to Fahrenheit degrees:

$$\Delta T_F = \frac{9\ \text{F}^\circ}{5\ \text{C}^\circ}\left(1.44\times10^{-3}\ \text{C}^\circ\right) = \boxed{2.6\times10^{-3}\ \text{F}^\circ}$$

**Insight:** This change in temperature was not measurable. Joule had to repeatedly raise and lower the weights to achieve a measurable temperature difference.

41. **Picture the Problem:** A hot object is immersed in water in a calorimeter cup. Heat transfers from the hot object to the cold water and cup, causing the temperature of the object to decrease and the temperature of the water and cup to increase.

**Strategy:** Since the heat only transfers between the water, cup, and object, we can use conservation of energy to calculate the heat given off by the object by summing the heats absorbed by the water and cup. Use the heat given off by the object and its change in temperature to calculate its specific heat.

**Solution: 1.** Let $\sum Q = 0$ and solve for $Q_{\text{Ob}}$:

$$0 = Q_{\text{Ob}} + Q_{\text{w}} + Q_{\text{Al}}$$

$$Q_{\text{Ob}} = -\left(Q_{\text{w}} + Q_{\text{Al}}\right) = -\left(m_{\text{w}}c_{\text{w}} + m_{\text{Al}}c_{\text{Al}}\right)\Delta T_{\text{w}}$$

**2.** Solve for the specific heat using equation 16-13:

$$c_{\text{Ob}} = \frac{Q_{ob}}{m_{\text{Ob}}\left(T - T_{\text{Ob}}\right)} = \frac{\left(m_{\text{w}}c_{\text{w}} + m_{\text{Al}}c_{\text{Al}}\right)\left(T_{\text{w}} - T\right)}{m_{\text{Ob}}\left(T - T_{\text{Ob}}\right)}$$

$$= \frac{\left(0.103\ \text{kg}\left[4186\ \text{J}/(\text{kg}\cdot\text{K})\right] + 0.155\ \text{kg}\left[900\ \text{J}/(\text{kg}\cdot\text{K})\right]\right)(20-22)\text{C}^\circ}{0.0380\ \text{kg}\,(22.0-100)\ \text{C}^\circ}$$

$$c_{\text{Ob}} = 385\ \text{J}/(\text{kg}\cdot\text{C}^\circ) = \boxed{385\ \text{J}/(\text{kg}\cdot\text{K})}$$

**3.** Look up the specific heat in Table 16-2:

The object is made of copper.

**Insight:** It is important to include the effect of the aluminum cup in this calculation. If the contribution of the cup were excluded, the specific heat of the object would have been calculated as 291 J/(kg K).

56. **Picture the Problem:** Two metal rods are connected in series. Heat flows from the high temperature source through the copper rod and then through the lead rod before reaching the cold temperature end.

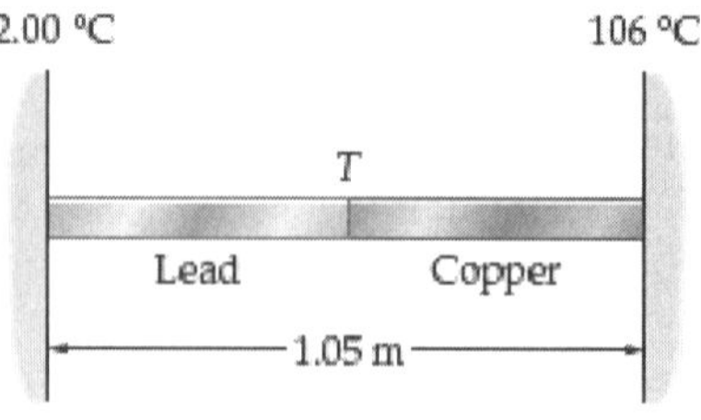


**Strategy:** Use the heat flow through the rods in equation 16-16 to calculate the temperature at the junction of the rods.

**Solution: 1. (a)** The temperature of the junction is greater than 54°C. Since lead has a smaller thermal conductivity than copper, it must have a greater temperature difference across it to have the same heat flow.

**2. (b)** Solve equation 16-16 for the junction temperature:

$$Q = k\,A\frac{T_2 - T_1}{L}t$$

$$T_1 = T_2 - \frac{QL}{k\,At} = 106^\circ\text{C} - \frac{(1.41\ \text{J})(0.525\ \text{m})}{\left[395\text{W}/(\text{m}\cdot\text{K})\right](0.015\ \text{m})^2(1\ \text{sec})} = \boxed{98^\circ\text{C}}$$

**Insight:** This problem can also be solved by finding the temperature difference across the lead:

$$T_2 = T_1 + \frac{QL}{k\,At} = 2\,°\text{C} + \frac{(1.41\text{ J})(0.525\text{ m})}{[34.3\text{W}/(\text{m}\cdot\text{K})](0.015\text{ m})^2(1\text{ sec})} = \boxed{98\,°\text{C}}$$

85. **Picture the Problem:** Heat conducts out of the body through the skin.

**Strategy:** Divide equation 16-16 by time to calculate the rate of heat transport through the skin. Use the thermal conductivity of water found in Table 16-3. The cross-sectional area is the surface area of the skin and the thickness of the conductor is the skin thickness.

**Solution: 1. (a)** Solve eq. 16-16 for the rate of heat transfer:

$$\frac{Q}{t} = kA\left(\frac{\Delta T}{L}\right) = [0.60\text{ W}/(\text{m}\cdot\text{K})](1.40\text{ m}^2)\left(\frac{3.0\text{ C}°}{0.012\text{ m}}\right) = \boxed{0.21\text{ kW}}$$

**2. (b)** Dropping the skin temperature to 31°C doubles the temperature difference, so the heat transport would also double: $2(210\text{ W}) = \boxed{0.42\text{ kW}}$

**Insight:** The rate of heat transport is proportional to the temperature difference.

## Answers to Practice Quiz

**1**. (e) **2**. (b) **3**. (c) **4**. (c) **5**. (a) **6**. (e) **7**. (d) **8**. (d) **9**. (c) **10**. (b) **11**. (b) **12**. (a)

# CHAPTER 17
# PHASES AND PHASE CHANGES

## Chapter Objectives

After studying this chapter, you should

1. understand the basic properties of an ideal gas.
2. understand how the kinetic theory of gases relates microscopic and macroscopic properties.
3. be able to work with stress and strain in the cases of elastic stretching (or compression), shear, and volume deformations of solids.
4. know the three phases of matter and be able to understand phase diagrams.
5. be able to calculate the heat required to change the phase of a substance.
6. know how to use the conservation of energy to keep track of heat transfer within a system.

## Warm-Ups

1. According to the kinetic theory model, what is the relation between the temperature of a gas and the mechanical properties of the moving gas particles?
2. Room temperature is taken to be 70 degrees Fahrenheit. Estimate how much the temperature would rise if the kinetic energy of every gas particle in the room doubled.
3. For the kinetic theory derivation of the gas law, why is it important that the particles interact only elastically?
4. According to kinetic theory, why does the gas law begin to fail as a gas approaches the boiling point temperature?

## Chapter Review

This is the second of three chapters on thermodynamics. In this chapter, we use the kinetic theory of gases to see how the microscopic properties of a system give rise to the macroscopic properties with which we are more familiar. We also explore the phenomenology of phase changes. The energy needed to bring about a phase change and the application of energy conservation to phase changes and other heat transfer processes within a system are also treated.

## 17–1 Ideal Gases

An **ideal gas** is one in which the gas particles (atoms or molecules) do not interact; that is, they move freely and independently of each other. The behavior of ideal gases is a close approximation to the behavior of most real gases. The state of the system of particles making up the gas is determined by the pressure, $P$, temperature, $T$, number of particles, $N$, and volume, $V$ of the gas. An equation that shows how these quantities depend on one another is called an **equation of state**. The detailed study of gases has shown that the equation of state for an ideal gas is

$$PV = NkT$$

where $k$ is a fundamental constant called the Boltzmann constant, which has the value

$$k = 1.38 \times 10^{-23}\ \text{J/K}$$

An alternative way to write the equation of state for an ideal gas uses the concept of the **mole** (mol). A mole is the amount of a substance that contains $6.022 \times 10^{23}$ entities; this value is called **Avogadro's number**, $N_A$. The masses of atoms (and molecules) are often quoted by stating the mass of one mole of atoms (or molecules), called the **atomic mass** (or molecular mass) of the substance. When moles are used in the equation of state, the number of particles in a gas is written as $N = nN_A$, where $n$ is the number of moles in the gas. The product of the two constants $N_A k$ is another constant called the gas constant, $R$, which has the value

$$R = 8.31\ \text{J/(mol}\cdot\text{K)}$$

Putting these factors together in the equation of state gives

$$PV = nRT$$

as a commonly used alternative version of the equation of state for an ideal gas; this latter version is often called the *ideal gas law*.

The ideal-gas equation of state contains all the information about how the relevant quantities relate. One relation is **Boyle's law**, which states that for *fixed N and T* the product of pressure and volume is constant:

$$P_i V_i = P_f V_f$$

Curves that are plotted under the condition of fixed temperature are called **isotherms**. Similarly, the ideal gas law also incorporates **Charles' law**, which states that for *fixed N and P*, the ratio $V/T$ is constant:

$$\frac{V_i}{T_i} = \frac{V_f}{T_f}$$

Notice that fixed $N$ also implies fixed $n$.

**Exercise 17–1 Inflating a Tire** An automotive worker needs to pump up an empty tire that has an inner volume of 0.0192 $m^3$. If the temperature in the manufacturing plant is 28.5 °C, what will be the gauge pressure in the tire, in psi, if 2.70 moles of air is pumped into it?

**Solution:** We are given the following information:

**Given:** $V = 0.0192\ \text{m}^3,\ T = 28.5\ °\text{C},\ n = 2.70\ \text{mol};$ **Find:** $P$ in psi

We can solve this problem by using the ideal-gas equation of state. Since we are given the number of moles of air, we shall use the form that contains $n$. Solving this equation for pressure we obtain

$$P = \frac{nRT}{V}$$

Since this expression assumes that $T$ is in Kelvin, we must remember to apply the conversion. Doing this conversion gives a pressure of

$$P = \frac{(2.70\ \text{mol})(8.31\ \text{J/mol}\cdot\text{K})\left[(28.5\ C^{\circ} + 273.15)\text{K}\right]}{0.0192\ \text{m}^3} = 3.525\times10^5\ \text{Pa}$$

The above result gives the absolute pressure in the tire. What we really want is the gauge pressure, which is obtained by subtracting atmospheric pressure. Thus,

$$P_g = P - P_{at} = 3.525\times10^5\ \text{Pa} - 1.013\times10^5\ \text{Pa} = 2.512\times10^5\ \text{Pa}\left(\frac{1.450\times10^{-4}\ \text{psi}}{1\ \text{Pa}}\right) = 36.4\ \text{psi}$$

which gives us the final result in pounds per square inch.

**Example 17–2 Compressed Air** If you must transfer air from a 3.75-$m^3$ container at atmospheric pressure to a 1.35-$m^3$ container at the same temperature, to what pressure must you compress the air?

**Picture the Problem** The sketch shows the larger container connected to the smaller container by a tube through which the air will pass. (The mechanism that compresses the air is not shown.)

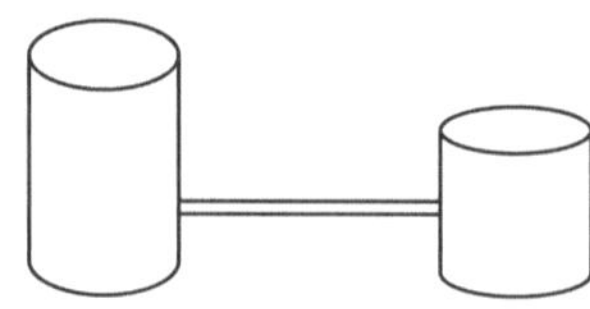

**Strategy** Both the amount of gas, $N$, and the temperature, $T$, remain fixed in this problem, so we can use Boyle's law to determine the final pressure.

**Solution**

1. Use Boyle's law to solve for the final pressure: $P_iV_i = P_fV_f \quad \Rightarrow \quad P_f = \dfrac{P_iV_i}{V_f}$

2. Obtain the numerical result: $P_f = \dfrac{(1.013\times10^5\ \text{Pa})(3.75\ \text{m}^3)}{1.35\ \text{m}^3} = 2.81\times10^5\ \text{Pa}$

**Insight** The fact that the pressure will increase if you squeeze the same gas into a smaller container should also appeal to your intuition. If it doesn't, think about it.

## Practice Quiz

1. For an ideal gas confined to a constant volume, if the temperature is increased by a factor of 2, what happens to the pressure?
   **(a)** The pressure becomes a factor of 2 smaller.
   **(b)** The pressure becomes a factor of 2 larger.
   **(c)** There is no change in the pressure.
   **(d)** The pressure goes to zero.
   **(e)** None of the above.

2. How many molecules are contained in 5.70 moles of a gas?
   **(a)** $6.02 \times 10^{23}$ **(b)** 6 **(c)** $3.43 \times 10^{24}$ **(d)** 22 **(e)** none of the above

3. You inflate your car tires to a gauge pressure of 2.15 atm at a temperature of 19.0 °C. After driving a couple of miles, the temperature of the tire increases by 27.0 C°. What is the new gauge pressure?
   **(a)** 2.35 atm **(b)** 2.84 atm **(c)** 1.72 atm **(d)** 3.06 atm **(e)** 2.77 atm

## 17–2 Kinetic Theory

The **kinetic theory of gases** uses the motion of the microscopic gas particles (atoms or molecules) to explain the macroscopic properties of gases. Specifically, we shall use kinetic theory to gain deeper insight into the origin of the pressure and temperature of a gas. In its simplest form, the kinetic theory of gases makes several reasonable assumptions: (a) a large number, $N$, of identical gas particles are in an impenetrable rigid container, (b) the particles move randomly within the container, and (c) the only interactions experienced by the particles are elastic collisions with each other and with the container.

Kinetic theory shows that the pressure exerted by a gas on a container of volume, $V$, is due to the collisions of the gas particles with the walls of the container. The speed of the gas particles (of mass $m$)

are not all the same but rather are distributed over a wide range of speeds according to the Maxwell speed distribution (see Figure 17-8 in the text). If you consider the change in momentum, $\Delta p$, of a typical gas particle upon collision with a wall of the container together with the amount of time between collisions with this wall, $\Delta t$, the force that this particle exerts on the wall, $F = \Delta p/\Delta t$, can be determined. Upon dividing this force by the area of the wall for all $N$ particles, the pressure works out to be

$$P = \frac{1}{3}\left(\frac{N}{V}\right)m\left(v^2\right)_{av}$$

In terms of the average translational kinetic energy of the particles, $K_{av} = \left(\tfrac{1}{2}mv^2\right)_{av}$, we get

$$P = \frac{2}{3}\left(\frac{N}{V}\right)K_{av}$$

This latter expression clearly shows that the pressure of a gas is directly proportional to the average translational kinetic energy of a gas particle.

If we compare the above results with the ideal-gas equation of state $PV = NkT$, we also find that the Kelvin temperature of a gas is directly proportional to the average kinetic energy of a gas particle. Specifically,

$$K_{av} = \tfrac{3}{2}kT$$

This result not only serves to give us a better idea of the physical meaning of the concept of temperature but also shows why the Kelvin temperature scale is a physically more appropriate scale to use in scientific work — it has the most direct connection to the energy of the particles in a system. Because the average kinetic energy is an important quantity, the speed associated with $K_{av}$ is also a useful measure of the behavior of the gas. This speed is the square root of $(v^2)_{av}$, which is called the root-mean-square speed (or *rms* speed):

$$v_{rms} = \sqrt{\left(v^2\right)_{av}} = \sqrt{\frac{3kT}{m}}$$

such that $K_{av} = \tfrac{1}{2}mv_{rms}^2$.

The total internal energy, $U$, of a gas whose particles consist of a single type of atom (a monatomic gas) is given by the average kinetic energy of a single particle times the number of particles:

$$U = \tfrac{3}{2}NkT = \tfrac{3}{2}nRT$$

where $n$ is the number of moles of the gas. If the gas particles consist of molecules, other contributions must be added to the internal energy. This latter case will not be treated in this chapter.

**Exercise 17–3 Kinetic Energy** If 15 moles of a gas are contained in a volume of 0.10 $m^3$ at atmospheric pressure, what is the average kinetic energy of a gas particle?

**Solution:** We are given the following information:

**Given:** $n = 15$ mol, $V = 0.10\ \text{m}^3$, $P = 101$ kPa; **Find:** $K_{av}$

The pressure is directly related to the average kinetic energy of a gas particle by kinetic theory. Solving this expression for the kinetic energy gives

$$K_{av} = \frac{3PV}{2N}$$

The only quantity we don't directly know at this point is the number of particles, $N$. However, we know that $N$ relates to the number of moles, $n$, according to $N = nN_A$. Therefore,

$$K_{av} = \frac{3PV}{2nN_A} = \frac{3\left(1.01\times10^5\ \text{Pa}\right)\left(0.10\ \text{m}^3\right)}{2\left(15\ \text{mol}\right)\left(6.022\times10^{23}\ \text{mol}^{-1}\right)} = 1.7\times10^{-21}\ \text{J}$$

You might look at this value and think that something must be wrong because it seems ridiculously small. Keep in mind, however, that most gas particles have very small mass, so this value is actually typical.

## Practice Quiz

4. If you pump energy into a gas such that the *rms* speed of the molecules triples, what happens to the temperature of the gas?
   **(a)** The temperature of the gas triples.
   **(b)** The temperature is reduced to one-third its previous value.
   **(c)** The temperature is reduced to one-ninth its previous value.
   **(d)** The temperature becomes nine times its previous value.
   **(e)** None of the above.

5. What is the average kinetic energy of an atom in a monatomic gas if its temperature is 20 °C?
   **(a)** $4.1\times10^{-22}$ J **(b)** $2.8\times10^{-22}$ J **(c)** $4.0\times10^{-21}$ J **(d)** $7.3\times10^{-21}$ J **(e)** $6.1\times10^{-21}$ J

## 17–3 Solids and Elastic Deformation

As with gases, the behavior of solids at the microscopic level helps us understand some of the macroscopic properties of solids. The intermolecular forces at work within a solid give rise to the "spring-like" behavior of solids when they are deformed by external forces. In this section, we treat the three cases of changing the length, shape, and volume of a solid.

When a force is applied perpendicularly to a side of an object of area, $A$, to stretch (or compress) the object, the amount that the length of the object increases (or decreases), $\Delta L$, is found to be directly proportional to the pressure, $F/A$, causing the deformation. Furthermore, the change in length of the object, for a given applied force per unit area, increases in direct proportion to the original length, $L_0$, of the object; therefore, we have $\Delta L \propto (F/A)L_0$. The proportionality constant that relates these quantities is called **Young's modulus**, $Y$. The final equality for the change in length of a solid is

$$F = Y\left(\frac{\Delta L}{L_0}\right)A$$

Dimensional analysis of the above equation shows that the SI unit of Young's modulus is $N/m^2$. While the above expression applies to both stretching and compression, you should be aware that some materials have different values of $Y$ for these two processes. Values of Young's modulus for different materials are listed in Table 17–1 in the text.

Another form of deformation (a *shear* deformation) occurs when the applied force is directed along the surface of the area over which it is applied; let's call it the top surface for convenience. In this case, if the opposite side of the object, the bottom surface, is held fixed, the shape of the object will change because the top surface will extend a distance $\Delta x$ beyond the bottom surface. If the distance between the top and bottom surfaces is $L_0$, then

$$F = S\left(\frac{\Delta x}{L_0}\right)A$$

where $S$ is called the **shear modulus**. This quantity plays the same role in shear deformation as Young's modulus plays in stretching. Values of $S$ for different materials are listed in Table 17–2 in the text.

Similarly, deformations that change the entire volume of an object, such as when fluid pressure is applied perpendicularly over the entire surface of a completely submerged object, obey a similar relation. If the pressure acting on an object changes by an amount $\Delta P$, the volume of the object will change in response by an amount $\Delta V$. The relationship between these two quantities is

$$\Delta P = -B\left(\frac{\Delta V}{V_0}\right)$$

where $V_0$ is the original volume. The quantity, $B$, is called the **bulk modulus**, which serves the same purpose as $Y$ and $S$ (and is also in the same units, $N/m^2$). Values of $B$ are listed in Table 17–3 in the text.

Each of the preceding three cases can be written such that we have an applied force per unit area that equals a substance-dependent factor (the modulus) times a quantity that represents how the material deforms. In general, the applied force per area is called a **stress** on the material, and the resulting deformation ($\Delta L/L_0$, $\Delta x/L_0$, $\Delta V/V_0$) is called the **strain** in the material. The preceding three cases all show that stress, ∝, strain, which is a direct result of the spring-like behavior of the material, that is in accordance with Hooke's law, to a good approximation. However, this behavior is true only within a limited range of stresses and strains known as the **elastic limit**. If a stress deforms an object beyond its elastic limit, the object no longer behaves like a spring. Instead, it becomes permanently deformed, and the preceding relations no longer hold.

---

**Example 17–4 Temperature-induced Stress** An aluminum cube has an edge length of 4.70 cm. If its temperature is raised from –2.00 °C to 26.0 °C, causing its volume to expand, what is the magnitude of the stress applied to the aluminum cube associated with this expansion?

**Picture the Problem** The diagram shows exaggerated views of an iron cube **(a)** before and **(b)** after the temperature has been raised.

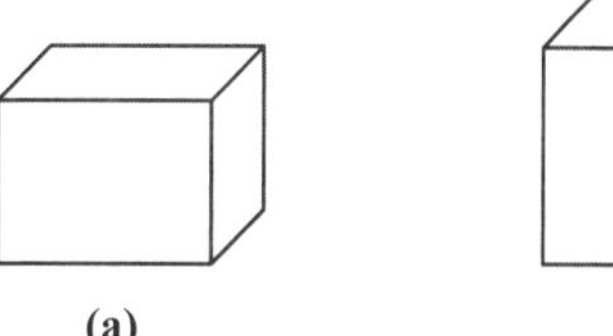


**Strategy** We can determine the stress, $|\Delta P|$, for this case of volume expansion by finding $\Delta V$ using the equation for volumetric thermal expansion.

**Solution**

1. Substitute $\Delta V$ from thermal expansion into the expression for the stress in volume expansion:

$$|\Delta P| = B\left(\frac{\beta V_0(\Delta T)}{V_0}\right) = B\beta(\Delta T)$$

2. Because no value of $\beta$ is given for aluminum in Table 16–1 of the text, use the result for $\alpha$:

$$\beta = 3\alpha = 3\left(24\times10^{-6}\ \text{K}^{-1}\right) = 72\times10^{-6}\ \text{K}^{-1}$$

3. Take the value of the bulk modulus for aluminum from Table 17–3 of the text, and calculate the final result:

$$|\Delta P| = B\beta(\Delta T) = \left(7.0\times10^{10}\ \text{N/m}^2\right)$$
$$\times\left(72\times10^{-6}\ \text{K}^{-1}\right)\left[26.0\ ^\circ C-\left(-2.00\ ^\circ\text{C}\right)\right]$$
$$= 1.4\times10^{8}\ \text{Pa}$$

**Insight** Here we calculated the stress as a magnitude because the pressure comes from within the solid rather than externally, as assumed by the minus sign in the expression. Also notice that the temperatures were not converted to Kelvin because we needed only the temperature difference, which is the same in Kelvin and Celsius.

---

## Practice Quiz

6. A pressure differential, $\Delta P$, applied to an object of original volume, $V_0$, decreases the volume of an object by an amount $\Delta V$. If, instead, the object's original volume was 2 $V_0$, its change in volume would be

   **(a)** $\Delta V/2$. **(b)** $2(\Delta V)$. **(c)** $4(\Delta V)$. **(d)** $\Delta V/4$. **(e)** $\Delta V$.

## 17–4 Phase Equilibrium and Evaporation

**Evaporation** is the release of molecules from the liquid phase to the gaseous phase. As with gases, molecules in a liquid have speeds that are distributed throughout a range of values. Some of the higher-speed molecules move sufficiently fast to spontaneously escape the liquid and enter the gaseous phase. If this liquid-gas mixture is in a closed container, then some of the gas molecules can also spontaneously enter the liquid phase. The result is that molecules flow between the two phases. When these flows become equal, so that the number of molecules in each phase remains constant, the molecules have reached a state of **phase equilibrium**. The pressure of the gas in phase equilibrium is called the **equilibrium vapor pressure**. A liquid boils at the temperature at which the equilibrium vapor pressure equals the external pressure on the liquid.

A plot of the equilibrium vapor pressure versus temperature produces the *vapor pressure curve*; on one side of this curve are the conditions (of temperature and pressure) that produce a gas, and on the other side are the conditions that produce a liquid. An analogous curve for the transition between the solid and liquid phases is the **fusion** curve and for the transition between the solid and gas phases is the **sublimation** curve. When these three curves are plotted on the same diagram, we have a **phase diagram** of the substance in question. This diagram provides graphical information on what properties of a substance correspond to what phase of matter.

## Practice Quiz

7. Besides fusion, a more common term for the phase transition from liquid to solid is

   **(a)** condensation. **(b)** boiling. **(c)** freezing. **(d)** melting. **(e)** liquefaction.

## 17–5 – 17–6 Latent Heats, Phase Changes, and Energy Conservation

Generally, heat is required to change the phase of a substance. During a phase change, there is no change in the temperature of the system undergoing the transformation, as all the heat is used to bring about the phase change. The amount of heat that is required to completely convert one kilogram of a substance from one phase to another is called the **latent heat**, $L$. Therefore, the amount of heat required to convert a mass, $m$, of a substance from one phase to another is given by

$$Q = mL$$

The SI unit of latent heat is J/kg. The value of $L$ depends on the type of phase change being considered. For phase changes between the liquid and gas phases, we use the latent heat of vaporization, $L_v$, and between the liquid and solid phases we use the latent heat of fusion, $L_f$. Values of $L_v$ and $L_f$ are listed in Table 17–4 of the text.

Often, instead of an external agent adding or removing heat to or from a system, we have heat exchanges within a given system. In these cases, it is useful to look at the energy balance between the parts of the system that lose energy and the parts that gain energy. In other words, by energy conservation, we set the magnitude of heat lost by one part of the system equal to the magnitude of heat gained by another part. When you perform this type of analysis, you must treat each step of the energy transfer separately (temperature changes should be handled separately from phase changes) and to recognize when a phase change will take place.

---

**Example 17–5 Boiling Ice** How much heat is required to convert 2.88 kg of ice at –3.00 °C to steam at 100 °C if the mixture is housed in an insulated container?

**Picture the Problem** In the sketch, the container on the left first has ice inside it, then it holds only liquid water, and finally it holds only steam.

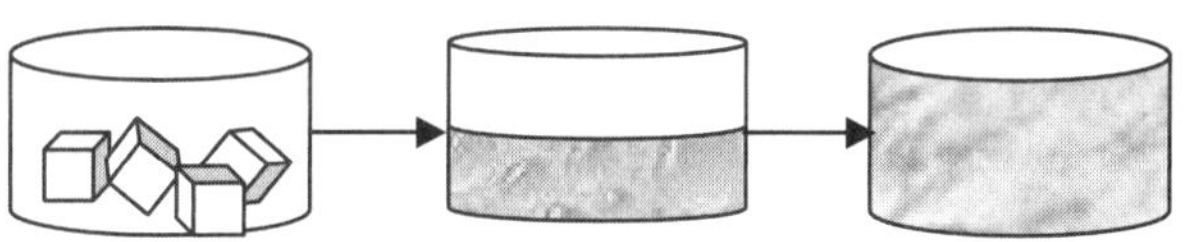

**Strategy** We divide the process into several steps. First, the ice warms to the melting point, then it melts into liquid water, then the liquid warms to the boiling point, and then it boils to become steam. In all phases, the mass remains equal to the mass of the ice.

**Solution**

1. The heat needed to warm the ice to the melting temperature 0 °C is: $Q_1 = m_{ice}c_{ice}\left(0\ ^\circ\text{C} - T_{0,ice}\right)$

2. The heat needed to melt the ice at 0 °C is: $Q_2 = m_{ice}L_f$

3. The heat needed to warm the liquid water from 0 °C to the boiling temperature 100 °C is:

$$Q_3 = m_{ice}c_w\left(100\ ^\circ\text{C} - 0\ ^\circ\text{C}\right)$$

4. The heat needed to convert the liquid to steam at 100 °C is:

$$Q_4 = m_{ice}L_v$$

5. Taking values of specific and latent heats from the appropriate tables in the text we can calculate the total heat required:

$$Q = m_{ice}c_{ice}\left(0\ ^\circ\text{C} - T_{0,ice}\right) + m_{ice}L_f + m_{ice}c_w\left(100\ ^\circ\text{C} - 0\ ^\circ\text{C}\right) + m_{ice}L_v$$

$$\therefore$$

$$Q = m_{ice}\left[c_{ice}\left(3.00\ ^\circ\text{C}\right) + L_f + c_w\left(100\ ^\circ\text{C}\right) + L_v\right]$$

$$= \left(2.88\ \text{kg}\right)\begin{bmatrix}\left(2090\ \text{J/kg}\cdot\text{C}^\circ\right)\left(3.00\ ^\circ\text{C}\right) + 33.5\times10^4\ \text{J/kg}\\ +\left(4186\ \text{J/kg}\cdot\text{C}^\circ\right)\left(100\ ^\circ\text{C}\right) + 22.6\times10^5\ \text{J/kg}\end{bmatrix}$$

$$= 8.70\times10^6\ \text{J}$$

**Insight** Approaching problems of these types in the step-by-step manner shown here is the best way to avoid mistakes. Also, notice that the units of specific heat are written with C° instead of K; remember, it doesn't matter which you use.

---

**Example 17–6 The Condensation and Melting of Water** How much steam at 100 °C must be placed into an insulated container that holds 0.75 kg of ice at –11 °C to finally produce liquid water at 12 °C?

**Picture the Problem** The sketch shows an ice–steam mixture that later becomes all liquid water.

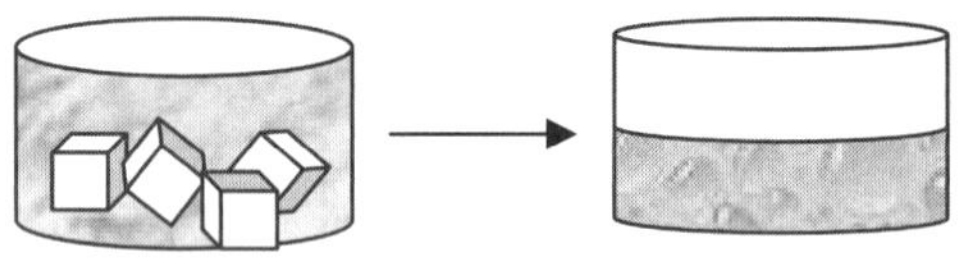

**Strategy** There will be heat lost by the water that starts as steam, and heat gained by the water that starts as ice. We can follow all the temperature changes and phase changes step-by-step and set the heat lost equal to the heat gained.

**Solution**

1. Heat is lost by the steam in condensing to liquid water at 100 °C:

$$Q_{lost,1} = m_{steam}L_v$$

2. Heat is subsequently lost by this liquid water in cooling to the equilibrium temperature: $Q_{lost,2} = m_{steam}c_w\left(100\,^\circ\text{C} - T_{equil}\right)$

3. Heat is gained by the ice to warm it to its melting temperature, 0 °C: $Q_{gain,1} = m_{ice}c_{ice}\left(0\,^\circ\text{C} - T_{0,ice}\right)$

4. Heat is gained by the ice to melt it at 0 °C: $Q_{gain,2} = m_{ice}L_f$

5. Heat is subsequently gained by this liquid water in warming to the equilibrium temperature: $Q_{gain,3} = m_{ice}c_w\left(T_{equil} - 0\,^\circ\text{C}\right)$

6. Setting the heat lost equal to the heat gained:

$$m_{steam}L_v + m_{steam}c_w\left(100\,^\circ\text{C} - T_{equil}\right) = m_{ice}c_{ice}\left(0\,^\circ\text{C} - T_{0,ice}\right) + m_{ice}L_f + m_{ice}c_w\left(T_{equil} - 0\,^\circ\text{C}\right)$$

7. Solving for the amount of steam gives:

$$m_{steam} = \frac{m_{ice}\left[c_{ice}\left(0\,^\circ\text{C} - T_{0,ice}\right) + L_f + c_w\left(T_{equil} - 0\,^\circ\text{C}\right)\right]}{L_v + c_w\left(100\,^\circ\text{C} - T_{equil}\right)}$$

8. Obtain the numerical result:

$$m_{steam} = \frac{(0.75\text{ kg})\left[\left(2090\ \tfrac{\text{J}}{\text{kg}\cdot\text{K}}\right)\left(11\text{ C}^\circ\right) + 33.5\times10^4\ \tfrac{\text{J}}{\text{kg}} + \left(4186\text{ J/kg}\cdot\text{K}\right)\left(12\text{ C}^\circ\right)\right]}{22.6\times10^5\text{ J/kg} + \left(4186\text{ J/kg}\cdot\text{K}\right)\left(88\text{ C}^\circ\right)} = 0.12\text{ kg}$$

**Insight** All the individual calculations of heat lost were calculated as positive quantities. Therefore, $\Delta T$ represents the magnitude of the temperature difference and is *not* always the final temperature minus the initial temperature.

## Practice Quiz

8. Does it require more or less heat transfer to fuse a certain amount of liquid water into ice than to convert the same amount of ice into liquid water? (Assume everything takes place at 0 °C.)

   **(a)** more **(b)** less **(c)** the same amount **(d)** the answer cannot be determined

9. When water vapor is transformed into liquid water,

   **(a)** heat is lost by the vapor. **(b)** heat is gained by the vapor.
   **(c)** no heat transfer is required. **(d)** the liquid water immediately freezes.
   **(e)** None of the above.

10. If 2.0 kg of ice at 0.0 °C is dropped into 3.82 kg of water at 7.0 °C, what percentage of the ice melts?

    **(a)** 100% **(b)** 2.0% **(c)** 28% **(d)** 17% **(e)** 81%

# Reference Tools and Resources

## I. Key Terms and Phrases

**ideal gas** a gas in which the gas particles do not interact except for elastic collisions

**equation of state** (for ideal gases) an equation that relates the temperature, pressure, volume, and number of particles of the gas

**mole** the amount of a substance that contains $6.022 \times 10^{23}$ entities

**Avogadro's number** the number of entities in a mole, equal to $6.022 \times 10^{23}$

**atomic (molecular) mass** the mass of one mole of atoms (molecules)

**isotherms** pressure-versus-volume curves plotted for fixed temperature and number of particles

**kinetic theory** relates the motion of the microscopic particles of a system to its macroscopic properties

**stress** the applied force per unit area that deforms a substance

**strain** the deformation that results from a stress applied to an object

**evaporation** the release of molecules from the liquid phase to the gaseous phase

**fusion** the freezing of a liquid to a solid

**sublimation** the direct transformation between the solid and gas phases

**phase diagram** a graph that shows the conditions (often of temperature and pressure) under which a substance will exist in different phases

**latent heat** heat required to completely change the phase of one kilogram of a substance

## II. Important Equations

| Name/Topic | Equation | Explanation |
|---|---|---|
| equation of state | $PV = NkT = nRT$ | Two forms of the equation of state for an ideal gas |
| kinetic theory | $P = \frac{2}{3}\left(\frac{N}{V}\right)K_{av}$<br>$K_{av} = \frac{3}{2}kT$ | Kinetic theory shows that both the pressure and Kelvin temperature of a gas are directly proportional to the average translational kinetic energy of a gas particle |

| | | |
|---|---|---|
| elastic deformation | $F = Y\left(\frac{\Delta L}{L_0}\right)A$<br>$F = S\left(\frac{\Delta x}{L_0}\right)A$<br>$\Delta P = -B\left(\frac{\Delta V}{V_0}\right)$ | Expressions describing the elastic deformation of solids for stretching, shearing, and volumetric stresses |
| latent heat | $Q = mL$ | The expression for the heat required to change the phase of an amount of mass, *m,* of a substance |

## III. Know Your Units

| Quantity | Dimension | SI Unit |
|---|---|---|
| the Boltzmann constant ( $k$ ) | $[M][L^2][T^{-2}][K^{-1}]$ | J/K |
| Mole | dimensionless | mol |
| Avogadro's number ( $N_A$ ) | dimensionless | molecules/mol |
| gas constant ( $R$ ) | $[M][L^2][T^{-2}][K^{-1}]$ | J/(mol·K) |
| Young's, shear, and bulk moduli ( $Y, S, B$ ) | $[M][L^{-1}][T^{-2}]$ | $N/m^2$ |
| Stress | $[M][L^{-1}][T^{-2}]$ | Pa (or $N/m^2$) |
| Strain | dimensionless | — |
| latent heat ( $L$ ) | $[L^2][T^{-2}]$ | J/kg |

# Puzzle

**VACUUM SUCKS!**

A vacuum chamber designed to be pumped down to a pressure of $10^{-4}$ Pa has stainless steel walls that are 3 mm thick. You must design a new chamber that can be pumped down to a pressure of $10^{-5}$ Pa. How thick must the walls of the new chamber be?

**a**. 0.3 mm **b**. 3 mm **c**. 1 cm **d**. 3 cm

## Answers to Selected Conceptual Questions

**4.** Yes. If the pressure and volume are changed in such a way that their product remains the same, it follows from the ideal-gas law that the temperature of the gas will remain the same. If the temperature of the gas is the same, the average kinetic energy of its molecules will not change.

**8.** If we look at the phase diagram in Figure 17-16, we can see that in order to move upward in the graph from the sublimation curve to the fusion curve, the pressure acting on the system must be increased.

**10.** No. Water is at 0 °C whenever it is in equilibrium with ice. The ice cube thrown into the pool will soon melt, however, showing that the ice cube–pool system is not in equilibrium. As a result, there is no reason to expect that the water in the pool is at 0 °C.

## Solutions to Selected End-of-Chapter Problems and Conceptual Exercises

14. **Picture the Problem:** A spherical balloon is filled with helium gas at a specified temperature and pressure. The number of atoms within the balloon is doubled, while the temperature and pressure are held constant. This increases the volume, and therefore the radius.

**Strategy:** Use the ideal gas law, equation 17-2, to calculate the number of atoms within the balloon. The volume of a sphere is related to the radius by $V = \frac{4}{3}\pi r^3$. Doubling the number of atoms at constant temperature and pressure will double the volume. Set the final volume equal to twice the initial volume and solve for the final radius in terms of the initial radius.

**Solution: 1. (a)** Solve the ideal gas law for the number of atoms:

$$PV = NkT \Rightarrow N = \frac{PV}{kT}$$

**2.** Write the volume in terms of the radius and insert the given values:

$$N = \frac{P\left(\frac{4}{3}\pi r^3\right)}{kT} = \frac{\left(2.4\times10^5 \text{ Pa}\right)\frac{4}{3}\pi\left(0.25 \text{ m}\right)^3}{\left(1.38\times10^{-23} \text{ J/K}\right)\left(273.15+18 \text{ K}\right)} = \boxed{3.9\times10^{24} \text{ atoms}}$$

**3. (b)** Set the final volume equal to twice the initial volume:

$$V_2 = 2V_1$$

$$\tfrac{4}{3}\pi r_2^3 = 2\left(\tfrac{4}{3}\pi r_1^3\right)$$

**4.** Solve for the final radius:

$$r_2 = 2^{1/3} r_1 = 1.26 r_1$$

The radius increases by a multiplicative factor of $\boxed{1.26}$.

**Insight:** Doubling the volume does not double the radius. To double the radius, the volume would need to increase by a factor of 8.

33. **Picture the Problem:** A uranium hexafluoride ($UF_6$) gas is made of molecules containing two different isotopes of uranium. Both types of molecules are at the same temperature, but because their masses differ slightly, the heavier isotope will have a smaller rms speed.

**Strategy:** Use equation 17-13 to calculate the ratio of rms speeds for the two isotopes. The molar mass of each molecule will be the mass of the uranium isotope plus 114 u (six times the molar mass of fluorine).

**Solution: 1.** Write the ratio of rms speeds using equation 17-13 and simplify:

$$\frac{v_{\text{rms, 238}}}{v_{\text{rms, 235}}} = \frac{\sqrt{3RT/M_{238}}}{\sqrt{3RT/M_{235}}} = \sqrt{\frac{M_{235}}{M_{238}}}$$

**2.** Insert the molar masses of both isotopes: $\dfrac{v_{\text{rms, 238}}}{v_{\text{rms, 235}}} = \sqrt{\dfrac{235\text{ u}+114\text{ u}}{238\text{ u}+114\text{ u}}} = \boxed{0.996}$

**Insight:** Even though the rms speeds differ by only 0.4%, this is still sufficient to separate out the U-235.

44. **Picture the Problem:** Two rods of differing materials but identical dimensions are placed end to end under a force of 8400 N. The force compresses each rod.

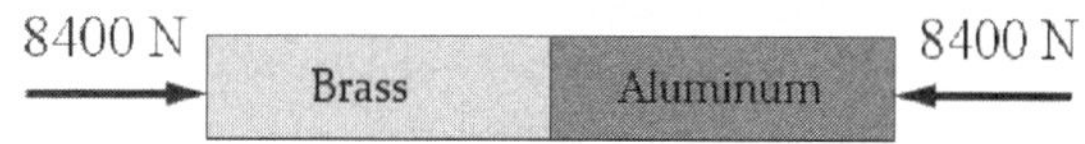


**Strategy:** The compressive force is felt equally by each rod. Therefore, use equation 17-17 to calculate the change in length of each rod separately. Sum the changes in length of both rods to determine to total change in length.

**Solution: 1. (a)** Solve equation 17-17 for the change in length:

$$F = Y\left(\frac{\Delta L}{L_0}\right)\left(\frac{\pi D^2}{4}\right) \Rightarrow \Delta L = \frac{4L_0 F}{Y\pi D^2}$$

**2.** Sum the changes in length for each rod and combine like terms:

$$\Delta L_{\text{total}} = \Delta L_{\text{Al}} + \Delta L_{\text{Br}} = \frac{4L_0 F}{Y_{\text{Al}}\pi D^2} + \frac{4L_0 F}{Y_{\text{Br}}\pi D^2} = \frac{4L_0 F}{\pi D^2}\left(\frac{1}{Y_{\text{Al}}} + \frac{1}{Y_{\text{Br}}}\right)$$

**3.** Solve numerically:

$$\Delta L_{\text{total}} = \frac{4(0.55\text{ m})(8400\text{ N})}{\pi(0.017\text{ m})^2}\left(\frac{1}{6.9\times10^{10}\text{ N/m}^2} + \frac{1}{9.0\times10^{10}\text{ N/m}^2}\right) = \boxed{0.52\text{ mm}}$$

**4. (b)** $\boxed{\text{Aluminum}}$ will have the larger change in length because it has the smaller Young's modulus.

**Insight:** The aluminum rod's change in length is 0.295 mm and the brass rod's change in length is 0.226 mm. As predicted, aluminum has the greater change in length.

52. **Picture the Problem:** The phase diagram for carbon dioxide shows the pressures and temperatures for which carbon dioxide is in the solid, liquid, and vapor phases. The lines separating the phases are the sublimation, fusion, and vapor pressure curves.

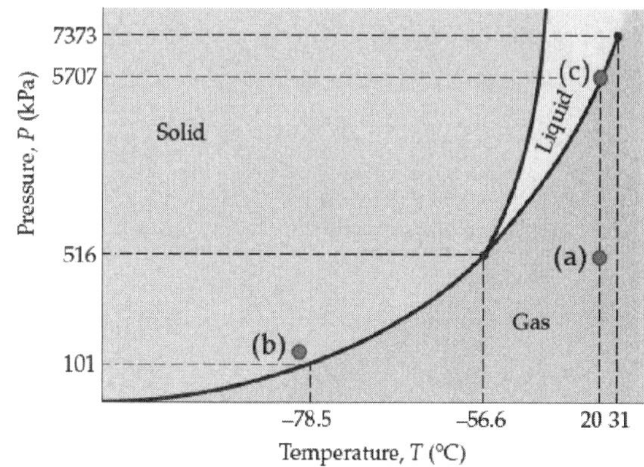


**Strategy:** Examine the graph to determine the phase at the given temperatures and pressures. Find the boiling pressure at 20°C.

**Solution: 1. (a)** It is in the $\boxed{\text{gas}}$ phase.

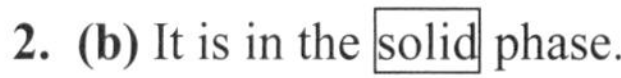

**2. (b)** It is in the $\boxed{\text{solid}}$ phase.

**3. (c)** The boiling pressure is $\boxed{5707\text{ kPa.}}$

**Insight:** Note that $CO_2$ does not have a liquid phase at atmospheric pressure.

63. **Picture the Problem:** Heat is added at a constant rate to ice initially at 0°C. Over time the heat melts the ice, raises the water temperature to the boiling point, and boils off the water.

**Strategy:** Calculate the time required for each step of the heating process by dividing the heat added in that process by the rate at which the heat is added.

**Solution: 1. (a)** Solve for the time to melt the ice:

$$t_{\text{AB}} = \frac{Q_{\text{AB}}}{\Delta Q/\Delta t} = \frac{mL_{\text{f}}}{\Delta Q/\Delta t} = \frac{1.000\text{ kg}\left(33.5\times10^4\text{ J/kg}\right)}{12,250\text{ J/s}} = \boxed{27.3\text{ s}}$$

**2. (b)** Solve for the time to heat the water to boiling:

$$t_{\text{BC}} = \frac{Q_{\text{BC}}}{\Delta Q/\Delta t} = \frac{mc\Delta T}{\Delta Q/\Delta t} = \frac{1.000\text{ kg}\left[4186\text{ J/}(\text{kg}\cdot\text{K})\right](100\text{ C}^\circ)}{12,250\text{ J/s}} = \underline{\underline{34.2\text{ s}}}$$

**3.** Add the time to melt the ice to the time to heat the water:

$$t_{\text{AC}} = t_{\text{AB}} + t_{\text{BC}} = 27.3\text{ s} + 34.2\text{ s} = \boxed{61.5\text{ s}}$$

**4. (c)** Solve for the time to boil off the water:

$$t_{\text{CD}} = \frac{Q_{\text{CD}}}{\Delta Q/\Delta t} = \frac{mL_{\text{v}}}{\Delta Q/\Delta t} = \frac{1.000\text{ kg}\left(22.6\times10^5\text{ J/kg}\right)}{12,250\text{ J/s}} = \underline{\underline{184.5\text{ s}}}$$

**5.** Add the time to boil the water to the previous two time periods:

$$t_{\text{AD}} = t_{\text{AB}} + t_{\text{BC}} + t_{\text{CD}} = 27.3\text{ s} + 34.2\text{ s} + 184.5\text{ s} = \boxed{246\text{ s}}$$

**6. (d)** Because 63 seconds is greater than 61.5 seconds and less than 246 seconds, $\boxed{\text{the water is boiling.}}$

**Insight:** The time required to boil off the water is significantly greater than any of the other time periods. This shows that the amount of heat required to boil water is significantly greater than the heat needed to melt the same amount of ice or to increase the temperature of the water from freezing to boiling.

71. **Picture the Problem:** A cold block is copper is dropped into an aluminum cup filled with water. The temperature of the copper rises as it absorbs heat from the aluminum and water. Eventually the temperatures of the copper, water, and cup come to equilibrium.

**Strategy:** Assume that the final temperature is greater than 0°C, such that no ice is formed. Use equation 16-13 to calculate the heat absorbed by the copper and the heat lost by the water and cup. Set the two heats equal and solve for the final temperature.

**Solution: 1. (a)** Set the heat gained equal to the heat lost:

$$m_{\text{Cu}}c_{\text{Cu}}\left[T_{\text{f}} - (-12°\text{C})\right] = m_{\text{w}}c_{\text{w}}\left(4.1°\text{C} - T_{\text{f}}\right) + m_{\text{Al}}c_{\text{Al}}\left(4.1°\text{C} - T_{\text{f}}\right)$$

**2.** Solve for the final temperature:

$$T_{\text{f}} = \frac{\left(m_{\text{w}}c_{\text{w}} + m_{\text{Al}}c_{\text{Al}}\right)(4.1°\text{C}) - m_{\text{Cu}}c_{\text{Cu}}(12°\text{C})}{m_{\text{w}}c_{\text{w}} + m_{\text{Al}}c_{\text{Al}} + m_{\text{Cu}}c_{\text{Cu}}}$$

**3.** Insert the given data:

$$T_{\text{f}} = \frac{\left(\begin{array}{c}\left\{(0.110\text{ kg})\left[4186\text{ J/}(\text{kg}\cdot\text{K})\right] + (0.075\text{ kg})\left[900\text{ J/}(\text{kg}\cdot\text{K})\right]\right\}(4.1°\text{C}) \\ -(0.048\text{ kg})\left[387\text{ J/}(\text{kg}\cdot\text{K})\right](12°\text{C})\end{array}\right)}{\left\{\begin{array}{c}(0.110\text{ kg})\left[4186\text{ J/}(\text{kg}\cdot\text{K})\right] + (0.075\text{ kg})\left[900\text{ J/}(\text{kg}\cdot\text{K})\right] \\ +(0.048\text{ kg})\left[387\text{ J/}(\text{kg}\cdot\text{K})\right]\end{array}\right\}}$$

$$= \boxed{3.6°\text{C}}$$

**4. (b)** Since $T_{\text{f}} > 0°\text{C}$, $\boxed{\text{no ice is present.}}$

**Insight:** If the calculation of the final temperature gave an answer less than zero, our assumption that all the ice melts would have been incorrect. In that case, we would assume the final temperature is 0°C and that some of the water would freeze to ice.

## Answers to Practice Quiz

**1.** (b) **2.** (c) **3.** (a) **4.** (d) **5.** (e) **6.** (b) **7.** (c) **8.** (c) **9.** (a) **10.** (d)

# CHAPTER 18
# THE LAWS OF THERMODYNAMICS

## Chapter Objectives

After studying this chapter, you should

1. understand the relationship between the first law of thermodynamics and the conservation of energy.
2. understand the behavior of gases that undergo constant-pressure, constant-volume, isothermal, and adiabatic processes.
3. know the difference between the specific heat at constant volume and the specific heat at constant pressure and how the two relate for a monatomic ideal gas.
4. know the basic functions of heat engines, refrigerators, and similar devices and how they function within the laws of thermodynamics.
5. develop a basic understanding of the concept of entropy and how it relates to order and disorder.
6. know the four laws of thermodynamics.

## Warm-Ups

1. Explain how it is possible to compress a gas to half its original volume without changing the pressure.
2. Estimate the change in temperature when one cubic meter of air is compressed to half its original volume adiabatically (start at room temperature and pressure).
3. Is melting an ice cube an irreversible process? How about freezing one?
4. Estimate the total work by a system taken around the cycle shown in the figure. Does the system have to be an ideal gas?

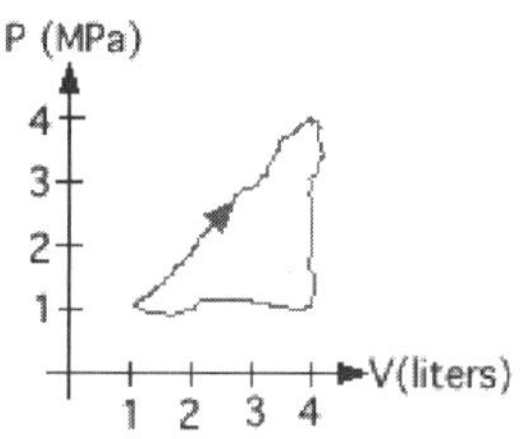

5. In your textbook, an illustration like this one is used to represent a heat engine. If this sketch is to represent your car, what parts of your car correspond to the parts of the diagram?

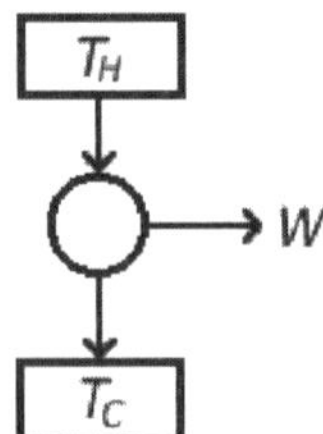


## Chapter Review

In this final chapter on thermodynamics, we bring all four laws together, hoping to provide a basic understanding of the importance of thermodynamics and many of its applications.

### 18–1 The Zeroth Law of Thermodynamics

The zeroth law of thermodynamics was introduced in Chapter 16; we review it here for the sake of completeness. The zeroth law of thermodynamics lays important groundwork for all of thermodynamics, especially for providing a working definition of temperature. The zeroth law says the following:

> *If system A is in thermal equilibrium with system B, and system C is also in thermal equilibrium with system B, then systems A and C will be in thermal equilibrium if brought into thermal contact.*

Temperature determines whether two systems will be in thermal equilibrium. When two systems are at the same temperature, they will be in thermal equilibrium if brought into thermal contact.

### 18–2 The First Law of Thermodynamics

The first law of thermodynamics is essentially just an application of the conservation of energy to situations involving heat. Any system has a certain amount of internal energy, $U$; this energy consists of all the potential and kinetic energy contained within the system. Changes in the internal energy result from either heat into (positive $Q$) or out of (negative $Q$) the system and/or work done by (positive $W$) or on (negative $W$) the system. The mathematical expression of this law is

$$\Delta U = Q - W$$

Notice that this expression is consistent with the intuitive notion that the internal energy will increase when either heat is added to the system and/or work is done on the system.

An important distinction between internal energy, heat, and work is that changes in internal energy depend only on the initial and final states of the system (which are determined by pressure, temperature, and volume) and $U$ is therefore called a **state function**. Both heat and work depend not only on the states involved but also on the process by which a system is changed from one state to another.

**Example 18–1 The First Law** If 4530 J of work is done on a 0.750-kg piece of copper while its temperature rises from 18.2 °C to 31.2 °C, what is the change in internal energy of the piece of copper? (Assume a constant pressure of 1 atm.)

**Picture the Problem** The sketch shows the piece of copper in question.

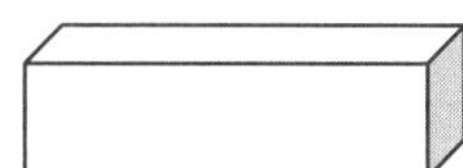

**Strategy** Because we are given the amount of work, $W$, if we can determine the heat flow, $Q$, we can use the first law to determine $\Delta U$.

**Solution**

1. Use the specific heat, Table 16–2 in the text, to determine $Q$:

$$Q = mc\Delta T = (0.750 \text{ kg})(390 \text{ J/kg}\cdot\text{C}^\circ)(13.0 \text{ C}^\circ) = 3802.5 \text{ J}$$

2. The work is done *on* the copper so:

$$W = -4530 \text{ J}$$

3. According to the first law of thermodynamics:

$$\Delta U = Q - W = 3802.5 \text{ J} - (-4530 \text{ J}) = 8330 \text{ J}$$

**Insight** Remember when using the first law of thermodynamics that it is important to keep track of whether the heat is into or out of the system and if the work is done on or by the system.

## Practice Quiz

1. The first law of thermodynamics is most closely related to
   **(a)** the conservation of energy.
   **(b)** the conservation of linear momentum.
   **(c)** the conservation of angular momentum.
   **(d)** Newton's second law.
   **(e)** None of the above.

2. If 13 J of work is done on a system to remove 9.0 J of heat, what is the change in the internal energy of the system?
   **(a)** – 4.0 J **(b)** 22 J **(c)** – 22 J **(d)** 4.0 J **(e)** none of the above

## 18–3 Thermal Processes

As mentioned at the end of the previous section, some quantities, such as $Q$ and $W$, depend on the process by which a system is altered. Because of this fact, we need to understand the basic types of processes used in the study and application of thermodynamics.

In this chapter, we consider only processes that are quasi-static and free from dissipative forces. Quasi-static means that a process takes place so slowly that we can consider the system to be in thermal equilibrium with its surroundings throughout the process. These conditions allow processes to be **reversible**, which means that both the system and its environment can be returned to their precise states at the beginning of the process. Processes for which these conditions do not apply are called **irreversible**.

### Constant Pressure and Constant Volume

For a gas that expands or contracts during a process while held at constant pressure (sometimes called an *isobaric* process), the work done by the gas during the process is found to be

$$W = P(\Delta V) \qquad \textit{(for constant pressure)}$$

where $\Delta V = V_f - V_i$ is the change in the volume of the gas. The above result can be interpreted graphically as the area under the curve of a pressure-versus-volume plot (see below). In fact, the area under the curve of a pressure-versus-volume plot for an expanding (or contracting) gas equals the work done by (or on) the gas for *any process*, not just processes at constant pressure.

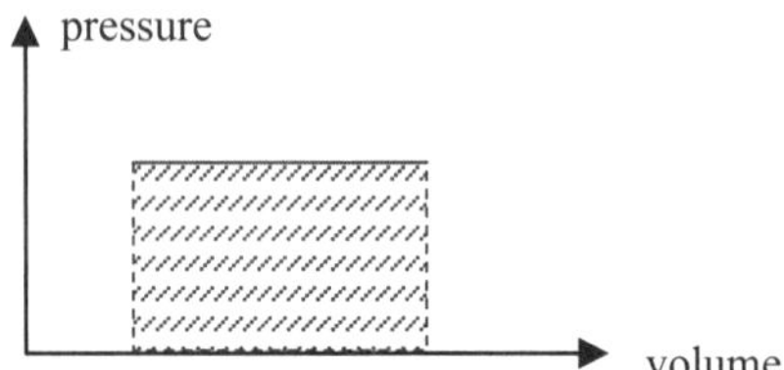


At constant volume (an *isochoric* process), no work is done by the gas during a reversible process. This fact is consistent with the preceding expression because $\Delta V = 0$ when the volume is held constant. Also, intuition suggests that there should be no work done because work results when a force acts through distance and, if the gas does not expand or contract through any distance, the net work done by the gas is zero. Thus,

$$W = 0 \qquad \textit{(for constant volume)}$$

### Isothermal Processes

An **isothermal process** is one that takes place at constant temperature. For an ideal gas, the relationship between the pressure and the volume during an isothermal process is

$$P = \frac{\text{constant}}{V}$$

where the constant is $NkT$. The work done by the gas during an isothermal process can be derived from the ideal-gas equation of state to be

$$W = NkT \ln\left(\frac{V_f}{V_i}\right) = nRT \ln\left(\frac{V_f}{V_i}\right) \qquad (\textit{for constant temperature})$$

**Adiabatic Processes**

If no heat flows during a process, the process is called **adiabatic**. This type of process occurs when a system is well insulated or when the process takes place so rapidly that heat doesn't have time to flow. During adiabatic processes, the pressure, volume, and temperature may all change. The pressure-versus-volume curve for a system undergoing an adiabatic process is called an *adiabat*.

---

**Example 18–2 Expanding Gas** A monatomic ideal gas consisting of 6.32 moles of atoms expands from a volume of 14.1 m$^3$ to 27.6 m$^3$. How much work is done by the gas if the expansion is **(a)** at a constant pressure of 133 kPa and **(b)** isothermal at $T$ = 303 K?

**Picture the Problem** The sketch shows the gas at its initial volume on the left and its final expanded volume on the right.

**Strategy** For both parts (a) and (b) we can use the above expressions to calculate the work done.

**Solution**

**Part (a)**

1. The work done at constant pressure gives: $W = P(\Delta V) = (133\times10^3 \text{ Pa})(13.5 \text{ m}^3) = 1.80\times10^6 \text{ J}$

**Part (b)**

2. Use the expression for the work done during an isothermal process:

$$W = nRT \ln\left(\frac{V_f}{V_i}\right)$$

$$= (6.32 \text{ mol})(8.31 \text{ J/mol}\cdot\text{K})(303 \text{ K})\ln\left(\frac{27.6 \text{ m}^3}{14.1 \text{ m}^3}\right)$$

$$= 1.07\times10^4 \text{ J}$$

**Insight** Recall that the natural logarithm, ln, is the same as log to the base $e$.

---

## Practice Quiz

3. During an isothermal process, a gas expands to twice its previous volume. The pressure in the gas
   **(a)** doubles.
   **(b)** triples.
   **(c)** becomes larger by a factor of $\sqrt{2}$.
   **(d)** stays the same.
   **(e)** None of the above.

4. The graph shown on the right most likely represents what type of process?
   **(a)** constant pressure **(b)** constant volume
   **(c)** isothermal **(d)** irreversible
   **(e)** none of the above

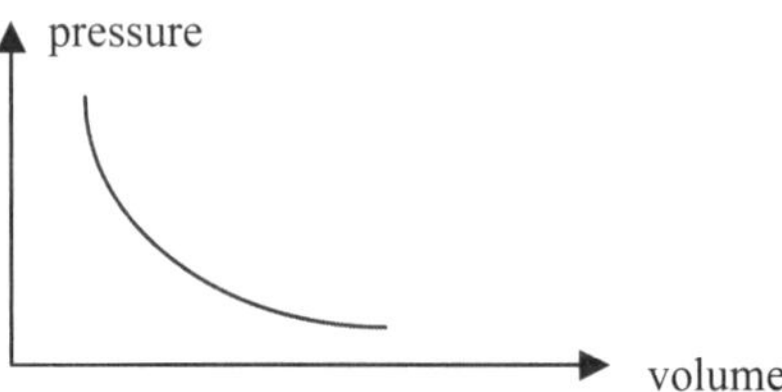


## 18–4 Specific Heats for an Ideal Gas: Constant Pressure, Constant Volume

It was mentioned previously that the amount of heat flow depends on the nature of the process; that is, heat is not a state function. This means that the specific heat is different for different processes. The relationship between the specific heats for constant pressure and constant volume processes is particularly useful. For constant volume, the specific heat, $c_v$ (lowercase $c$), is defined by

$$Q_v = mc_v\Delta T = nC_v\Delta T$$

where $C_v$ is called the **molar specific heat** (with a capital $C$) representing the heat needed to change the temperature of 1 mole of a substance by 1 Celsius degree. Similarly, the molar specific heat for constant pressure is determined by

$$Q_p = nC_p\Delta T$$

For a monatomic ideal gas, the molar specific heats are

$$C_v = \tfrac{3}{2}R \quad \text{and} \quad C_p = \tfrac{5}{2}R$$

which further implies that $C_p - C_v = R$. This last relation is true for all ideal gases, not just monatomic ones. Given this information we are now ready to write down a pressure-volume relation for adiabatic processes, which we could not do in the previous section. If we define the quantity $\gamma = C_p/C_v$, then, for an ideal gas undergoing an adiabatic process we have

$$PV^\gamma = \text{constant} \qquad (\textit{for adiabatic processes})$$

For a monatomic gas, $\gamma = 5/3$.

**Exercise 18–3 An Adiabatic Process** During an adiabatic process of a monatomic ideal gas containing 100 moles, the pressure in the gas increases from 105 kPa to 115 kPa. If the original volume of the gas was 2.25 m$^3$, **(a)** what is the new volume of the gas and **(b)** what is its final temperature?

**Solution:** We are given the following information:

**Given:** $n = 100$ mol, $P_i = 105$ kPa, $P_f = 115$ kPa, $V_i = 2.25\text{ m}^3$; **Find:** (a) $V_f$, (b) $T_f$

**Part (a)**

Since the gas is a monatomic ideal gas, we know that $\gamma = 5/3$ and we can use the result $PV^\gamma =$ constant.

$$P_i V_i^{5/3} = P_f V_f^{5/3} \quad \Rightarrow \quad V_f = V_i \left(\frac{P_i}{P_f}\right)^{3/5} = \left(2.25\text{ m}^3\right)\left(\frac{105\text{ kPa}}{115\text{ kPa}}\right)^{3/5} = 2.13\text{ m}^3$$

**Part (b)**

Now that we know the final pressure and volume, we can use the equation of state of an ideal gas to determine the temperature:

$$PV = nRT \quad \Rightarrow \quad T_f = \frac{P_f V_f}{nR} = \frac{\left(1.15\times10^5\text{ Pa}\right)\left(2.130\text{ m}^3\right)}{\left(100\text{ mol}\right)\left(8.31\text{ J/mol}\cdot\text{K}\right)} = 295\text{ K}$$

## Practice Quiz

5. If the volume of a monatomic ideal gas expands adiabatically such that its volume triples, what factor times the initial pressure equals the final pressure of the gas?

   **(a)** 6.2 **(b)** 0.16 **(c)** 3 **(d)** 0.33 **(e)** 1.9

## 18–5 The Second Law of Thermodynamics

In Chapter 16, we noted that temperature is the quantity that indicates the existence and direction of the flow of heat from one system to another when the systems are in thermal contact. The second law of thermodynamics states this direction explicitly:

> *When two systems at different temperatures are brought into thermal contact, the spontaneous flow of heat is always from the system at the higher temperature to the system at the lower temperature.*

## 18–6 – 18–7 Heat Engines, Carnot Cycle, Refrigerators, Air Conditioners, and Heat Pumps

A device that converts heat to work is called a **heat engine**. Heat engines require a high-temperature region to supply energy to the system, a low-temperature region to exhaust wasted heat, an engine that operates cyclically, and a working substance. If we take the *magnitude* of the heat supplied to the engine as $Q_h$ and the *magnitude* of the heat exhausted to the low-temperature region as $Q_c$, the work done by the heat engine is given by

$$W = Q_h - Q_c$$

Heat engines are often characterized by their **efficiency**, $e$, which is the fraction of the heat supplied that appears as work:

$$e = \frac{W}{Q_h} = 1 - \frac{Q_c}{Q_h}$$

The maximum possible efficiency of a heat engine that operates from a single hot reservoir at temperature $T_h$ and a single cold reservoir at $T_c$ results when all processes in the cycle are reversible – this statement is known as **Carnot's theorem**. Analysis of engines of this type gives the maximum efficiency to be

$$e_{\text{max}} = 1 - \frac{T_c}{T_h}$$

Notice that the maximum possible efficiency depends only on the temperatures of the reservoirs and not on any details of the engine. Furthermore, this theorem shows that no engine can be perfectly efficient. Thus, we can see that the maximum work that a particular heat engine can perform is

$$W_{\text{max}} = e_{\text{max}} Q_h = \left(1 - \frac{T_c}{T_h}\right) Q_h$$

Although it may not be obvious, Carnot's theorem is equivalent to the statement of the second law of thermodynamics given previously and is often considered to be an alternative version of it.

---

**Example 18–4 The Efficiency of an Engine** A heat engine generates 4.1 kJ of work at 35% efficiency. How much heat is exhausted in the cold reservoir?

**Picture the Problem** The diagram shows a schematic of the heat engine.

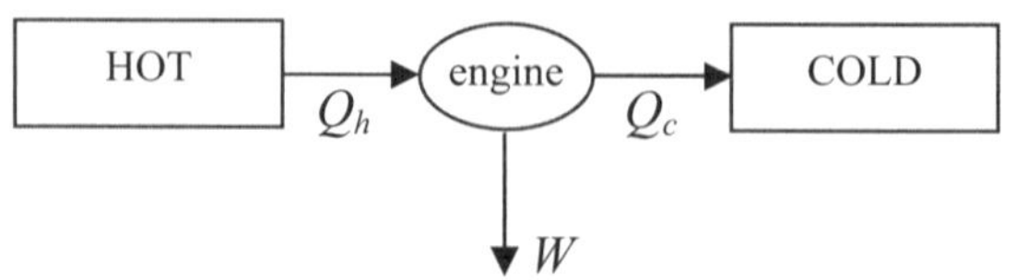


**Strategy** Because both the work done and the efficiency depend on $Q_h$, we can combine those two expressions to eliminate $Q_h$ and solve for $Q_c$.

**Solution**

1. The expressions for $W$ and $e$ are: $W = Q_h - Q_c$ and $e = \dfrac{W}{Q_h} \Rightarrow Q_h = \dfrac{W}{e}$

2. Substitute $Q_h$ into the expression for the work: $W = \frac{W}{e} - Q_c$

3. Solve this equation for $Q_c$: $Q_c = \frac{W}{e} - W = W\left(\dfrac{1-e}{e}\right)$

4. Obtain the numerical result: $Q_c = \left(4.1\times10^3\ \text{J}\right)\left(\dfrac{1-0.35}{0.35}\right) = 7.6\ \text{kJ}$

**Insight** Notice here that the efficiency is used as a decimal number in the solution even though it is quoted as a percentage in the problem. It may seem strange that more heat is exhausted than is converted to work, but that should be expected for an efficiency of less than 50%.

---

Recall that a heat engine uses the natural tendency for heat to flow from hot to cold as a means of generating work. A **refrigerator** is almost the precise reverse of a heat engine. In a refrigerator, work is the input that ultimately leads to a cooler region (the refrigerator) becoming even cooler and a warmer region (the room) becoming even warmer. The effectiveness of a refrigerator is indicated by its **coefficient of performance**, COP, which compares the heat forced from the cold reservoir, $Q_c$, to the amount of work required to do it, $W$:

$$\text{COP}_{\text{ref}} = \frac{Q_c}{W}$$

An **air conditioner** is the same as a refrigerator, except that the room being cooled is the cold reservoir and the outdoor air is the warm region.

A **heat pump** is the reverse of an air conditioner. This device uses work to remove energy from the cold reservoir (the outdoor air) and pump it into the warm reservoir (the room being heated). In an ideal (reversible) heat pump, the amount of work required to pump an amount of heat, $Q_h$, into a room is given by the same relationship as for a Carnot heat engine: $W = Q_h(1 - T_c/T_h)$. The COP for a heat pump is:

$$\text{COP}_{\text{hp}} = \frac{\text{Q}_\text{h}}{\text{W}}$$

## Practice Quiz

6. If a heat engine requires 4500 J of heat to produce 1200 J of work, what is its efficiency?

   **(a)** 0.27  **(b)** 100%  **(c)** 0.73  **(d)** 0.36  **(e)** 0%

7. In terms of process and function, which pair of devices is most alike?
   **(a)** heat engine and refrigerator
   **(b)** heat engine and heat pump
   **(c)** refrigerator and air conditioner
   **(d)** air conditioner and heat pump
   **(e)** air conditioner and heat engine

8. If heat flow to the hot reservoir increases over time for a refrigerator (assuming $Q_c$ remains constant), its COP
   **(a)** increases.
   **(b)** stays the same.
   **(c)** decreases.

## 18–8 – 18–9 Entropy, Order, and Disorder

The fact that heat flows naturally only from systems of higher temperature to systems of lower temperature is only one part of a larger "directionality" to the laws of thermodynamics. The basic quantity in physics that encompasses that directionality is called **entropy**, $S$. The entropy of a system is a state function, and the change in entropy is defined to be

$$\Delta S = \frac{Q}{T}$$

where it is understood that the heat, $Q$, is transferred reversibly at a fixed absolute temperature $T$.

The concept of entropy becomes particularly important when we consider the total change in entropy of a system *and its surroundings* during a process. It is found that, although the entropy of an individual system can decrease, this total entropy never decreases. For reversible processes, the change in the total entropy is zero, and for all other processes the total entropy increases. Remembering that reversibility is an idealization and that all real processes are irreversible, we see that *the total entropy of the universe is always increasing*. This last point is the sense in which entropy provides us with directionality to the laws of physics — an arrow of time. The statement that the entropy of the universe is increasing is another equivalent way to state the second law of thermodynamics. In fact, the second law of thermodynamics is very commonly called "the law of increase of entropy."

One can gain some intuitive insight into the physical meaning of entropy by thinking of it as a measure of the amount of disorder in the universe. When processes occur, the universe always comes away more disordered than it was before. An example is the natural flow of heat from a hot reservoir to a cold reservoir. In the cold reservoir, the molecules are moving more slowly and scattering off each other at a slower rate. If the system is cold enough they may even clump together and solidify, forming a crystal lattice. When heat from the hot reservoir comes in, it increases the speed of the molecules, causing them to move about more randomly and increasing the disorder (and therefore the entropy) of the system. Of course, the disorder (entropy) of the hot reservoir decreases, but it can be shown that the increase in entropy of the cold reservoir is always greater in magnitude than the decrease in entropy of the hot reservoir. The total disorder always increases.

---

**Example 18–5 Ice and Steam** A mass of 2.88 kg of ice at 0.00 °C is mixed with 0.500 kg of steam at 100 °C. Estimate the change in entropy only up to the point when all the ice has melted and all the steam has condensed.

**Picture the Problem** The sketch on the left shows an initial mixture of ice and steam that ultimately becomes all liquid water.

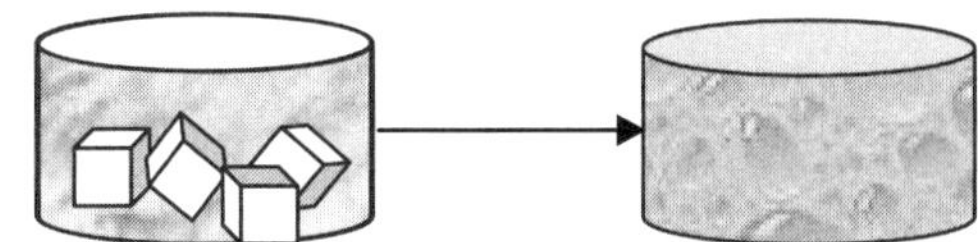

**Strategy** To estimate the total change in entropy, we need to calculate the heat flow for all the processes that take place until we reach the point of having all liquid water.

**Solution**

1. Use Table 17–4 in the text to obtain the latent heat of fusion for water, the heat needed to melt the ice at 0 °C:

$$Q_i = m_i L_f = (2.88\text{ kg})(33.5\times10^4\text{ J/kg}) = 9.648\times10^5\text{ J}$$

2. Also, using Table 17–4, find the heat needed to condense the steam:

$$Q_s = m_s L_v = (0.500\text{ kg})(22.6\times10^5\text{ J/kg}) = 1.130\times10^6\text{ J}$$

3. Since $Q_s > Q_i$, the melted ice warms up; the amount of heat $\Delta Q$ that warms up the liquid is:

$$\Delta Q_w = Q_s - Q_i = 1.130\times10^6\text{ J} - 9.648\times10^5\text{ J} = 1.652\times10^5\text{ J}$$

4. Using Table 16–2 for the specific heat, the change in temperature of the melted ice is:

$$\Delta Q_w = m_i c_w \Delta T_{wat} \quad\Rightarrow\quad \Delta T_w = \Delta Q_w / m_i c_w$$

$$\therefore\quad \Delta T_w = \frac{1.652\times10^5\text{ J}}{(2.88\,kg\times4186\text{ J/kg}\cdot\text{K})} = 13.7\text{ K}$$

**5**. The change in entropy for the melting ice is: $\Delta S_i = \dfrac{Q_i}{T_i} = \dfrac{9.648\times10^5\text{ J}}{273.15\text{ K}} = 3532\text{ J/K}$

**6**. The change in entropy to condense the steam is: $\Delta S_s = \dfrac{-Q_s}{T_i} = \dfrac{-1.130\times10^6\text{ J}}{373.15\text{ K}} = -3028\text{ J/K}$

**7**. The approximate change in entropy to warm the water is:

$$\Delta S_w \approx \frac{\Delta Q_w}{T_{w,av}} = \frac{\Delta Q_w}{\left[T_i + \left(T_i + \Delta T_w\right)\right]/2}$$
$$= \frac{1.652\times10^5\text{ J}}{\left[(273.15+286.85)/2\right]\text{K}} = 590\text{ J/K}$$

**8**. The total change in entropy is:

$$\Delta S_{tot} \approx \Delta S_i + \Delta S_s + \Delta S_w$$
$$= (3532 - 3028 + 590)\text{ J/K} = 1090\text{ J/K}$$

**Insight** This result is only an estimate because we had to estimate $\Delta S_w$ due to the fact that there was a temperature change. To handle this change in temperature, we used the average temperature during the change. As long as the change is comparatively small, this estimate is reasonable. Note also that the temperatures were converted to kelvin because the change in entropy is defined in terms of the kelvin temperature. Also keep in mind that, as the problem requests, this result gives the total entropy change only up to the point when the system is all liquid and does not include the change in entropy for the system to reach an equilibrium temperature.

## Practice Quiz

9. Can the entropy of a system ever decrease?

   **(a)** No, because entropy always increases.

   **(b)** Yes, but only in some reversible processes.

   **(c)** No, because the change in entropy is zero for any system.

   **(d)** Yes, because entropy always decreases.

   **(e)** Yes, if there's a net heat flow out of the system.

10. When the total entropy increases during a process, the systems involved become

   **(a)** more orderly.

   **(b)** less energetic.

   **(c)** crystallized.

   **(d)** more disordered.

   **(e)** more energetic.

**11**. If 2.09 kg of ice at temperature $T' = 0$ °C is melted into water at the same temperature by placing it in contact with a reservoir at temperature $T = 24.4$ °C, what is the change in the entropy of the universe?

**(a)** 210 J/K
**(b)** 0 J/K
**(c)** 2350 J/K
**(d)** 2560 J/K
**(e)** $\infty$

### 18–10 The Third Law of Thermodynamics

This chapter ends with the statement of the third law of thermodynamics. This law states the following:

*It is impossible to lower the temperature of a system to absolute zero in a finite number of steps.*

Perhaps the simplest way to accept the impossibility of absolute zero is to try to find a process that would produce the first system to be cooled all the way to this temperature. It would require either bringing the system into thermal contact with a cooler system (which would have to reach absolute zero first) or pumping heat out of the system, which would have to be isolated from everything warmer (including the pump) to prevent thermal conduction from increasing the temperature. Again, this is a heuristic discussion designed to appeal to your intuition – think about it.

## Reference Tools and Resources

### I. Key Terms and Phrases

**state function** a quantity that only depends on the thermodynamic state of a system (from $P$, $V$, and $T$)

**reversible process** a process that allows a system to return precisely to a previous state

**irreversible process** a process that is not reversible

**isothermal process** a process that takes place at constant temperature

**adiabatic process** a process during which no heat is transferred

**molar specific heat** the heat needed to change the temperature of 1 mol of a substance by 1 C°

**heat engine** a device that converts heat into work

**Carnot's theorem** states the conditions that give the maximum efficiency of a heat engine

**refrigerator** a device that uses work to cause heat to flow from a cooler region to a warmer region

**entropy** the ratio of the heat that flows at a fixed temperature to that temperature for a reversible process. It measures the amount of disorder in a system.

## II. Important Equations

| Name/Topic | Equation | Explanation |
|---|---|---|
| the first law of thermodynamics | $\Delta U = Q - W$ | The conservation of energy for systems with heat flow |
| constant pressure processes | $W = P(\Delta V)$ | The work done by a gas expanding under constant pressure |
| constant temperature processes | $W = NkT \ln\left(\frac{V_f}{V_i}\right) = nRT \ln\left(\frac{V_f}{V_i}\right)$ | The work done by a gas expanding under constant temperature |
| specific heats | $C_v = \frac{3}{2}R, \quad C_p = \frac{5}{2}R$ | The molar specific heat at constant volume and at constant pressure |
| adiabatic processes | $PV^{\gamma} = \text{constant}$ | The behavior of an ideal gas during an adiabatic process |
| heat engines | $e = \frac{W}{Q_h} = 1 - \frac{Q_c}{Q_h}$ | The efficiency of a heat engine |
| entropy | $\Delta S = \frac{Q}{T}$ | The change in entropy of a system at fixed temperature under a reversible process |

## III. Know Your Units

| Quantity | Dimension | SI Unit |
|---|---|---|
| molar specific heat ( $C$ ) | $[\mathrm{M}][\mathrm{L}^2][\mathrm{T}^{-2}][\mathrm{K}^{-1}]$ | J/(mol·K) |
| efficiency ( $e$ ) | dimensionless | — |
| coefficient of performance (COP) | dimensionless | — |
| entropy ( $S$ ) | $[\mathrm{M}][\mathrm{L}^2][\mathrm{T}^{-2}][\mathrm{K}^{-1}]$ | J/K |

## Puzzle

**PROCESS PROBLEMS**

The figure shows five different processes (labeled a–e). Each leads from an initial state ($P$ = 3 atm, $V$ = 1 m$^3$) to a final state ($P$ = 1 atm, $V$ = 3 m$^3$). Please answer each of the following questions. For which process is $W$ the largest? The smallest? For which process is $Q$ the largest? The smallest?

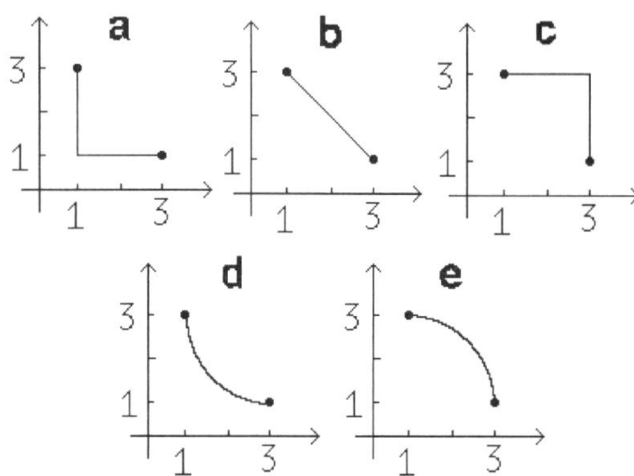


## Answers to Selected Conceptual Questions

**2.** **(a)** Yes. Heat can flow into the system, if at the same time the system expands, as in an isothermal expansion of a gas. **(b)** Yes. Heat can flow out of the system, if at the same time the system is compressed, as in an isothermal compression of a gas.

**4.** No. The heat might be added to a gas undergoing an isothermal expansion. In this case, there is no change in the temperature.

**8.** The final temperature of an ideal gas in this situation is $T$; that is, there is no change in temperature. The reason is that as the gas expands into the vacuum is does no work – it has nothing to push against. The gas is also insulated, so no heat can flow into or out of the system. It follows that the internal energy of the gas is unchanged, which means that its temperature is unchanged as well.

## Solutions to Selected End-of-Chapter Problems and Conceptual Exercises

9. **Picture the Problem**: As a car operates, its engine converts the internal energy of the gasoline into both work and heat.

**Strategy:** Use the first law of thermodynamics to calculate the heat given off, $Q_{\text{rel}}$.

**Solution: 1. (a)** Solve the first law for the heat released.

$$\Delta U = \left(-Q_{\text{rel}}\right) - W$$

$$Q_{\text{rel}} = -\Delta U - W = -\left(\frac{-1.19\times10^{8}\ \text{J/gal}}{25.0\ \text{mi/gal}}\right) - 5.20\times10^{5}\ \text{J/mi} = \boxed{4.24\ \text{MJ/mi}}$$

**2. (b)** Increasing miles per gallon improves efficiency, resulting in a $\boxed{\text{decrease}}$ of heat released to the atmosphere.

**Insight:** The gas mileage of a car is a measure of how much work (miles transported) the car can do for a given amount of available energy (in gallons of gas). The maximum amount of work would occur if all of the energy was converted to work and no heat was ejected. However, the Second Law of Thermodynamics says this is not possible.

22. **Picture the Problem**: A monatomic ideal gas expands at constant temperature.

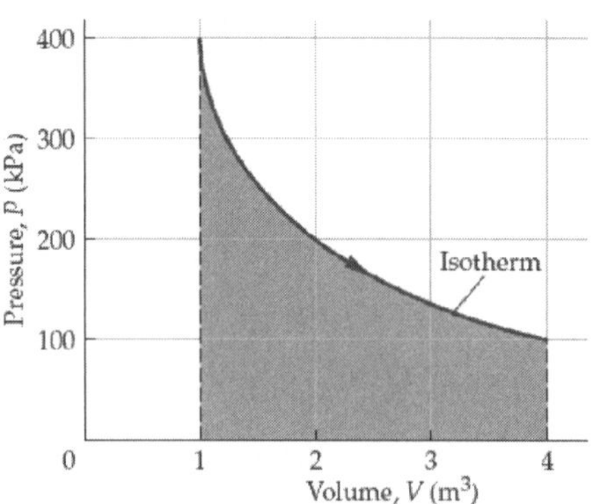


**Strategy:** Use equation 18-5 and the first law of thermodynamics to calculate the heat between any two volumes of the isothermal expansion.

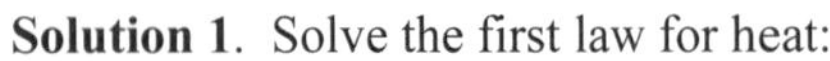

**Solution 1**. Solve the first law for heat: $Q = \Delta U + W$

**2.** Set $\Delta U = 0$ and use equation 18-5 for *W*: $Q = 0 + nRT\ln\left(\frac{V_f}{V_i}\right)$

**3. (a)** Because the gas expands ($V_f > V_i$) heat must be positive, so heat enters the system.

**4. (b)** Heat input equals the work done; work done is equal to the area under the *PV* plot. The area from $1.00\text{ m}^3$ to $2.00\text{ m}^3$ is greater than the area from $3.00\text{ m}^3$ to $4.00\text{ m}^3$.

**5. (c)** Insert the numeric values for the expansion from $1.00\text{ m}^3$ to $2.00\text{ m}^3$: $Q = 400\text{ kPa}\left(1.00\text{ m}^3\right)\ln\left(\frac{2.00\text{ m}^3}{1.00\text{ m}^3}\right) = \boxed{277\text{ kJ}}$

**6. (d)** Insert the numeric values for the expansion from $3.00\text{ m}^3$ to $4.00\text{ m}^3$: $Q = 100\text{ kPa}\left(4.00\text{ m}^3\right)\ln\left(\frac{4.00\text{ m}^3}{3.00\text{ m}^3}\right) = \boxed{115\text{ kJ}}$

**Insight:** The heat absorbed in the expansion from $1.00\text{ m}^3$ to $2.00\text{ m}^3$ is equal to the heat absorbed in the expansion from $2.00\text{ m}^3$ to $4.00\text{ m}^3$, because the heat absorbed is a function of the relative volume expansion. In both of these cases the volume doubles.

26. **Picture the Problem**: A monatomic ideal gas undergoes a process in which its volume increases while the pressure remains constant.

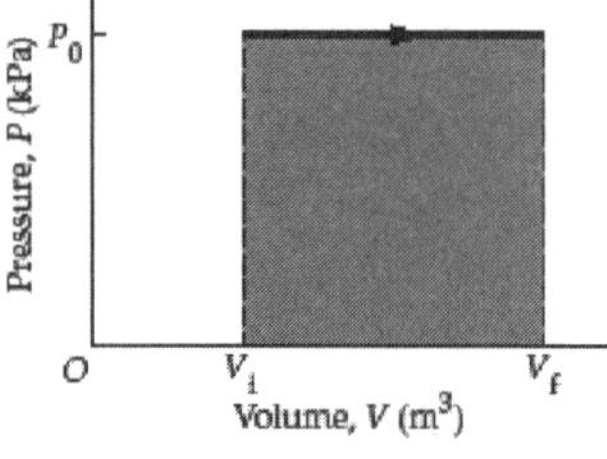


**Strategy:** Use the area under the *PV* plot to find the work done by the gas. Calculate the initial and final temperatures from the ideal gas law. Use the change in temperature to calculate the change in internal energy. Finally, calculate the heat added to the system using the first law of thermodynamics.

**Solution: 1. (a)** Multiply *P* by $\Delta V$: $W = P\Delta V = 210\text{ kPa}\left(1.9\text{ m}^3 - 0.75\text{ m}^3\right) = \boxed{2.4\times10^5\text{ J}}$

**2. (b)** Solve the ideal gas law for *T*: $T = \frac{PV}{nR}$

**3.** Insert initial conditions: $T_i = \frac{210\times10^3\text{ Pa}\left(0.75\text{ m}^3\right)}{49\text{ mol}\left[8.31\text{ J/}\left(\text{mol}\cdot\text{K}\right)\right]} = \boxed{3.9\times10^2\text{ K}}$

**4.** Insert the final conditions: $T_f = \frac{210\times10^3\text{ Pa}\left(1.9\text{ m}^3\right)}{49\text{ mol}\left[8.31\text{ J/}\left(\text{mol}\cdot\text{K}\right)\right]} = \boxed{9.8\times10^2\text{ K}}$

**5. (c)** Use equation 17-15 to solve for $\Delta U$: $\Delta U = \frac{3}{2}nR\Delta T = \frac{3}{2}\left(49\text{ mol}\right)\left[8.31\text{ J/}\left(\text{mol}\cdot\text{K}\right)\right]\left(980\text{ K} - 390\text{ K}\right)$

$= \boxed{3.6\times10^5\text{ J}}$

**6. (d)** Solve the first law for heat: $Q = \Delta U + W = 360\text{ kJ} + 240\text{ kJ} = \boxed{6.0\times10^5\text{ J}}$

**Insight:** The work done by the gas is proportional to the pressure. The temperature, and thus the change in internal energy, is also proportional to the pressure. Therefore the heat absorbed must also be proportional to the pressure. If the initial pressure was 420 kPa (double that given in the problem), then the work done would be 480 kJ, the change in internal energy 720 kJ, and the heat 1200 J.

30. **Picture the Problem**: A monatomic ideal gas is adiabatically compressed.

**Strategy:** Use the ideal gas law to relate the initial and final conditions of the gas. Then use equation 18-9 to eliminate the unknown volumes from the equation.

**Solution: 1. (a)** Doing work on the system must $\boxed{\text{increase}}$ the internal energy and therefore the temperature when no heat flows into or out of the system.

**2. (b)** Write the ideal gas law in terms of the initial and final states:

$$PV = nRT \Rightarrow \frac{P_\text{i}V_\text{i}}{T_\text{i}} = nR = \frac{P_\text{f}V_\text{f}}{T_\text{f}}$$

**3.** Solve for the final temperature:

$$T_\text{f} = \left(\frac{P_\text{f}}{P_\text{i}}\right)\left(\frac{V_\text{f}}{V_\text{i}}\right)T_\text{i}$$

**4.** Now solve equation 18-9 for the ratio of volumes:

$$P_iV_i^{\gamma} = P_fV_f^{\gamma} \Rightarrow \frac{V_\text{f}}{V_\text{i}} = \left(\frac{P_\text{i}}{P_\text{f}}\right)^{\frac{1}{\gamma}} = \left(\frac{P_\text{f}}{P_\text{i}}\right)^{-\frac{1}{\gamma}}, \text{ where } \gamma = \tfrac{5}{3}$$

**5.** Insert this ratio into the equation for $T_\text{f}$:

$$T_\text{f} = T_\text{i}\left(\frac{P_\text{f}}{P_\text{i}}\right)\left(\frac{P_\text{f}}{P_\text{i}}\right)^{-\frac{1}{\gamma}} = T_\text{i}\left(\frac{P_\text{f}}{P_\text{i}}\right)^{1-\frac{1}{\gamma}} = (280\text{ K})\left(\frac{140\text{ kPa}}{110\text{ kPa}}\right)^{(1-\frac{3}{5})} = \boxed{310\text{ K}}$$

**Insight:** As predicted, when work was done on the gas in an adiabatic process the temperature increased.

40. **Picture the Problem**: A thermally isolated monatomic ideal gas is compressed causing its pressure and temperature to increase.

**Strategy:** Use equation 18-9 to calculate the final volume in terms of the initial and final pressures. Then combine equation 18-9 and the ideal gas law to derive an equation for the final volume in terms of the temperatures.

**Solution: 1. (a)** Solve equation 18-9 for the final volume:

$$P_\text{i}V_\text{i}^{\gamma} = P_\text{f}V_\text{f}^{\gamma} \Rightarrow V_\text{f} = \left(\frac{P_\text{i}}{P_\text{f}}\right)^{\frac{1}{\gamma}} V_\text{i} = \left(\frac{105\text{ kPa}}{145\text{ kPa}}\right)^{3/5}\left(0.0750\text{ m}^3\right) = \boxed{0.0618\text{ m}^3}$$

**2. (b)** Combine the Ideal Gas Law with equation 18-9 and solve for final volume:

$$P_\text{i}V_\text{i}^{\gamma} = P_\text{f}V_\text{f}^{\gamma}$$

$$\left(\frac{nRT_\text{i}}{V_\text{i}}\right)V_\text{i}^{\gamma} = \left(\frac{nRT_\text{f}}{V_\text{f}}\right)V_\text{f}^{\gamma}$$

$$V_\text{f} = V_\text{i}\left(\frac{T_\text{f}}{T_\text{i}}\right)^{\frac{1}{1-\gamma}} = \left(\frac{295\text{ K}}{317\text{ K}}\right)^{1/(1-\frac{5}{3})}\left(0.0750\text{ m}^3\right) = \boxed{0.0835\text{ m}^3}$$

**Insight:** When a monatomic ideal gas is compressed adiabatically, its volume decreases while its temperature and pressure increase. Because the temperature in part (b) is less than the initial temperature, the gas had to expand as it cooled. This is seen in the final volume being greater that the initial volume.

51. **Picture the Problem**: The efficiency of a Carnot engine is increased by lowering the cold temperature reservoir.

**Strategy:** Use equation 18-13 to solve for the temperature of the cold temperature reservoir using the initial efficiency and then the new efficiency.

**Solution: 1. (a)** Solve equation 18-13 for $T_c$: $T_c = T_h(1 - e_{max}) = 545 \text{ K}(1 - 0.300) = \boxed{382 \text{ K}}$

**2. (b)** The efficiency of a heat engine increases as the difference in temperature of the hot and cold reservoirs increases. Therefore, the temperature of the low temperature reservoir must be $\boxed{\text{decreased}}$.

**3. (c)** Insert the new efficiency: $T_c = 545 \text{ K}(1 - 0.400) = \boxed{327 \text{ K}}$

**Insight:** The efficiency of a Carnot engine is increased when the temperature difference between the two reservoirs increases. When the hot temperature reservoir is fixed, the efficiency can be increased by lowering the temperature of the cold reservoir.

59. **Picture the Problem**: An air conditioner extracts heat from a room at a rate of 11 kW.

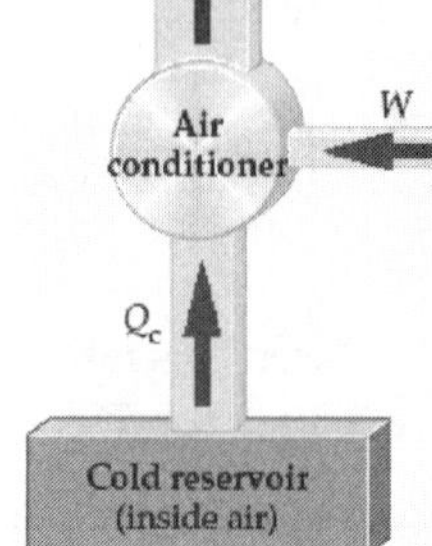


**Strategy:** Modify equation 18-10 and the Carnot relation to be rate equations by dividing each energy term by time. Then use the resulting equations to determine the mechanical power.

**Solution: 1.** Solve equation 18-10 for $Q_h/t$: $Q_h/t = W/t + Q_c/t$

**2.** Write the Carnot relation as a rate equation: $\dfrac{Q_h/t}{Q_c/t} = \dfrac{T_h}{T_c}$

**3.** Substitute the expression for $Q_h/t$ from step 1: $\dfrac{W/t + Q_c/t}{Q_c/t} = \dfrac{T_h}{T_c}$

**4** Solve for mechanical power: $P = W/t = (Q_c/t)\left(\dfrac{T_h}{T_c} - 1\right)$

**5.** Insert numerical values: $P = 11 \text{ kW}\left(\dfrac{273.15 + 32\,°\text{C}}{273.15 + 21°\text{C}} - 1\right) = \boxed{0.41 \text{ kW}}$

**Insight:** The coefficient of performance for the air conditioner can be found by dividing the rate of heat flow by the mechanical power giving $COP = 26.8$.

71. **Picture the Problem**: Imperfect insulation in a house allows heat $Q_c$ to leak into the house from the outside. An air conditioner extracts that same heat from the house and expels slightly more heat outside, where the temperature is warmer. The entropy change of the universe takes is the sum of the changes in entropy due to the heat entering the house and the air conditioner expelling the heat outside.

**Strategy:** Use equation 18-18 to sum the entropy changes for inside and outside the house. The resulting equation can be divided by time to determine the rate of entropy change. Use equation 18-10 to determine the rate that the air conditioner expels heat to the external world.

**Solution: 1. (a)** The entropy of the universe will $\boxed{\text{increase}}$. The air conditioner has the efficiency of a Carnot engine, so it does not increase the entropy of the universe, but the heat leaking into the house does produce a net entropy increase.

**2. (b)** Sum the inside and outside changes in entropy due to the air conditioner and the leaking insulation:

$$\Delta S = \left[\left(\frac{-Q_c}{T_c}\right)_{\text{in}} + \left(\frac{Q_h}{T_h}\right)_{\text{out}}\right]_{\text{AC}} + \left[\left(\frac{Q_c}{T_c}\right)_{\text{in}} + \left(\frac{-Q_c}{T_h}\right)_{\text{out}}\right]_{\text{Leak}}$$

**3.** Eliminate the indoor changes in entropy because they are equal and opposite:

$$\Delta S = \left(\frac{Q_{\text{h}}}{T_{\text{h}}}\right) + \left(\frac{-Q_{\text{c}}}{T_{\text{h}}}\right) = \frac{Q_{\text{h}} - Q_{\text{c}}}{T_{\text{h}}}$$

**4.** Use equation 18-10 to eliminate $Q_{\text{h}}$, and divide by time to obtain the rate of entropy increase:

$$\Delta S = \frac{Q_{\text{h}} - Q_{\text{c}}}{T_{\text{h}}} = \frac{W}{T_{\text{h}}} \Rightarrow \frac{\Delta S}{t} = \frac{W/t}{T_{\text{h}}} = \frac{0.41\text{ kW}}{32 + 273.15\text{ K}} = \boxed{1.3\text{ W/K}}$$

**Insight:** The heat exchange through the air conditioner does not change the entropy of the universe, but it does increase due to the heat flow of 11 kW into the house through the windows and doors (see problem 69).

## Answers to Practice Quiz

**1**. (a) **2**. (d) **3**. (e) **4**. (c) **5**. (b) **6**. (a) **7**. (c) **8**. (c) **9**. (e) **10**. (d) **11**. (a)